M000240277

Concrete Pavement Design, Construction, and Performance

Also available from Taylor & Francis

Reynolds's Reinforced Concrete Designer Handbook 11th edn
T. Threlfall *et al.*

Hb: ISBN 978–0–419–25820–9
Pb: ISBN 978–0–419–25830–8

Reinforced Concrete 3rd edn
P. Bhatt *et al.*

Hb: ISBN 978–0–415–30795–6
Pb: ISBN 978–0–415–30796–3

Concrete Bridges
P. Mondorf

Hb: ISBN 978–0–415–39362–1

Concrete Mix Design, Quality Control and Specification 3rd edn
K. Day

Hb: ISBN 978–0–415–39313–3

Binders for Durable and Sustainable Concrete
P.-C. Aïtcin

Hb: ISBN 978–0–415–38588–6

Aggregates in Concrete
M. Alexander *et al.*

Hb: ISBN 978–0–415–25839–5

Concrete Pavement Design Guidance Notes
G. Griffiths *et al.*

Hb: ISBN 978–0–415–25451–9

Information and ordering details

For price availability and ordering visit our website **www.sponpress.com**
Alternatively our books are available from all good bookshops.

Concrete Pavement Design, Construction, and Performance

Norbert Delatte

Taylor & Francis
Taylor & Francis Group

LONDON AND NEW YORK

First published 2008
by Taylor & Francis
2 Park Square, Milton Park, Abingdon, Oxon OX14 4RN

Simultaneously published in the USA and Canada
by Taylor & Francis
270 Madison Ave, New York, NY 10016, USA

Reprinted 2010

Taylor & Francis is an imprint of the Taylor & Francis Group, an informa business

© 2008 Norbert Delatte

Typeset in Sabon by
Integra Software Services Pvt. Ltd, Pondicherry, India
Printed and bound in Great Britain by
Cpod, a division of The Cromwell Press Group, Trowbridge, Wiltshire

All rights reserved. No part of this book may be reprinted or reproduced or utilised in any form or by any electronic, mechanical, or other means, now known or hereafter invented, including photocopying and recording, or in any information storage or retrieval system, without permission in writing from the publishers.

The publisher makes no representation, express or implied, with regard to the accuracy of the information contained in this book and cannot accept any legal responsibility or liability for any efforts or omissions that may be made.

British Library Cataloguing in Publication Data
A catalogue record for this book is available from the British Library

Library of Congress Cataloging-in-Publication Data
Delatte, Norbert J.
Concrete pavement design, construction, and performance / Norbert Delatte.
p. cm.
Includes bibliographical references and index.
ISBN 978-0-415-40970-4 (hardback : alk. paper) 1. Pavements, Concrete.
I. Title.
TE278.D45 2007
625.8′4--dc22
2007001787

ISBN10: 0–415–40970–5 (hbk)
ISBN10: 0–203–96108–0 (ebk)

ISBN13: 978–0–415–40970–4 (hbk)
ISBN13: 978–0–203–96108–7 (ebk)

I dedicate this book to my late father, Norbert Delatte Sr., who taught me to love books.

Contents

Figures

Tables

Acknowledgements

There are too many people to thank for helping me with this book, but I will try anyway. To begin with, the members of American Concrete Institute Committees 325, Concrete Pavements, Committee 327, Roller Compacted Concrete Pavements, Committee 330, Concrete Parking Lots and Site Paving, and Committee 522, Pervious Concrete, have helped me considerably with discussions during committee meetings and over beers afterward. James Shilstone, Sr. (Jim) and James Shilstone, Jr. (Jay) supplied material for Chapter 5, and Juan Pable Covarrubias supplied material for Chapter 9. Tim Smith of the Cement Association of Canada, Dale Crowl of the Ohio Department of Transportation, and George Palko of The Great Lakes Construction Company supplied photographs. In addition to my colleagues, I'd like to thank my students in courses at Cleveland State University and at the University of Alabama at Birmingham over the years, because teaching them challenged me to research and synthesize the material. Finally, I need to thank my wife Lynn and our children, Isabella and Joe, for their patience while I was drafting this manuscript.

Figures 2.1, 2.2, 2.3, and 18.7 are reprinted by permission of the American Concrete Pavement Association, Skokie, IL.

Figures 3.1, 3.5, 14.1–14.11, 15.1–15.3, 17.1–17.6, and Table 15.1 are reprinted from course materials developed by the American Concrete Pavement Association and used in National Highway Institute (NHI) training. NHI is a part of the Office of Professional and Corporate Development, Federal Highway Administration.

Chapter I

Introduction

Concrete pavements have been used for highways, airports, streets, local roads, parking lots, industrial facilities, and other types of infrastructure. When properly designed and built out of durable materials, concrete pavements can provide many decades of service with little or no maintenance. "Concrete generally has a higher initial cost than asphalt but lasts longer and has lower maintenance costs" (Hoel and Short 2006: 26).

In some cases, however, design or construction errors or poorly selected materials have considerably reduced pavement life. It is therefore important for pavement engineers to understand materials selection, mixture proportioning, design and detailing, drainage, construction techniques, and pavement performance. It is also important to understand the theoretical framework underlying commonly used design procedures, and to know the limits of applicability of the procedures.

The beginnings

The first concrete pavement was built in Bellefontaine, Ohio, in 1891, by George Bartholomew. He had learned about cement production in Germany and Texas and found pure sources of the necessary raw materials, limestone and clay, in central Ohio. Because this was the first concrete pavement, the city council required him to post a $5,000 bond that guaranteed the pavement would last 5 years. Over 100 years later, part of his pavement was still in use. Details of the history of the project are provided by Snell and Snell (2002).

The American Concrete Pavement Association (ACPA) "100 Years of Innovation" website (http://www.pavement.com/PavTech/AbtConc/History/Introduction.html) notes that the pavement "was an immediate success. Local businessmen petitioned to have the entire block around the Square paved with concrete. In 1893, Court Avenue and Opera Street were paved. Columbus Avenue and the remainder of Main Street followed in 1894." At that time, the term "concrete" was not yet in general use, so the material was called "artificial stone" and mixed by hand in 1.5 m (5 ft) square forms. Other early concrete pavements included Front Street in Chicago, which was built in 1905 and lasted 60 years, and Woodward Avenue in Detriot (1909) which was the first mile of concrete pavement.

There may have been even earlier concrete pavement experiments. Pasko (1998) notes "according to Blanchard's American Highway Engineers' Handbook of 1919, in 1879 in Scotland, a concrete was used with portland cement for binding. 'The surface was very good, but when the road commenced to break, it went to pieces very fast.' Blanchard goes on to say that the first portland cement concrete (PCC) pavement in the United States was put down in 1893 on South Fitzhugh Street in Rochester, N.Y., by J.Y. McClintock, Monroe county engineer. This was a section of portland cement grouted macadam, a forerunner of the modern concrete pavement of the Hassam type. The cost of this pavement was $1 per square yard (per 0.84 square meters). However, the pavement soon deteriorated." In spite of this possible earlier history, it is clear the Bellefontaine was the first successful, long-lasting concrete pavement.

Wider availability of automobiles led to increasing demand for paved roads. In 1913, 37 km (23 miles) of concrete pavement was built near Pine Bluff, Arkansas, at a cost of one dollar per linear foot. It became known as the "Dollarway." The pavement was 2.7 m (9 ft) wide and 125 mm (5 in) thick. The remains of Dollarway are preserved as a rest area along US 6. This was followed, in 1914, by 79 kms (49 miles) of concrete pavement for rural county roads in Mississippi, and by the end of 1914, a total of 3,778 km (2,348 miles) of concrete pavement had been built in the United States (ACPA 2006).

Despite the growing importance of the automobile, it was in fact a bicyclists' association that was organized and effective enough to press for the passage of the first Federal-Aid Highway Act in 1916. In the same year, the Portland Cement Association was organized to promote the use of portland cement and concrete. The concrete industry paved "seedling miles" with the hope that the public would demand they be linked together with more concrete pavement.

Early road tests

To supplement theory in the quest to develop design procedures for concrete pavements, many road tests were held over the years. It is believed that the first controlled evaluation of concrete pavement performance was conducted in 1909 by the Public Works Department of Detroit. Steelshod shoes and heavy iron wheels were mounted at opposite ends of a pole, revolving around a circular track, to simulate horse and wagon traffic of the day. Test sections included concrete, granite, creosote block, and cedar block. Based on this study, Wayne County, Michigan paved Woodward Avenue with concrete and then paved sixty more miles of concrete roads in the following 2 years (ACPA 2006).

After 1916, concrete roads were being built 125–225 mm (5–9 in) thick, but little was known about thickness requirements. During 1912–1923, the State of Illinois conducted the Bates Road Test. Using old World War I trucks with wheel loads from 454 to 5,900 kg (1,000 to 13,000 lbs), traffic was applied to test sections of different materials and thicknesses. The concrete sections of uniform cross-section were 100–225 mm (4–9 in) thick while thickened-edge sections were 229-125-229 mm and 229-152-229 mm (9-5-9 and 9-6-9 in), some with edge bars. The results showed that of the 22 brick, 17 asphalt, and 24 concrete sections in the test, one brick, three asphalt, and 10 concrete sections satisfactorily withstood the truck loadings. As a result, several design formulations were developed for Illinois to use in building its first state highway system (ACPA 2006). The Bates Road Test provided basic data that was used by design engineers for many years (Yoder and Witczak 1975: 4).

Until 1922, many pavements had been built with no joints and a thickened center section in an attempt to prevent the formation of an erratic longitudinal crack that developed in 4.9–5.5 m (16–18 ft) wide pavements. Based on the results of the Bates Road Test, center joints to eliminate longitudinal cracking were adopted (ACPA 2006).

Other road tests followed. The Pittsburg, California Road Test of 1921–1923 compared reinforced concrete pavement to plain concrete (ACPA 2006). This test showed no particular advantage to reinforcement, but the test pavements were built without joints and did not reflect modern practice (Ray 1981: 5).

> In 1950 and 1951, the Bureau of Public Roads (now FHWA) with the Highway Research Board (now the Transportation Research Board), several states, truck manufacturers, and other highway-related industries conducted Road Test One – MD just south of Washington, D.C. An existing 1.1 mi (1.8 km) of two-lane highway was carefully inventoried, instrumented, and traversed by 1,000 trucks per day. The results showed the value of good load transfer between slabs, the effects of speed and axle weights, and the problems caused by pumping. It produced the first dynamic wheel equivalence factors.
>
> (Pasko 1998)

This road test evaluated a section of US 301 consisting of two 3.66 m (12 ft) wide lanes with a 229-178-229 mm (9-7-9 in) thickened edge cross-section. The pavement had been built in 1941. The slabs were reinforced with wire mesh, and had 12.2 m (40 ft) contraction joint spacing and 36.6 m (120 ft) expansion joint spacing (Yoder and Witczak 1975: 597).

Truck loads of 80 and 100 kN (18,000 and 22,400 lb) single axles and 142 and 200 kN (32,000 and 44,800 lb) tandem axles were applied between June and December 1950 (Huang 2004: 19–20). The primary aim of this road test was not road design, but to determine proper tax allocation between trucking and railroads.

These road tests applied accelerated traffic to pavements, simulating a lifetime of traffic within a few months. However, it was not possible to accelerate environmental effects in the field. Pasko (1998) notes "One also needs to look at the effects of uncontrolled variables (environment) on pavement performance. A good example is the Road Test One – MD where controlled testing during July and August produced negligible damage. In September, the area had very heavy rains. In August, eight joints were pumping compared to 20 and 28 in September and October, respectively. The edge pumping was 162 ft (50 m) in August, 462 ft (140 m) in September, and 380 ft (116 m) in October after the heavy rain."

At the Maryland Road Test, progression of cracking was definitely linked to the development of pumping. Pumping was found to be greatest at expansion joints. Pumping occurred on plastic clay soils, but not on granular subgrades with low percentages of silt and clay. Deflections due to temperature curling and warping were also measured, and found to be significant (Yoder and Witczak 1975: 598).

The first concrete airfield pavement was built in 1928 at Ford Field in Dearborn, Michigan. Lunken Field in Cincinnati, Ohio, was constructed a year later. Like many early highway pavements, these airfields used a thickened-edge section, with the pavement at the edge 50 mm (2 in) thicker than at the centerline. Throughout the 1930s, the predominant aircraft was the relatively light DC-3, so the trucks that delivered fuel to the aircraft controlled the design. Early airfield concrete pavements were generally 152 mm (6 in) thick. The advent of the B-29 Superfortress bomber in World War II required 305 mm (12 in) pavements, so existing pavements were overlaid with 152–178 mm (6–7 in) of new concrete. With larger postwar bombers, pavement thickness increased to 508–686 mm (20–27 in) (ACPA 2006).

The work of the US Army Corps of Engineers in concrete pavement development was significant, particularly for airfields. With the B-29 in late World War II and progressively heavier postwar long range bombers, the Corps faced the difficult task of designing pavements around the world for very heavy loadings. In 1975, Yoder and Witczak noted that the "Corps. . . has for the past 20 years conducted extensive research programs on prototype pavements as well as pavement test sections" (Yoder and Witczak 1975: 4).

Ray (1981: 4–8) summarized the advances in pavement design and performance that were made between 1945 and 1980.

- Based on observations of pavement deformation due to expansion or shrinkage of highly expansive soils in the late 1930s, treatments for these soils with cement or lime were developed.

- Mechanisms of frost heave were identified and analyzed, and methods to address frost heave through careful subgrade construction and relatively thin granular subbase layers were adopted.
- Pumping was identified as a potential failure mechanism for concrete pavement joints, and soils and conditions susceptible to pumping were identified. Pumping may be controlled through use of granular subbases and doweled joints.
- Inadequate drainage or nonuniform compaction can lead to problems with thin granular subbases.
- Cement-treated bases (CTBs) or subbases provide excellent support and prevent pumping.
- Under the right conditions, thin concrete pavements can provide satisfactory service. Some 150 mm (6 in) airfield pavements lasted 30 years or more as long as the design aircraft wheel loads were not exceeded. Similarly, Iowa rural farm-to-market roads only 110 mm (4½ in) thick carried traffic for 30 years.
- Even if slab stresses are not high, excessive deflections can lead to pavement performance problems by aggravating curling and warping and erosion.
- Most agencies switched exclusively to jointed plain concrete pavement (JPCP) from jointed reinforced concrete pavement (JRCP) due to spalling, blowups, and broken reinforcement at intermediate cracks that form in JRCP slabs.
- Problems with undoweled JPCP joints led to wider adoption of dowels for highway and airport pavements.
- JPCP with short joint spacing on the order of 4.6 m (15 ft) provides the lowest life cycle costs for most applications.
- Concrete shoulders tied to mainline concrete pavements reduce deflections and stresses and improve safety and surface drainage and reduce shoulder maintenance.
- Properly designed and built overlays – either bonded, unbonded, or whitetopping – can provide many years of service.

The AASHO road test

The Dwight D. Eisenhower Federal-Aid Highway Act of 1956 created the 66,000 km (41,000 mile) interstate highway system. Sixty percent of the system was paved with concrete. There was obviously a need for research to support this construction effort. The AASHO Road Test, a $27-million study, was undertaken near Ottawa, Illinois. Six different test loops were built and loaded around the clock for 2 years. Twelve different combinations of axles and many thicknesses of asphalt and concrete were evaluated in order to establish performance histories and trends (ACPA 2006).

Pasko (1998) notes that

> At the AASHO Road Test, there were two distinctive failure modes for concrete pavements. The very thin pavements failed with continuous edge pumping that caused edge-cracking that coalesced into a longitudinal edge crack. The thicker pavements failed by joint pumping that caused transverse cracking starting particularly in the traffic leave side of the joints. The data from both were averaged together in the road test analysis to develop a performance equation. Even so, of the 84 pavement test sections greater than 8 in (200 mm) in thickness, only seven sections had a serviceability index of less than 4.0 at the end of the testing. In fact, only three sections could actually be considered as having failed. Hence, one can conclude that even though the AASHO data is the best that we have, it hardly predicts failure of the thicknesses of pavement that are now being built (greater than 8 in). Additionally, at the road test, there were no punchthroughs (shear failure) such as those produced at the Pittsburg Road Test under steel wheels, nor were there other types of environmentally induced failures such as blow-ups, CRCP punchouts, and so forth.

Early concrete pavements often suffered distress due to freezing and thawing cycles, or scaling due to deicing salts, or subgrade pumping. These problems were addressed during the 1930s, with air entrainment of concrete to attack the durability problems. The conditions likely to lead to pumping were identified – subgrade soil with fines that could go into suspension, water in the subgrade, and frequent heavy axle loads. Where these conditions exist, subbase layers between the concrete pavement and the subgrade are used to prevent pumping.

During this period, improvements to concrete paving equipment had been relatively minor, with motorized mixers replacing hand mixers. Pavements were placed using relatively expensive side forms. The slipform paver was developed in Iowa between 1946 and 1949 by two state highway engineers, James W. Johnson and Bert Myers. The innovation was first used in 1949 to pave a 2.7 m (9 ft) wide, 150 mm (6 in) thick county road. By using two pavers side by side, a county road could be paved in one pass. By 1955, the Quad City Construction Company had developed an improved, track propelled paver that could place 7.3 m (24 ft) wide slabs 250 mm (10 in) thick, and over time even larger pavers were developed. During this time, sawn joints began to replace hand-tooled joints and improve pavement smoothness.

Construction practices also improved. "The 1960's and Interstate construction also saw a number of advances in concrete pavement construction technology. Electronic controls were added to slipform pavers. Subgrade trimmers were introduced for better grade control. Tied concrete shoulders, first tested in Illinois in 1964, were found to add significant structural value to concrete pavements. Concrete saws were increased in size and capability."

(ACPA 2006). Central mixing plants increased production capacity by a factor of 10 over the previous dry batch process. The net result of this work was dramatic improvements in speed of construction.

Elements of modern slipform roadway paving are shown in Figures 1.1 through 1.3. In Figure 1.1, dowel and tie bars have been placed and secured in baskets in order to provide load transfer across sawn joints. Figure 1.2

Figure 1.1 Dowel and tie bar baskets placed in preparation for slipform paving (photo courtesy of The Great Lakes Construction Company, Hinckley, Ohio).

Figure 1.2 (a) and (b) Slipform pavers (photo courtesy of The Great Lakes Construction Company, Hinckley, Ohio).

Figure 1.3 Finishing slipformed pavement around a banked curve (photo courtesy of The Great Lakes Construction Company, Hinckley, Ohio).

illustrates the extrusion of concrete from the slipform paver itself, along with finishing operations. Figure 1.3 shows how slipforming can be used around a banked curve.

Evolution of design

Yoder and Witczak note that in the early development of concrete pavements, they were built directly on subgrade regardless of subgrade type or drainage conditions. Typical slab thicknesses were 152–178 mm (6–7 in). With increasing traffic after World War II, pumping became an increasingly important phenomenon, although it had been described as early as 1932 (Yoder and Witczak 1975: 596).

Thickened edge sections were common in the 1930s and 1940s. For example, a pavement would be built 152 mm (6 in) thick in the center and 203 mm (8 in) thick along the edges, which was described as an 8-6-8 design. Pavements were typically only 5.5–6.1 m (18–20 ft) wide (Yoder and Witczak 1975: 596–597).

As designs evolved, pavements were built over granular subbases to prevent pumping. In northern climates, thick bases were also used for frost protection. Pavement thickness was increased to 229–254 mm (9–10 in) for highways (Yoder and Witczak 1975: 596–597). In modern practice, thicker designs are often used for heavily traveled highways.

The long term pavement performance program

The last and greatest road test of all began in 1987 and continued for the next two decades (and in fact continues today). This was the Federal Highway Administration long term pavement performance (FHWA LTPP) program, which is part of the strategic highway research program (SHRP). The LTPP experiment encompassed rigorous field tests of more than 2,400 500 m (1,640 ft) asphalt and concrete pavement sections across the United States and Canada (FHWA LTPP 2006).

The program consisted of both pavements that were in service at the beginning of the test, termed general pavement studies (GPS) sections, and pavement sections that were constructed specifically for the LTPP, termed specific pavement studies (SPS) sections. SPS sections were constructed to a specific experimental design within each experiment. In the original program, there were 777 GPS sites and 234 SPS sites (Huang 2004: 26). An additional experiment was the seasonal monitoring program (SMP) that focused on environmental effects.

Within the GPS program, GPS-3, GPS-4, and GPS-5 analyzed JPCP, JRCP, and continuously reinforced concrete pavement (CRCP), respectively. Two other experiments analyzed concrete overlays on concrete pavements, with GPS-8 for bonded overlays and GPS-9 for unbonded overlays (Huang 2004: 25–26). However, the GPS-8 bonded concrete overlay program was not pursued.

Five of the nine SPS programs were relevant to rigid pavements. These were SPS-2, structural factors, SPS-4, preventive maintenance effectiveness, SPS-6, rehabilitation, SPS-7, bonded concrete overlays, and SPS-8, study of environmental effects in the absence of heavy loads (Huang 2004: 26). Because few states were willing to construct bonded concrete overlay test sections, only four sites were built, and the SPS-7 experiment was terminated early.

All sites had original inventory information and materials samples collected. LTPP pavement sites are visited periodically by contractors for data collection. Data collected includes traffic, profile, smoothness, distress, friction, and falling weight deflectometer (FWD) readings. Detailed climate information is also gathered (Huang 2004: 25–26).

The LTPP program offered many advantages over previous road tests. The previous tests were limited to a single geographical location, and generally used accelerated traffic loading. Accelerated traffic tends to obscure the importance of environmental effects. In contrast, the LTPP program is nearly two decades old, and realistic traffic and environmental conditions have been monitored over that time.

The LTPP program produced a large number of research reports and design and analysis tools, including the spreadsheet-based 1998 American Association of State Highway and Transportation Officials (AASHTO) rigid pavement design procedure (AASHTO 1998). However, the greatest importance of the LTPP program was in the development and calibration of models for the new AASHTO Mechanistic-Empirical Pavement Design Guide (M-EPDG).

In order to make LTPP data more accessible to users, the FHWA published several versions of DataPave software. For several years, the LTPP data and DataPave software were used for a student contest (Delatte 2002). The software has been superseded by DataPave Online, a website that allows users to access and analyze LTPP data. "The LTPP DataPave Online is a major effort to make the LTPP data more accessible to worldwide transportation community" (FHWA DataPave 2006). The Transportation Research Board (TRB) also has a program for LTPP studies (TRB 2006).

Economy, service life, and life cycle costs

Highway agencies have reported service lives for concrete pavements of as much as 25–40 years, or generally $1^1/_2$–2 times the service life of asphalt pavements designed and built to similar standards. In general, concrete is selected for heavily trafficked pavements and typically carries four times as much daily truck traffic as asphalt pavements. Respective life cycle costs of pavement designs depend greatly on material costs at the time of construction, but concrete pavements often have significantly lower maintenance costs although initial construction costs may be higher (Packard 1994).

Challenges for the future

The interstate highway system was a monumental and historic undertaking. However, in some respects, the engineers and builders of the early highways and the interstate highway system had an easier task than those of today. As the highway system has become built out, it must be maintained, repaired, and reconstructed while, in many cases, allowing traffic to continue to use the roadway. Much tighter environmental controls are now in place, and quality raw materials for use in highway construction are becoming increasingly scarce in many areas. Concrete overlays of various types have an important part to play in the upgrading and maintaining of the overall network. Accelerated or "fast track" paving materials and techniques have been developed to get traffic onto pavements more quickly.

Maintenance and rehabilitation

Even the best designed and built concrete pavements will eventually wear out and require maintenance or rehabilitation. Concrete pavement performance and distress are addressed in Chapter 3. Maintenance techniques and maintenance management are discussed in Chapter 16, and rehabilitation techniques in Chapter 17. It is important to recognize that timely maintenance, or pavement preservation, can substantially extend pavement life and delay the need for rehabilitation. When major work becomes necessary, overlays, as discussed in Chapter 18, may be used.

An effective maintenance and rehabilitation program, however, also requires a deeper understanding of how pavements behave. The basic design aspects discussed in Chapter 7 can provide insight as to whether a particular pavement distress may be due to inadequate pavement thickness, an unstable subbase material, or excessive joint spacing.

Meeting congestion and safety challenges

"Rehabilitation of today's highways requires traffic management, contracting and construction techniques that are quick and efficient and that result in a pavement with superior rideability and service life" (ACPA 2000c: 1). Pavement maintenance and rehabilitation are particularly challenging for high volume pavements, particularly in congested urban areas. The American Concrete Pavement Association (ACPA) has published *Traffic Management – Handbook for Concrete Pavement Reconstruction and Rehabilitation* (ACPA 2000c) to address some of these issues. Some important points made in that document include:

- On many urban freeway reconstruction projects, the road user delay costs may exceed $50,000 per day.
- Concrete pavement does not require a curing time of 5–14 days before opening to traffic – mixtures are available that allow traffic in 6–8 hours.
- Limiting contractors to single lane closures or nighttime work hours may be unwise for both safety and productivity.
- The public generally prefers a short project with major disruptions to a long project with smaller disruptions. The slogan is "get in, get out, stay out."
- Issues that should be addressed include scope of the project, traffic management, safety, construction requirements, innovative bidding, constructability, emergency planning, and public information and coordination.
- Innovative bidding and contracting practices, such as incentives/disincentives and A+B bidding (bidding both price and time),

motivate and challenge contractors to finish projects and open them to public traffic more quickly.

- Bridge abutments and clearances may constrict the range of available work zone configuration options, and should be considered early in the planning process.
- Recycling paving materials in place, as discussed later in this chapter, reduces project duration.
- Availability and capacity of detour routes and access for work vehicles are key considerations. Pavements and bridges on detour routes that will carry significant increases in traffic require structural evaluation.
- It is important to communicate with the public about the impacts, future benefits, and progress of the project to establish and maintain support.

Non-destructive testing technologies such as maturity or ultrasonic pulse velocity can estimate in-place pavement strength for opening to traffic. ACPA (2000c: 28) provides a table for determining the strength necessary to open the concrete pavement to public traffic.

Safety in US highway construction work zones is addressed in Part VI of the *Manual on Uniform Traffic Control Devices* (FHWA 2003). Congestion and capacity may be analyzed using *Highway Capacity Manual 2000* (AASHTO 2000).

Accelerated concrete pavement construction and reconstruction are also addressed by the American Concrete Institute (ACI) committee report *Accelerated Techniques for Concrete Paving, ACI 325.11R-01* (ACI Committee 325 2001). It is important to note that accelerated paving does not always require high early strength concrete. With materials and mixtures in use today, conventional paving concrete generally achieves sufficient strength to carry traffic in 2–4 days, so if opening to traffic can be delayed that long there is no need to use high early strength concrete. Small weekend closure projects such as urban intersections can be built with conventional paving concrete. If all concrete is placed by Saturday evening, it will have a day and a half to cure for a Monday morning opening.

One interesting case study is the reconstruction of heavily trafficked Interstate Highway 15 in Southern California. The concrete truck lanes were badly deteriorated, and a 4.5 km (2.8 mile) section was rebuilt in only two nine day closures. Because this is the main highway between Los Angeles and Las Vegas, it has heavy traffic on weekends. The project is discussed in detail by Lee et al. (2005). Some of the key features of the project included:

- The pavement was rebuilt in nine day closures as opposed to overnight closures, because the longer closures had been found to yield much higher contractor productivity.

- Longer closures reduce the overall disruption to the traveling public, provide greater life expectancy for the pavement, improve safety, and significantly reduce construction costs.
- The existing pavement of 200 mm (8 in) of concrete over 100 mm (4 in) of CTB over 300 mm (12 in) of aggregate base was replaced with 290 mm (11½ in) of concrete over 150 mm (6 in) of asphalt over half of the reused aggregate base. The pavement grade was not changed. The outer truck lane was widened by 600 mm (2 ft) to reduce edge-loading stresses.
- Caltrans went to considerable effort to communicate to the public how the selected strategy would reduce the overall project duration and minimize disruption to the traveling public. An automated work zone information system gave the public travel time and detour information on changeable message signs. Public outreach materials included comprehensive project brochures and flyers, a construction advisory electronic bulletin, and a project information hotline. A project website was used to communicate information to the public and also to gather feedback. Several public meetings were held for local communities.
- Rapid strength gain concrete with type III cement allowed opening the pavement to traffic in 12 hours. About 50 percent of the project was able to use a more conventional concrete with type II cement, as long as 24 hours of curing was available.
- Project completion time incentives and disincentives and late opening penalties were specified in the contract.

On larger projects, the bulk of the project may be paved with regular concrete with small amounts of high early strength concrete used for small final closure strips. This approach was adopted for critical taxiway tie-ins during new runway construction at Cleveland Hopkins International Airport, although it was found that with proper planning the conventional concrete achieved sufficient early strength and it was not necessary to use high early strength concrete (Peshkin et al. 2006b).

Case studies of similar projects can provide valuable insights into available techniques and strategies. Peshkin et al. (2006a,b) review a number of airport case studies. The ACPA report *Traffic Management – Handbook for Concrete Pavement Reconstruction and Rehabilitation* (ACPA 2000c: iii, 50) discusses highway and street case studies:

- Reconstruction of the Eden's Expressway in north suburban Chicago required the Illinois DOT to use a wide range of accelerated paving techniques and provided much of the information for the ACPA Handbook.
- In Denver, Colorado, 75 days were cut from a 200-day schedule for construction of an urban arterial.

- In Johnson County, Kansas, a major intersection was repaved over a weekend with a whitetopping overlay.

If high early strength concrete is used, it is important to consider durability. Some mixtures that achieve strength in 6–8 hours have shown durability problems, although not in all cases. Durability of high early strength concrete is discussed in Chapter 6 and by Van Dam et al. (2005). It is particularly frustrating to agencies and users when repairs fail quickly and need to be repeated.

Recycling and reuse of materials

The construction of highways, bridges and buildings has been increasing from the beginning of the past century, especially in areas of high population density. These facilities need to be repaired or replaced with the passing of time because their end of service life is reached or the original design no longer satisfies the needs due to the growth in population or traffic. These facts have generated two important issues. First, a growing demand for construction aggregates and, second, an increase in the amount of construction waste. Two billion tons of aggregate are produced each year in the United States. Production is expected to increase to more than 2.5 billion tons per year by the year 2020. This has raised concerns about the availability of natural aggregates and where we will find new aggregate sources. On the other hand, the construction waste produced from building demolition alone is estimated to be 123 million tons per year. Historically, the most common method of managing this material has been through disposal in landfills. As cost, environmental regulations and land use policies for landfills become more restrictive, the need to seek alternative uses of the waste material increases. This situation has led state agencies and the aggregate industry to begin recycling concrete debris as an alternative aggregate.

(FHWA 2004: 5–6)

Pavements with steel reinforcement may also be recycled, and equipment has been developed to economically remove reinforcement and dowels without the need for hand labor. Continuous steel or mesh is usually removed at the demolition site, while dowels and tie bars may be removed at a plant where the material is processed (ACPA 1993c). Any reinforcing steel in the concrete has to be removed.

Existing concrete can be crushed and recycled into concrete pavement, either as new concrete pavement or as a base or subbase course under concrete pavement. Crushing and processing may occur at or very near to the construction site. The recycled concrete aggregate (RCA) offers a number of advantages:

- reduces the need for virgin aggregates;
- reduces the volume of waste materials brought to landfills;
- reduces transportation costs as well as the number of trucks that must maneuver in and out of the work zone, which can reduce project durations as well as costs;
- residual cementitious properties as well as the angularity of crushed concrete particles can enhance the stability and stiffness of a recycled concrete base or subbase;
- when old concrete is crushed to a smaller size, susceptibility to D-cracking is reduced.

There are also some risks and cautions associated with the re-use of existing concrete:

- Existing concrete often has unknown properties. Construction debris may contain unwanted bricks, wood, steel, ceramics, glass, or other materials. As a result, some agencies only allow concrete from their previous agency or DOT projects to be recycled for their new projects.
- Use of RCA in new concrete can create workability problems unless the moisture content of the material is controlled.
- Some RCA needs to be washed before being used in new concrete, which adds to cost.
- Fines in unwashed RCA may reduce concrete compressive strength.
- RCA generally has a higher absorption and lower specific gravity than the original aggregate used in the concrete, due to the inclusion of mortar made of cement, water, and air. It is best to keep stockpiles in saturated surface dry (SSD) condition.
- When used as a base or subbase, the RCA should be kept wet to limit dust and excess working should be avoided to prevent segregation. The material should be compacted in a saturated condition.
- Concerns about leaching remain when RCA is used in a drainage layer or near a water source.

The Federal Highway Administration Recycling Team performed a national review of the state of the practice in *Transportation Applications of Recycled Concrete Aggregate*. The study identified 41 states recycling concrete aggregate, with 38 using it as aggregate base and 11 as an aggregate for PCC. The report reviewed and summarized state practices in Texas, Virginia, Michigan, Minnesota and California. Texas, Minnesota, and California reported that RCA performed better than virgin aggregate as a base. One value engineering proposal in Michigan saved $114,000 on a $3 million project through use of RCA. Michigan has also used recycled concrete as backfill for edge-drains. Two developments that make the use of RCA more attractive are the use of mobile crusher units that can be moved to stockpiles of construction debris, and mobile units that crush pavement in place into base or subbase material (FHWA 2004). The FHWA has established a website at http://www.fhwa.dot.gov/pavement/recycling/index.cfm to provide information on recycling.

An important thing to remember is that concrete often fails for a reason. If the concrete failed due to fatigue, there is little risk involved in recycling the material. On the other hand, if the failure was due to a material-related distress such as alkali-silica reaction (ASR), recycling the material into new concrete simply sets the stage for a future failure. One cautionary tale on the problems that can occur when crushed concrete is recycled in an environment prone to sulfate attack is discussed in detail by Rollings et al. (2006) and reviewed briefly in Chapter 5. It may be prudent to carefully examine the existing concrete for ASR or D-cracking problems to determine whether recycling is an option (ACPA 1993c).

When the concrete pavement has a thin asphalt overlay, it is possible to crush the two together if it is to be used as a base or a subbase, but not in new concrete. The asphalt in the material is not harmful, and it is not necessary to mill off the overlay. Minnesota allows up to 3 percent asphalt cement or approximately 50 percent reclaimed asphalt pavement (RAP) in the crushed base material. California goes further and allows any combination of RCA and RAP (FHWA 2004: 26).

The ACPA has published *Recycling Concrete Pavement, Technical Bulletin TB–014.P*, which discusses concrete pavement recycling from old pavement removal, to concrete crushing, to replacing the new pavement. The bulletin covers critical aspects of mix design for RCA in fresh concrete. It may be necessary to determine the optimum rate of recycled concrete fines in the mixture, and to pay special attention to control of water content. A guide specification is included in the Bulletin (ACPA 1993c).

Environmental impacts

Road and airport construction and the production of concrete and other construction materials have environmental impacts. For the most part, the impacts of development on the landscape are similar regardless of the pavement type, with the exception of pervious pavements. Also, both asphalt and concrete pavements require large quantities of aggregate, so their impact through aggregate mining and production are similar.

The production of portland cement impacts the environment due to the large amounts of energy required and the release of CO_2 into the atmosphere. Concrete production requires approximately 3.4 GJ per cubic meter

(Mindess et al. 2003: 5). "As a major part of the world economy, the concrete industry must play an active role in sustainable development. Because there are few new technologies on the horizon that can reduce CO_2 emissions from the manufacturing of portland cement, however, the answer to reducing CO_2 emissions lies in minimizing the output of cement clinker" (Malhotra 2006: 42).

The total worldwide annual production of fly ash is about 900 million tonnes (990 million tons), principally in China, India, and the US. China alone produces two-thirds of the worldwide amount. Use of coal in power plants, and therefore production of fly ash, is expected to increase substantially in these countries over the next few decades. These countries currently use approximately 10–15 percent of the fly ash in concrete. In the United States, fly ash is separately batched at concrete plants, whereas overseas the fly ash is often blended into the cement at a rate of about 20 percent (Malhotra 2006: 42–43).

Not all fly ash meets specifications for use in concrete. Most of the fly ash that does is type F, which is a by-product of combustion of anthracite or bituminous coal, and has pozzolanic but not cementitious properties. High calcium fly ash, type C, which has cementitious properties as well as pozzolanic properties, is produced from lignite or sub-bituminous coal. "Technologies are available, however, that can beneficiate fly ash that fails to meet fineness and carbon content requirements – the two most important parameters of fly ash used in concrete. These technologies include removal of carbon by electrostatic and floatation methods. Grinding and air classification methods have also been used to produce fly ash with high fineness" (Malhotra 2006: 43).

Malhotra (2006: 45) suggests reducing CO_2 emissions using the following strategies:

- "Using less portland cement;
- Using more supplementary cementitious materials;
- Using less unit water content by using more water-reducing admixtures and HRWRAs [high-range water reducing admixtures];
- Incorporating recycled aggregates in concrete; . . .
- Where possible, specifying strength acceptance criteria at 56 or 91 days instead of 28 days."

For paving concrete, then, the strategies for reducing cement use and thus CO_2 emissions are:

- maximize the aggregate structure as discussed in Chapter 6 to minimize the overall paste volume;

- use water reducers to reduce the amount of water and thus the cementitious material content required to maintain the same water/cementitious materials (w/cm) ratio; and
- replace as much cement as practicable with fly ash or ground granulate blast furnace slag (GGBFS).

Malhotra (2006: 43–44) has developed high volume fly ash (HVFA) concrete where as much as half or more of the cement is replaced by fly ash. To date, HVFA concrete has not been used for pavement, to the author's knowledge.

Some other supplementary cementitious materials that could be used in paving concrete are GGBFS, natural pozzolans, rice-husk ash, silica fume, and metakaolin. GGBFS is already extensively used in paving concrete, but availability is limited. Only about 25 million tonnes (28 million tons) are produced a year. Rice-husk ash is not yet commercially available, but the estimated worldwide potential is 20 million tonnes (22 million tons). Silica fume and metakaolin are only available worldwide in relatively small quantities (Malhotra 2006: 44). To date, silica fume and metakaolin have not been widely used for pavements. Silica fume has been used for dense concrete bridge deck overlays, and could be used for pavement bonded concrete overlays.

Sustainability

Sustainable development is of increasing interest, and was written into the American Society of Civil Engineers Code of Ethics in 1996. This was added as item 1e – "Engineers should seek opportunities to be of constructive service in civic affairs and work for the advancement of the safety, health and well-being of their communities, and the protection of the environment through the practice of sustainable development" (ASCE 2006).

Sustainable development is defined as "the challenge of meeting human needs for natural resources, industrial products, energy, food, transportation, shelter, and effective waste management while conserving and protecting environmental quality and the natural resource base essential for future development" (ASCE 2006). Another term for sustainable development is "green building."

Pavements represent an important potential application area for sustainability concepts. "As the world's non-renewable resources such as fossil fuels and roadway aggregates decrease in availability, it is important for all levels of government to begin considering paving structures on a sustainability basis rather than just a first cost basis" (Smith and Jolly 2005: 585).

Some of the environmental and sustainability benefits of concrete pavements include:

- Reductions of 2.4–30 percent in energy requirements for pavement construction and maintenance, with higher savings for more heavily trafficked highways.
- Reduced heavy vehicle fuel consumption and greenhouse gas production (particularly CO_2) because concrete pavement has less rolling resistance, therefore requiring lower vehicle power. Fuel reductions of up to 20 percent have been observed.
- Concrete pavements use less granular material or aggregate throughout the pavement structure because base layers are not needed. Asphalt pavements may use twice as much. These materials are growing scarcer, and hauling aggregates represents a significant fraction of the environmental impact of highway construction.
- Supplementary cementitious materials – fly ash, ground granulated blast furnace slag, and silica fume – are industrial by-products and their use in concrete reduces volumes of waste.
- Concrete pavements reduce the heat island effect. Urban areas tend to be 5 °C (9 °F) warmer than the surrounding countryside because dark pavements and roofs retain heat. Concrete pavement has a lighter color and reflects light, reducing the heat island effect. Black surfaces can be 21 °C (38 °F) warmer than the most reflective white surfaces. Reducing the heat island effect can also reduce smog and ozone in urban areas.
- Because concrete is more reflective than asphalt, lighting requirements are lower for concrete parking lots, reducing energy consumption (Smith and Jolly 2005: 585–606).

An US Environmental Protection Agency (EPA) website on the heat island effect states "For millions of Americans living in and around cities, heat islands are of growing concern. This phenomenon describes urban and suburban temperatures that are 2 to 10 °F (1 to 6 °C) hotter than nearby rural areas. Elevated temperatures can impact communities by increasing peak energy demand, air conditioning costs, air pollution levels, and heat-related illness and mortality" (EPA 2006). Lighter colored and porous paving materials are recommended by the EPA for reducing the heat island effect.

One way of addressing sustainability in constructed facilities is through the leadership in energy and environmental design (LEED) program. Although LEED primarily applies to buildings, it provides principles that are relevant to sustainability in general, and applies to parking facilities and industrial pavements. Under LEED, buildings are eligible for points that may be used to achieve different levels of certification. Potential LEED points applicable to ready-mixed concrete include storm water management, landscape paving, managing construction waste, use of recycled materials, and reductions in the use of portland cement (RMC Research Foundation 2005).

Possible LEED strategies include:

- use of pervious concrete pavement, discussed in Chapter 2;
- concrete parking lots to reduce the heat island effect;
- diversion of construction debris from landfills through recycling or salvage, which primarily applies to overlays, discussed in Chapter 18;
- use of by-products as supplementary cementitious materials, such as fly ash, GGBFS, silica fume, or rice-husk ash. Use of these materials in concrete usually reduces the amount of portland cement needed;
- use of local materials to reduce transportation costs (RMC Research Foundation 2005).

The Environmental Council of Concrete Organizations (ECCO) (www. ecco.org) notes

As a nearly inert material, concrete is an ideal medium for recycling waste or industrial byproducts. Many materials that would end up in landfills can be used instead to make concrete. Blast furnace slag, recycled polystyrene, and fly ash are among materials that can be included in the recipe for concrete and further enhance its appeal. Waste products such as scrap tires and kiln dust are used to fuel the manufacture of cement. And even old concrete itself can be reborn as aggregate for new concrete mixtures.

Another environmental plus for concrete is energy efficiency. From manufacture to transport to construction, concrete is modest in its energy needs and generous in its payback. The only energy intensive demand is in the manufacture of portland cement, typically a 10–15% component of concrete. Since the materials for concrete are so readily available, concrete products and ready-mixed concrete can be made from local resources and processed near a jobsite. Local shipping minimizes fuel requirements for handling and transportation.

(ECCO 2006)

Sources for information, research reports, guidelines, and standards

A number of organizations publish references that are useful to the concrete pavement engineer. Many of these references were consulted during the development of this book and are listed in the bibliography. Although many of these organizations are in the United States, their documents are often used and referenced worldwide.

The websites will be useful to readers who would like to pursue the topics discussed. Also, readers who wish to keep up with the latest developments or contribute to the state of the practice may wish to participate in one or more of these organizations.

One of the oldest organizations to develop information for concrete pavements is the Portland Cement Association (PCA), website www.cement.org. The PCA developed a number of important pavement design guides up through the 1980s. The American Concrete Pavement Association (ACPA), website www.pavement.com, subsequently has assumed most of the PCA's concrete pavement work. At present, the ACPA publishes design and construction documents for highways, parking lots, streets and local roads, airports, and industrial pavements. The ACPA and PCA also sell a number of concrete pavement design computer programs. The PCA continues to publish literature on roller-compacted concrete (RCC) pavements as well as cement-stabilized base and subbase layers, as well as a key publication on *Design and Control of Concrete Mixtures* (Kosmatka et al. 2002).

The National Ready Mixed Concrete Association (NRMCA), website http://www.nrmca.org/, is at present the primary source for information on pervious concrete pavements. NRMCA also provides a number of certifications for concrete suppliers and contractors. The PCA, ACPA, NRMCA are all industry associations that are supported and financed by their members. All three of these organizations employ engineers to draft their technical documents and also contract for outside research.

The American Concrete Institute (ACI), website www.concrete.org, works to advance concrete knowledge through a large number of technical committees. The four technical committees most directly relevant to concrete pavement are 325 Concrete Pavements, 327 Roller Compacted Concrete Pavements, 330 Concrete Parking Lots and Site Paving, and 522 Pervious Concrete. The technical committee membership includes members of industry associations, government agency representatives (e.g. FHWA, FAA, and USACE, discussed below), academics, and consulting engineers. Documents are produced through a rigorous consensus process, which means they take longer to develop but are often more authoritative than those from other sources. The documents of the four pavement committees, as well as many others with direct application to pavements (e.g. mixture proportioning, testing, and curing), are contained in the six volume Manual of Concrete Practice and may also be purchased separately. ACI also has a number of certification programs for testing laboratories and contractors.

ACI, APCA, and PCA publications are particularly relevant for concrete pavement types that are not addressed by governmental organizations such as FHWA, AASHTO, or PCA. These include parking lots, streets and local roads, RCC, pervious concrete, and industrial pavements.

The American Society for Testing and Materials (ASTM), website www.astm.org, works through technical committees in a similar manner to ACI. There is also some overlap of membership between the two organizations. ASTM publishes testing protocols and materials standards.

The Transportation Research Board (TRB) is a division of the National Research Council. The TRB website is www.trb.org. The National Research Council is jointly administered by the National Academy of Sciences, the National Academy of Engineering, and the Institute of Medicine. TRB is a very large organization with international membership that covers a broad array of transportation topics, including concrete and concrete pavement. Concrete pavement research results are often presented at the TRB annual meeting, and papers are published in Transportation Research Record (TRB). Like ACI and ASTM, TRB works through constituent committees. TRB was formerly the Highway Research Board.

The International Society for Concrete Pavements (ISCP), website http://www.concretepavements.org/, is a relatively young organization whose main activity has been to organize the International Conferences on Concrete Pavements. Eight of these conferences have been held every 4 years, and the conference proceedings are a valuable source of information. Many ISCP members also participate in TRB and ACI.

Highways

Under the United States Department of Transportation, the Federal Highways Administration (FHWA), website http://www.fhwa.dot.gov/, is responsible for highways. FHWA pavement research reports may be downloaded free from http://www.fhwa.dot.gov/pavement/pub_listing.cfm. The FHWA also has a computer program for pavement drainage design, which may also be downloaded free and is discussed in detail in Chapter 4.

The American Association of State Highway and Transportation Officials (AASHTO), website http://www.transportation.org/, sells pavement design guides and materials and test standards. AASHTO standards are often similar or identical to ASTM standards. Many AASHTO documents, including the pavement design guides, are developed under contract research through the National Cooperative Highway Research Program (NCHRP), which is under TRB. NCHRP online publications may be downloaded free from http://www4.trb.org/trb/onlinepubs.nsf/web/crp.

Many US state Departments of Transportation have websites with research reports and standards. There are obviously too many to list here.

Airports

In the United States, civil aviation is governed by the Federal Aviation Administration (FAA), website http://www.faa.gov/. The FAA's design and construction standards are published in the Series 150 Advisory Circulars and may be downloaded free from http://www.faa.gov/airports_airtraffic/ airports/resources/ recent_advisory_circulars/. The FAA also has free design computer programs that may be downloaded. These are discussed in Chapter 10.

The Innovative Pavement Research Foundation (IPRF), website www.iprf.org, has published a number of research reports on airfield pavements. IPRF reports may be downloaded free from the website.

For US military facilities, standards and guidelines are developed by the US Army Corps of Engineers (USACE) and published as unified facilities criteria (UFC) documents on the web at http://65.204.17.188/ report/doc_ufc.html. These include UFC 3-250-02 (2001a) "Standard Practice Manual for Rigid Pavements," and UFC 3-260-02 (2001b) "Pavement Design for Airfields."

Chapter 2

Types of concrete pavements

There are a number of different types of concrete pavements that have been built. However, for the most part, they have two features in common. First, they resist traffic loads through flexure of the concrete. If reinforcement is used, it is used for crack control and not to carry load. The second element is that concrete pavements contract due to drying shrinkage of the concrete, and expand and contract due to thermal effects, and these movements must be dealt with. Different types of pavements use joints, reinforcing steel, or both.

The term "conventional concrete pavements" is generally taken to mean either jointed plain, jointed reinforced, or continuously reinforced concrete pavements (the first three categories described below) but not other types. Design and detailing of joints is important for these pavements. All three conventional pavement types have been used as overlays, although jointed plain overlays are most common.

Prestressed and precast concrete pavements are used for similar applications as conventional concrete pavements, but have been used infrequently. Other types of concrete pavement include roller compacted concrete (RCC) and pervious or porous concrete, which are generally used for specialized industrial or parking lot applications.

Jointed plain concrete pavement

Jointed plain concrete pavement, or JPCP, consists of unreinforced concrete slabs 3.6–6.0 m (12–20 ft) in length with transverse contraction joints between the slabs. The joints are spaced closely enough together so that cracks should not form in the slabs until late in the life of the pavement. Therefore, for JPCP, the pavement expansions and contractions are addressed through joints. JPCP is illustrated in Figure 2.1.

One important performance issue with JPCP is load transfer across the joints. If joints become faulted, then drivers encounter bumps at the joints and experience a rough ride. Two methods are used to provide load transfer across JPCP joints – aggregate interlock and dowels.

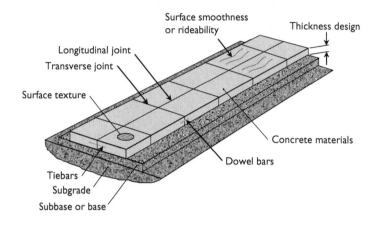

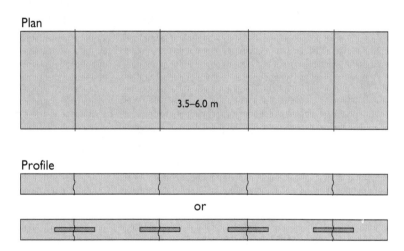

Figure 2.1 Jointed plain concrete pavement (JPCP) (courtesy: ACPA).

Aggregate interlock joints are formed during construction by sawing 1/4–1/3 of the way through the pavement to create a plane of weakness. A crack then propagates through the remaining thickness of the pavement as the concrete contracts. This crack has a rough surface because it propagates around the aggregates through the green cement paste, and as long as it remains narrow the joint can transfer load from one slab to another through bearing stress of the aggregate particles against each other across the crack. Load transfer is compromised if the joint opens too widely or if the aggregates wear away. The quality and erosion resistance of the material supporting the slab at the joint also affect load transfer.

When the pavement carries heavy vehicle traffic, particularly at high speeds, aggregate interlock will break down over time and will not prevent faulting over the life of the pavement. In this case, dowels are provided across the joint for load transfer. Dowels are smooth rods, generally plain- or epoxy-coated steel, which are usually greased or oiled on side to allow the joints to open and close without resistance.

JPCP is the most commonly used type of concrete pavement because it is usually the cheapest to construct. A 1999 survey by the American Concrete Pavement Association identified 38 US states as constructing JPCP (ACPA 1999c). It is economical because there is no need to pay for any reinforcing steel in the slabs or for labor to place the steel. In most regions, also, contractors will have more familiarity with JPCP than with other types of concrete pavement. In those regions where corrosion of steel is a problem, the absence of steel reinforcement means an absence of steel corrosion issues, although the steel dowels can still corrode. To address dowel corrosion, a variety of different dowel materials and coatings have been investigated (Snyder 2005).

JPC pavements, like other conventional concrete pavements, often use tie bars to connect adjacent traffic lanes. Tie bars are deformed reinforcing steel and, unlike dowels, are not intended to allow the joints to open and close. Tie bars are used to separate lanes for highway pavements. In contrast, airfield JPC pavements generally use dowels at all joints.

Most of this book's design and construction guidelines primarily address JPCP because it is the predominant concrete pavement type by far. Special considerations for transitions between pavements and bridges and CRCP details are discussed in Chapter 12.

Key performance issues of JPCP include:

- initial pavement smoothness, which is a function of construction practices;
- adequate pavement thickness to prevent mid-slab cracking;
- limiting the joint spacing, also to prevent mid-slab cracking; and
- adequate joint design, detailing, and construction.

Jointed reinforced concrete pavement

Jointed reinforced concrete pavement, or JRCP, is distinguished from JPCP by longer slabs and light reinforcement in the slabs. This light reinforcement is often termed temperature steel. JRCP slab lengths typically range from 7.5 to 9 m (25–30 ft), although slab lengths up to 30 m (100 ft) have been used. With these slab lengths, the joints must be doweled. The slab steel content is typically in the range of 0.10–0.25 percent of the cross-sectional area, in the longitudinal direction, with less steel in the transverse direction. Either individual reinforcing bars or wire fabrics and meshes may be used. Because the steel is placed at the neutral axis or midpoint of the slab, it

Plan

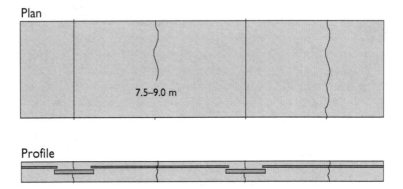

7.5–9.0 m

Profile

Figure 2.2 Jointed reinforced concrete pavement (JRCP) (courtesy: ACPA).

has no effect on the flexural performance of the concrete and serves only to keep cracks together. JRCP is illustrated in Figure 2.2.

Although JRCP was widely used in the past, it is less common today. The only advantage that JRCP has over JPCP is fewer joints, and this is outweighed by the cost of the steel and the poor performance of the joints and the cracks. Because the joints are spaced further apart than JPCP, they open and close more, and load transfer suffers as joints open wider. JRCP joints always use dowels. Furthermore, even though the slabs are longer, the cracks still form at the same interval as JPCP, and therefore JRCP slabs generally have one or two interior cracks each. The light steel reinforcement across these cracks is generally not enough to maintain load transfer, and therefore the cracks fault as well as the joints. As a result, the latest proposed AASHTO M-EPDG procedure does not have provisions for JRCP. A 1999 survey by the American Concrete Pavement Association identified only nine US states as constructing JRCP, and several of those are states that have small highway networks (ACPA 1999c).

Continuously reinforced concrete pavement

Continuously reinforced concrete pavement, or CRCP, is characterized by heavy steel reinforcement and an absence of joints. Much more steel is used for CRCP than for JRCP, typically on the order of 0.4–0.8 percent by volume in the longitudinal direction. Steel in the transverse direction is provided in a lower percentage as temperature steel. CRCP is illustrated in Figure 2.3.

Cracks form in CRCP approximately 0.6–2 m (2–6 ft) apart. The reinforcement holds the cracks tightly together and provides for aggregate interlock and shear transfer. CRC pavements require anchors at the beginning and end of the pavement to keep the ends from contracting due to shrinkage, and to help the desired crack pattern develop. Special CRCP

Plan

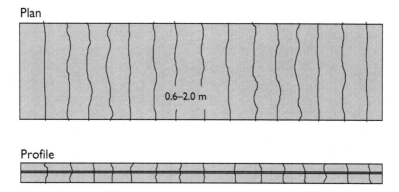

0.6–2.0 m

Profile

Figure 2.3 Continuously reinforced concrete pavement (CRCP) (courtesy: ACPA).

design considerations, including determination of the proper reinforcement percentage, are discussed in Chapter 12.

Use of CRCP dates back to the 1921 Columbia Pike experimental road in Virginia, with over 22,500 km (14,000 miles) of two-lane highway built in the United States by 1982. CRCP has also been used for major airports around the world (CRSI 1983: 1).

Because of the steel reinforcement, CRCP costs more than JRCP, and is thus used less frequently in most regions. However, it provides a smoother ride and a longer life than any other type of pavement, and is therefore a preferred type in Texas and Illinois. A 1999 survey by the American Concrete Pavement Association identified only eight US states as constructing CRCP (ACPA 1999c).

In many regions, the performance of CRCP has been excellent. A 2000 study of concrete pavement performance in the Southeastern US used the LTPP database to investigate 14 CRCP sections in the states of Alabama, Florida, Georgia, Mississippi, North Carolina, and South Carolina. At the time of the study the sections were 21–30 years old and had carried heavy traffic, but were generally in very good to excellent condition. With three exceptions, the pavements had serviceability indices of 4 or better, despite the fact that they had already exceeded their 20-year design lives (Delatte et al. 2000: 4-3–4-6).

Key performance considerations for CRCP include:

- initial pavement smoothness;
- adequate pavement thickness to prevent excessive transverse cracking; and
- adequate reinforcing steel to hold cracks together and prevent punchouts. Punchouts are a distress mechanism distinct to CRCP and are discussed in Chapter 3.

Conventional pavement joints

Conventional pavements (JPCP, JRCP, and CRCP) make use of several types of transverse and longitudinal joints. Transverse contraction joints are used in JPCP and JRCP, usually with dowels. At the end of each daily paving operation, or for a significant delay in paving, transverse construction joints are placed, generally at the location of a planned contraction joint for JPCP or JRCP. Transverse expansion or isolation joints are placed where expansion of the pavement would damage adjacent bridges or other drainage structures. Longitudinal contraction joints are created where two or more lane widths or shoulders are paved at the same time. In contrast, longitudinal construction joints are used between lanes or shoulders paved at different times (ACPA 1996a: VII-9–VII-10).

> The performance of concrete pavements depends to a large extent upon the satisfactory performance of the joints. Most jointed concrete pavement failures can be attributed to failures at the joint, as opposed to inadequate structural capacity. Distresses that may result from joint failure include faulting, pumping, spalling, corner breaks, blowups, and mid-panel cracking. Characteristics that contribute to satisfactory joint performance, such as adequate load transfer and proper concrete consolidation, have been identified through research and field experience. The incorporation of these characteristics into the design, construction, and maintenance of concrete pavements should result in joints capable of performing satisfactorily over the life of the pavement. Regardless of the joint sealant material used, periodic resealing will be required to ensure satisfactory joint performance throughout the life of the pavement. Satisfactory joint performance also depends on appropriate pavement design standards, quality construction materials, and good construction and maintenance procedures.
>
> (FHWA 1990a)

Transverse contraction joints

A transverse construction joint is defined as "a sawed, formed, or tooled groove in a concrete slab that creates a weakened vertical plane. It regulates the location of the cracking caused by dimensional changes in the slab, and is by far the most common type of joint in concrete pavements" (FHWA 1990a).

Lightly loaded jointed pavement contraction joints may rely only on aggregate interlock across joints. More heavily loaded pavements almost always use load transfer dowels in the joints. Dowels prevent vertical movement or faulting between slabs, but allow the joint to open and close to relieve stress buildup due to moisture and temperature changes in the concrete pavement. A dowel basket assembly with corrosion resistant

Figure 2.4 Dowel basket assembly with corrosion resistant epoxy-coated dowels (photo by author).

epoxy-coated dowels is shown in Figure 2.4. Dowel baskets are also shown in Figure 1.1.

For jointed highways, transverse joints (usually doweled) are used perpendicular to traffic, and longitudinal joints are used parallel to traffic, between traffic lanes. Airport pavements are much wider, and jointed airport pavements often use square or nearly square panels with dowels around all four sides. Figure 2.5 shows a cross-section view of a doweled transverse joint.

> The primary purpose of transverse contraction joints is to control the cracking that results from the tensile and bending stresses in concrete slabs caused by the cement hydration process, traffic loadings, and the environment. Because these joints are so numerous, their performance significantly impacts pavement performance. A distressed joint typically exhibits faulting and/or spalling. Poor joint performance frequently leads to further distresses such as corner breaks, blowups, and mid-panel cracks. Such cracks may themselves begin to function as joints and develop similar distresses. The performance of transverse contraction joints is related to three major factors: joint spacing, load transfer across the joint, and joint shape and sealant properties.
>
> (FHWA 1990a)

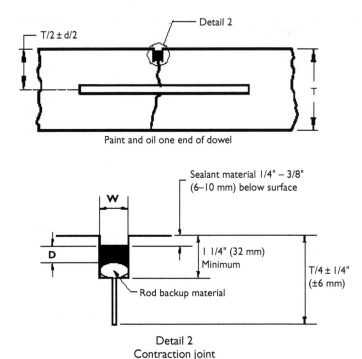

Paint and oil one end of dowel

Detail 2
Contraction joint

Figure 2.5 Doweled joint (FAA 2004: 86, 86–1).

Pavement distresses, including joint distresses, are discussed in Chapter 3. Joint spacing and load transfer are discussed in Chapter 7, and joint shape and sealants are discussed in Chapter 15.

Longitudinal joints

A longitudinal joint is defined as

> a joint between two slabs which allows slab warping without appreciable separation or cracking of the slabs . . . Longitudinal joints are used to relieve warping stresses and are generally needed when slab widths exceed [4.6 m] 15 feet. Widths up to and including [4.6 m] 15 feet have performed satisfactorily without a longitudinal joint, although there is the possibility of some longitudinal cracking. Longitudinal joints should coincide with pavement lane lines whenever possible, to improve traffic operations. The paint stripe on widened lanes should be at [3.7 m] 12 feet and the use of a rumble strip on the widened section

is recommended... Load transfer at longitudinal joints is achieved through aggregate interlock.

(FHWA 1990a)

To aid load transfer, tie bars are often used across longitudinal joints. Tie bars are thinner than dowels, and use deformed reinforcing bars rather than smooth dowel bars. Design of longitudinal joints and tie bars is discussed in Chapter 7. Tie bars in basket assemblies between lanes are shown in Figures 1.1 and 2.6, along with dowel baskets. The tie bars are longer and thinner than the dowel bars, and are also epoxy coated for corrosion protection. Figure 2.7 shows a cross-section view of a longitudinal joint with tie bars.

Longitudinal joints may be sawn or built as construction joints. If they are sawn, basket assemblies as shown in Figures 1.1 and 2.6 are used and the joint is sawn in a manner similar to transverse contraction joints. If built as construction joints, tie bars are used to connect the new and old concrete.

For crowned pavements, it is important to provide a properly detailed longitudinal joint at the crown. Otherwise, a crack is almost certain to form there (Rollings 2001, 2005).

Figure 2.6 Tie bar basket assemblies with corrosion resistant epoxy-coated tie bars – dowel baskets are also shown (photo by author).

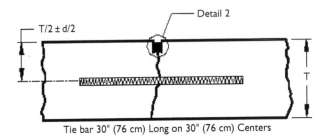

Figure 2.7 Longitudinal joint (FAA 2004).

Construction joints

A construction joint is defined as "a joint between slabs that results when concrete is placed at different times. This type of joint can be further broken down into transverse and longitudinal joints" (FHWA 1990a). A header and dowel basket for a transverse construction joint are shown in Figure 2.8.

After paving up to the header, the header will be removed. The next paving day will start with the new concrete butted up against the old concrete.

Figure 2.8 Header and dowel basket for a transverse construction joint (photo by author).

Transverse construction joints should normally replace a planned contraction joint. However, they should not be skewed, as satisfactory concrete placement and consolidation are difficult to obtain. Transverse construction joints should be doweled... and butted, as opposed to keyed. Keyed transverse joints tend to spall and are not recommended. It is recommended that transverse construction joints be sawed and sealed. The reservoir dimensions should be the same as those used for the transverse contraction joints.

(FHWA 1990a)

Longitudinal construction joints with tie bars are discussed below.

It is essential that the tiebars be firmly anchored in the concrete. Tiebars should be either mechanically inserted into the plastic concrete or installed as a two-part threaded tiebar and splice coupler system. It is recommended that periodic pullout tests be conducted to ensure the tiebars are securely anchored in the concrete... Bending of tiebars is not encouraged. Where bending of the tiebars would be necessary, it is recommended that a two-part threaded tiebar and splice coupler system be used in lieu of tiebars. If tiebars must be bent and later straightened during construction, Grade 40 [less than 276 MPa or 40 ksi yield strength] steel should be used, as it better tolerates the bending. It may be necessary to reapply a corrosion-resistant coating to the tiebars after they have been straightened. When pullout tests are performed, they should be conducted after the tiebars have been straightened. It is recommended that longitudinal construction joints be sawed and sealed. The reservoir dimensions should be the same as those used for the longitudinal joints.

(FHWA 1990a)

Keyed longitudinal joints have been used in the past but are now less common. These joints use a sort of tongue and groove configuration where one slab has a formed slot and the adjacent slab has a key that fits in the slot to transfer shear. "The decision to use keyed longitudinal construction joints should be given careful consideration. The top of the slab above the keyway frequently fails in shear. For this reason, it is recommended that keyways not be used when the pavement thickness is less than [250 mm] 10 inches. In these cases, the tiebars should be designed to carry the load transfer" (FHWA 1990a).

Expansion joints

An expansion joint is defined as "a joint placed at a specific location to allow the pavement to expand without damaging adjacent structures or the pavement itself" (FHWA 1990a). These are generally needed at bridge abutments and at embedded utility structures in streets. For example, Figure 2.4 shows a utility structure blockout adjacent to a dowel basket. The embedded utility structure will require expansion joints.

Early pavement designs used transverse expansion joints as well as contraction joints, but performance was poor. One early design combined expansion joints at 28 m (90 ft) intervals supplemented with contraction joints at 9 m (30 ft) intervals in between. The expansion joints closed and allowed the contraction joints to open too widely (Ray 1981: 6).

> Good design and maintenance of contraction joints have virtually eliminated the need for expansion joints, except at fixed objects such as structures. When expansion joints are used, the pavement moves to close the unrestrained expansion joint over a period of a few years. As this happens, several of the adjoining contraction joints may open, effectively destroying their seals and aggregate interlock. The width of an expansion joint is typically [19 mm] 3/4 inch or more. Filler material is commonly placed [19 to 25 mm] 3/4 to 1 inch below the slab surface to allow space for sealing material. Smooth dowels are the most widely used method of transferring load across expansion joints. Expansion joint dowels are specially fabricated with a cap on one end of each dowel that creates a void in the slab to accommodate the dowel as the adjacent slab closes the expansion joint.
>
> (FHWA 1990a)

This detail is shown in Figure 2.9.

At bridges, expansion joints are very important, because an expanding pavement can build up considerable force and damage bridge superstructures and abutments. Expansion joints at bridges are discussed in Chapter 12.

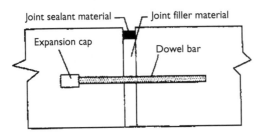

Figure 2.9 Expansion joint detail (FHWA 1990a).

Pressure relief joints are intended to serve the same purpose as expansion joints, except that they are installed after initial construction to relieve pressure against structures and to alleviate potential pavement blowups. Pressure relief joints are not recommended for routine installations. However, they may be appropriate to relieve imminent structure damage or under conditions where excessive compressive stresses exist.

(FHWA 1990a)

Overlays

Concrete pavements may also be used as overlays for either existing asphalt or concrete pavements. For each of the two existing pavement types, there are two overlay classifications based on whether the overlay is bonded to the existing pavement, or whether the bond is either ignored or prevented, and thus not considered in the design. A complete discussion of concrete overlays is provided by Smith et al. (2002) and ACI Committee report *ACI 325.13R-06, Concrete Overlays for Pavement Rehabilitation* (ACI Committee 325 2006). Overlays are discussed in detail in Chapter 18.

The oldest type of concrete overlay over an existing asphalt pavement is termed whitetopping. Generally, no special measures are taken to either achieve or prevent bond between the new concrete and the old asphalt. Many of these were constructed in the United States when existing asphalt highways were added to the Interstate Highway system and upgraded. For all practical purposes, these are designed and built as conventional concrete pavements, using the existing asphalt as a high quality base. Special details may be needed when the new concrete must be wider than the existing asphalt. Traditionally, whitetopping overlay designs do not consider any reduction in flexural stress in concrete due to bond.

However, for some time it has been recognized that the concrete bonds to asphalt, and thinner overlays have been developed to take advantage of that fact. Ultra-thin whitetopping (UTW) was developed in the United States in the early 1990s, with overlays between 50 and 100 mm (2 and 4 in) in thickness. In order to reduce curling stresses in such thin pavements, they are typically cut into squares 0.6–2 m (2–6 ft) on a side. Conventional whitetopping overlays are typically at least 200 mm (8 in) thick.

Subsequently, thin whitetopping overlays have been developed to fill the gap between light traffic UTW and conventional whitetopping. These are generally 100–200 mm (4–8 in) thick, with 1.2–2 m (4–6 ft) squares. As with UTW, thin whitetopping overlays rely on bond to reduce the flexural stress in the concrete. A complete discussion of thin and ultrathin whitetopping is provided by Rasmussen and Rozycki (2004).

Unbonded concrete overlays are constructed over existing concrete pavements with specific measures taken to prevent bonding between the two layers. Most commonly, the bond breaker is a thin layer of hot-mix asphalt. The reason for the bond breaker is to keep cracks and other damage in the existing pavement from propagating up through the new pavement – this propagation is termed "reflective cracking." With the bond breaker layer, the existing pavement functions as a high quality base of support for the new pavement, so there are similarities to conventional whitetopping.

Bonded concrete overlays have also been constructed. For these overlays, the existing concrete pavement is carefully prepared to facilitate bond. These measures are discussed in detail by Delatte et al. (1996a,b). These can be as thin as 50 mm (2 in) because of the composite action with the existing pavement. However, because any damage in the base concrete pavement will reflect through the overlay, bonded concrete overlays are limited to existing pavement in good condition. This is probably the main reason that these overlays remain rare – they are only suitable for pavements in relatively good condition, and agencies are reluctant to invest in these pavements and tend to allocate funds to pavements in worse condition. The other reason they are rare is that if bond is not achieved, the thin overlays will fail rapidly.

A third type of concrete overlay on concrete pavement, the partially bonded overlay, has been used in the past. For these overlays, no special measures are taken to either prevent or achieve bond – the new concrete is simply placed on top of the old. Some of these overlays were US Army Corps of Engineers airfields, with thick overlays placed over thin existing pavements. As a result, the contribution of the existing pavement may have been very small. There seems to be very little recent construction of partially bonded concrete overlays. In theory, at least, partially bonded overlays would seem to be susceptible to reflective cracking.

Prestressed and precast concrete pavement

All conventional concrete pavements rely on the flexural strength of the concrete to resist traffic loads over time. By using prestressing tendons to induce a net compressive force in the pavement section it is possible to considerably decrease the thickness of the pavement, because the traffic loads must overcome the compressive stress before inducing a net tensile stress and flexural fatigue into the pavement.

In addition to prestressed pavements for original construction or overlays, precast concrete sections with either conventional or prestressed reinforcement have been used as full depth patches. Precast sections may be left in place as a permanent pavement, or may be temporary to allow traffic until a permanent full depth patch is placed.

Prestressed pavement

Pasko (1998) stated

> prestressed concrete was introduced in the late 1940s and was used first in airport pavements. About 1959, two-way prestressed slabs were used

at Biggs military airfield in Texas. The 24-in (610-mm) plain pavement was replaced with 9-in (230-mm) post-tensioned slabs. Unfortunately, the fear of the unknown, the need to use more skilled labor, and the reluctance of mile-a-day slipform contractors to embrace this unproven technology have held this concrete-saving technology back. About a dozen highways with prestressed concrete pavements of various designs were built in the United States between 1970 and 1990.

The first known prestressed test highway pavement was a short section in Delaware in 1971. Other demonstration projects were at Dulles Airport, along with two in Pennsylvania (Huang 2004: 16).

Another demonstration project, an unbonded overlay at Chicago O'Hare International Airport, on Runway 8R-27L, was documented in the FAA Engineering Brief No. 24. The base pavement was 305 mm (12 in) thick CRCP on a 457 mm (18 in) thick crushed limestone base. The prestressed overlay was 203 mm (8 in) thick on the western half and 228 mm (9 in) thick on the eastern half, based on Boeing 747 aircraft. The dimensions of the overlay were 244 m (800 ft) by 46 m (150 ft). The pavement was prestressed in both the longitudinal and transverse directions (FAA 1981).

The 1993 AASHTO Pavement Design Guide discusses the design of prestressed pavements. Prestressed pavement's potential advantages include more efficient use of construction materials (due to reduced pavement thickness) and fewer joints and cracks, with reduced maintenance and longer pavement life. Items addressed include:

- The pavement may be prestressed only longitudinally (with the pavement unreinforced or conventionally reinforced transversely), prestressed both longitudinally and transversely, or prestressed at an angle.
- Although prestressed pavement should work with low strength support, fairly high strength subbases of 54 MPa/m (200 psi/in) or more have generally been used.
- Much longer slabs may be used than conventional pavements – generally on the order of 122 m (400 ft) although slabs as long as 300 m (1,000 ft) have been built in Europe.
- Prestress levels are typically 689–2,070 kPa (100–300 psi) in the longitudinal direction and 0–1,380 kPa (0–200 psi) in the transverse direction.
- Typical tendons are 15 mm (0.6 in) strand stressed to 80 percent of yield, spaced at two to four times slab thickness longitudinally and three to six times slab thickness transversely.
- Because little is known about fatigue of prestressed pavements, and because little warning may precede failure, designers should use conservative fatigue safety factors.

- Prestressed pavement thickness is on the order of 40–50 percent of conventional concrete pavement thickness, or about 100–150 mm (4–6 in) for highways.
- Subgrade friction restraint is an important factor, and friction reducing layers such as sand over building paper or polyethylene sheeting are often used.
- Prestress losses on the order of 15–20 percent should be considered in the design (AASHTO 1993: II-65–II-67).

The FAA notes that

> Prestressed concrete pavements have been used in airport applications in Europe and to a limited extent in the United States. Prestressed concrete airport pavements are usually post-tensioned with high strength steel strands. These pavements are usually considerably thinner than plain, jointed reinforced, or continuously reinforced concrete pavements yet provide high load carrying capacity. Slab lengths on the order of 400–500 ft (120–150 m) are generally used.
>
> (FAA 2004: 102)

Some of these projects had prestressing strand in only one direction, and tended to have cracks form parallel to the strand due to lack of prestress in the transverse direction. A 150 mm (6 in) pavement with prestressing in both directions, 1.6 km (1 mile) long, was built on I-35 in Texas and was still in excellent condition after 17 years (Merritt et al. 2002: 5).

Recently there has been renewed interest in precast and prestressed pavement systems. Tyson and Merritt (2005) described the following projects:

- A Texas Department of Transportation project, completed in 2002, near Georgetown, TX, to demonstrate the viability of constructing a pavement using precast prestressed concrete. TxDOT placed approximately 700 meters (2,300 feet) of two-lane pavement (plus shoulders) on the frontage road along Interstate 35.
- The California Department of Transportation (Caltrans) built a demonstration project in April 2004. This was built at night on an interstate with a low tolerance for lane closure. The agency placed approximately 76 m (250 ft) of two-lane (plus shoulder) pavement on Interstate 10 in El Monte, CA. This project is discussed in detail in Merritt et al. (2005).

At the time the article was written in 2005, additional demonstration projects were planned for Missouri and Indiana.

The system described by Tyson and Merritt (2005) consists of the following features:

- Individual precast, prestressed pavement panels approximately 200 mm (8 in) thick are constructed ahead of time. These are roughly equivalent in traffic capacity to conventional pavement 355 mm (14 in) thick. The panels are cast at the full pavement width.
- Three types of panels are used – joint, central stressing, and base panels.

> Joint panels are located at the ends of each posttensioned section and contain dowelled expansion joints to absorb horizontal slab movements. The central stressing panels are placed at the middle of each posttensioned section and contain large pockets or blockouts where the posttensioning strands are fed into the ducts and stressed. The base panels make up the majority of the pavement, placed between the joint panels and stressing panels. During a single construction operation, workers must place at least one complete section of panels from joint panel to joint panel.
>
> (Tyson and Merritt 2005)

- Panels were placed over a hot-mix asphalt leveling course.
- Shear keys along the edges of the panels facilitate vertical alignment during construction and ensure satisfactory ride quality by preventing joint faulting.

Precast pavement panels

Precast concrete panels have been used in two ways for concrete pavement rehabilitation. These are:

- temporary replacement for removed panels until concrete can be placed during the next scheduled closure; and
- permanent pavement (selective panel replacement).

Potential advantages of precast concrete panels include higher quality concrete, better curing, less risk of weather disruption, and reduced delay before opening to traffic (Lane and Kazmierowski 2005: 771). Important issues include leveling the panels to avoid bumps at panel edges, and load transfer between precast panels or between precast panels and existing pavement. Precast panels are generally reinforced with mild steel, primarily to prevent damage during transportation and handling.

Precast panels were used for temporary closure during runway intersection reconstruction at Charleston (SC) International Airport in 1990 and

Savannah/Hilton Head International Airport in 1996. Both pavement intersections were reconstructed during multiple eight-hour overnight closures, using concrete with proprietary high early strength cement. During each closure, two 7.6 by 7.6 m (25 by 25 ft) sections were excavated and two were filled with concrete. Precast panels approximately 3.8 m (12½ ft) square (four per section) or 3.8 m by 2.4 m (12½ by 8 ft) (six per section) were cast before construction, to fulfill several functions:

- Producing the precast panels allowed the contractor to gain experience with and verify the properties and workability of the high early strength concrete.
- Panels could be used to open the runway to traffic in the event of an emergency.
- Panels were used to speed production – following each overnight closure, one section was filled with precast panels. On the next closure, the panels were removed and the new concrete was placed as two more panels were removed.

The contractor for the Savannah project was able to observe the Charleston project and made improvements to the process (Drinkard 1991, McGovern 1998, Peshkin et al. 2006b).

Precast panels were evaluated on 12-lane Highway 427 in Toronto, Canada. This is a major north–south commuter route through Toronto. All replacement slabs were reinforced with epoxy-coated reinforcement. Three different methods were tested, including two patented methods. All three methods performed well, despite a relatively inexperienced contractor (Lane and Kazmierowski 2005).

In October and November 2003, 157 distressed slabs were rehabilitated with precast panels at 18 locations along I-25 north of Denver, Colorado. The slabs were stabilized and slab jacked to match the elevation of the existing pavement using high density polyurethane foam, and the joints were backfilled with a joint bonding material. Fiber glass thin tie bars were used instead of dowels across panel joints, but some panels cracked next to the tie bars. Productivity increased substantially as the workers became more familiar with the technology. It proved to be necessary to diamond grind most of the panels to produce an acceptable ride (Buch et al. 2005). Overall, precast slabs were found to be a satisfactory way to accommodate short lane closures, in the same manner as the Charleston and Savannah airport projects.

An innovated compression joint system has been developed in Japan and used to construct 800 square meters (8,600 square feet) of taxiway pavement at the Sendai Airport. The individual slabs are 14.5 m long by 2.5 m wide by 240 mm thick (47.6 ft by 8.2 ft by 9.4 in) and are prestressed. The joints are connected with prestressing tendons which are tensioned to provide load transfer. The existing pavement was badly distressed asphalt. All of the work was completed during a 9 pm–7 am overnight closure (Hachiya et al. 2001).

Roller compacted concrete

Roller compacted concrete (RCC) pavements are described in detail by ACI Committee 325 in report ACI 325.10R (ACI 1995). RCC is a very dry mixture that can be produced as ready-mixed concrete but is often produced in a pugmill. RCC is also extensively used for dam construction and rehabilitation, although the mixtures and construction methods are different. It is a low to no slump mixture that is closer in some respects to cement treated aggregate base than a conventional, flowing concrete.

The construction process resembles that of hot-mix asphalt pavements. The material is delivered by dump trucks, placed into an asphalt paver, and then rolled with steel wheel rollers. The RCC is then cured. The pavement may be allowed to crack naturally, or joints may be cut. Because RCC shrinks less than conventional concrete, the joints or cracks are further apart than those for JPCP.

Traditionally, the construction process has left RCC pavements with rough surfaces which are not suitable for high speed traffic. Unsurfaced RCC pavements have many industrial applications – multimodal freight transfer facilities, automobile plants, and military facilities. Recent large scale industrial projects have included Honda and Mercedes facilities in Alabama (Delatte et al. 2003).

RCC pavements have also been constructed with hot-mix asphalt riding surfaces. These have been widely used for new subdivision construction in Columbus, Ohio, and for streets and roads in Quebec province, Canada. Figure 2.10 illustrates RCC pavement construction in Columbus for a new subdivision.

Some small state demonstration projects have been conducted in South Carolina and Tennessee. The state of Georgia also used RCC to replace deteriorated asphalt shoulders on Interstate 285 around Atlanta (Bacon 2005). Information on these and other projects has been made available by the Southeast Cement Promotion Association on the web at http://www.rccpavement.info/.

RCC pavements have proven to be very economical to construct. Compared to JPCP pavements, RCC pavements do not require forms, dowels or tie bars, or labor for texturing and finishing, so construction costs are lower. Maintenance costs also tend to be lower than other pavement types. The flexural strength of RCC is typically equal to or better than that of conventional concrete. A recent review of RCC in service documented

Figure 2.10 RCC pavement construction – Columbus, Ohio (photo by author).

excellent long-term performance (Piggott 1999). For high speed applications, the use of bonded concrete overlays on RCC has been investigated, and a trial pavement has been built at an industrial facility in Michigan (Kreuer 2006).

Pervious/porous concrete

All of the concrete pavement types discussed above provide impervious cover. Except for small amounts of infiltration through joints and cracks, precipitation runs off of the pavement surface and must be handled by a separate drainage system.

However, in many regions developers are limited in the amount of impervious cover allowed on a site. This has led to increasing interest in pervious or porous pavements – pavements that allow rapid water flow into and through the pavement structure. Different types of porous pavements, including pervious concrete, are described by Ferguson (2005). Figure 2.11 shows the permeability of a pervious concrete pavement demonstration project constructed in Cleveland, Ohio, in August 2005.

Pervious concrete generally has much lower strength than conventional concrete. Niethelath et al. (2005) discussed the potential benefits of pervious

Figure 2.11 Permeability of a pervious pavement demonstration project – Cleveland, Ohio (photo by author).

concrete for reducing pavement noise, but did not address the structural aspects of producing a durable mainline highway pavement. The solution would most likely be a conventional concrete pavement structure with a pervious pavement overlay.

Procedures for design and construction of pervious concrete pavements have recently been published by ACI Committee 522 (ACI Committee 522 2006). Information about pervious concrete has been made available by the Southeast Cement Promotion Association on the web at http://www.pervious.info/.

Because pervious concrete has a much lower flexural strength than conventional concrete, it has been most widely used for parking lots and light-traffic streets and roads. For these applications, a pavement thickness of 150 mm (6 in) has generally been found to be adequate. In the future, however, the use of pervious concrete pavement will probably be expanded to heavier traffic. Pervious pavement parking lots may be used as part of building construction projects to accrue LEED points (RMC Research Foundation 2005: 6–13).

Chapter 3

Performance

Concrete pavement engineering is the selection of design, materials, and construction practices to ensure satisfactory performance over the projected life of the pavement. Pavement users are sensitive to the functional performance of pavements – smoothness and skid resistance – rather than structural performance. Pavements, as a general rule, develop distresses gradually over time under traffic loading and environmental effects. An exception is when poor material choices or construction practices cause defects before or shortly after the pavement is put into service.

As part of the LTPP Program, the FHWA has developed a Distress Identification Manual (Miller and Bellinger 2003). This manual provides descriptions and photographs to identify the different distress types and classify their severity.

This manual lists the following distress types for JPCP and JRCP:

- Cracking – divided into corner breaks, durability ("D") cracking, longitudinal cracking, and transverse cracking.
- Joint deficiencies – joint seal damage (transverse or longitudinal), and joint spalling (transverse or longitudinal).
- Surface defects – divided into map cracking, scaling, polished aggregates, and popouts.
- Miscellaneous distresses – classified as blowups, faulting of transverse joints and cracks, lane-to-shoulder dropoff, lane-to-shoulder separation, patch deterioration, and water bleeding and pumping.

This manual lists the following distress types for CRCP:

- Cracking – as above, except CRCP cannot have corner breaks.
- Surface defects – as above.
- Miscellaneous distresses – as above, with the addition of punchouts, transverse construction joint deterioration, and longitudinal joint seal damage. Also, CRCP does not have joints, so joint faulting does not occur.

Another manual for concrete condition assessment is provided by the ACI Committee 201 report 201.1R-92 *Guide for Making a Condition Survey of Concrete in Service* (ACI Committee 201 1992).

Cracking

Cracks may form in concrete pavements due to a one time overload or due to repeated fatigue loading. The exception is tight, closely spaced transverse cracks formed intentionally in CRCP.

Corner breaks

Corner breaks only occur at corners of JPCP or JRCP. A triangular piece of concrete, from 0.3 m (1 ft) to half the width of the slab, breaks off (Miller and Bellinger 2003: 36). These are more likely with longer slabs, because as the slabs warp or curl upward the slab corners may become unsupported and break off when heavy vehicles travel across them. Huang (2004: 378) notes that "load repetitions combined with loss of support, poor load transfer across the joint, and thermal curling and moisture warping stresses usually cause corner breaks." Corner breaks may be avoided by limiting slab lengths, particularly with stiffer subbases, and by providing load transfer to adjacent slabs through dowels and tie bars. A moderate severity corner break is shown in Figure 3.1.

Figure 3.1 Corner breaks (ACPA 1996a: VIII-16).

Durability ("D") cracking

Durability or "D" cracking occurs near joints, cracks, and free edges, and is manifested as a "closely spaced crescent-shaped hairline cracking pattern" (Miller and Bellinger 2003: 37). The cracks are often darker than the remaining uncracked concrete. Yoder and Witczak (1975: 636) note that the phenomenon is regional and due to use of non-durable materials and/or severe climatic conditions. It is a progressive failure mechanism than may eventually result in nearly total disintegration of the slab.

According to Huang (2004: 387), it is caused by freeze-thaw expansion of some types of coarse aggregate. Mindess et al. (2003: 503) note that the problem occurs frequently with limestones in midwestern states in the United States. There is also a critical size for many aggregates below which they are not susceptible to fracture (Mindess et al. 2003: 502). Therefore, D-cracking may be prevented by avoiding susceptible aggregates or, if only susceptible aggregates are available, by using smaller maximum coarse aggregate size. For some poorly consolidated sedimentary rocks, this critical dimension may be as small as 12–25 mm (1/2–1 in) (Mindess et al. 2003: 140). High severity D-cracking is shown in Figure 3.2.

Longitudinal cracking

Longitudinal cracks are defined as those parallel to the pavement centerline (Miller and Bellinger 2003: 38). Huang (2004: 380) suggests that longitudinal cracks are caused by a combination of heavy load repetitions, loss of foundation support, and curling and warping stresses, or by improper construction of longitudinal joints. If longitudinal cracks are not in vehicle wheel paths and do not fault appreciably, the effect on pavement performance may not be significant. Longitudinal cracks are also likely to occur at the crowns of crowned pavements if longitudinal joints are not provided (Rollings 2001, 2005). A high severity longitudinal crack is shown in Figure 3.3.

Transverse cracking

Transverse cracks are defined as those perpendicular to the pavement centerline (Miller and Bellinger 2003: 40). They are a key JPCP concrete pavement performance measure, because once a transverse crack forms its faulting and deterioration leads to severe roughness. JPCP does not have steel across the crack to hold it together. The cracking can progress and lead to a shattered slab, requiring slab replacement (Hoerner et al. 2001: 70). Huang (2004: 384) notes that transverse cracks are "usually caused by a combination of heavy load repetitions and stresses due to temperature gradient, moisture gradient, and drying shrinkage." A high severity transverse crack in JRCP is shown in Figure 3.4.

Figure 3.2 High severity D-cracking (Miller and Bellinger 2003: 37).

Figure 3.3 High severity longitudinal crack (Miller and Bellinger 2003: 39).

Figure 3.4 High severity transverse crack in JRCP (Miller and Bellinger 2003: 41).

Yu et al. (1998) found that transverse cracking risk is higher in drier climates with longer joint spacings, due to the higher temperature gradients that develop in these climates. Overall, increased joint spacing greatly increases transverse cracking. Determining proper joint spacing is discussed in detail in Chapter 7.

In JRCP and CRCP, transverse cracks are expected to form. The key difference is how the cracks perform. With CRCP, there is enough steel to keep cracks tightly closed unless they progress to punchouts. With JRCP, there is not enough steel to prevent crack deterioration and faulting.

Joint deficiencies

As only JPCP and JRCP have joints, these pavement types alone can have joint deficiencies. These are classified as seal damage or spalling.

Joint seal damage (transverse or longitudinal)

Joint seals are used to keep incompressible materials and water from penetrating joints. Incompressible materials can lead to stress concentrations when open pavement joints close, causing some of the concrete to spall off. Water leads to deterioration in the pavement and underlying layers.

Typical types of joint seal damage include extrusion (seal coming up out of joint), hardening, adhesive failure (loss of bond), cohesive failure (splitting), complete loss of sealant, intrusion of foreign material, or weed growth in the joint (Miller and Bellinger 2003: 44). Joints must be periodically cleaned out and resealed, and this type of damage usually indicates a need to maintain the joints.

Transverse joint seal damage is only an issue with JPCP or JRCP. Longitudinal joint seal damage may occur with any type of conventional concrete pavement.

Joint spalling (transverse or longitudinal)

Joint spalling is defined as "cracking, breaking, chipping, or fraying of slab edges within 0.3 m (1 foot) from the face of the joint" (Miller and Bellinger 2003: 45–46). Spalls are a surface phenomenon and are generally caused by incompressible materials creating stress concentrations in joints as they close due to slab expansion or traffic loading. They may also be caused by "poorly designed or constructed load transfer devices" (Huang 2004: 382). Therefore, the best way to avoid spalls is to properly maintain joints. Spalls may also be caused by poor construction practices, such as failing to properly cure pavement joints after saw cutting. High severity transverse joint spalling is shown in Figure 3.5.

Figure 3.5 High severity transverse joint spalling (ACPA 1996a: VIII-47).

Surface defects

Unlike cracking and joint deficiencies, surface defects are usually unrelated to design. They are due to either poor materials selection or poor construction practices, or both.

Map cracking

Map cracking is defined as "a series of cracks that extend only into the upper surface of the slab. Larger cracks frequently are oriented in the longitudinal direction of the pavement and are interconnected by finer transverse or random cracks" (Miller and Bellinger 2003: 47). According to Huang (2004: 387), it is usually caused by overfinishing of concrete. Mindess et al. (2003: 507) note that map cracking can either be caused by excessive bleeding and plastic shrinkage from finishing too much or too early, which leads to fine cracks, or by ASR, which leads to coarse cracks. Map cracking is shown in Figure 3.6.

It is useful to distinguish between map cracking due to finishing problems, which is unlikely to progress further, and map cracking due to ASR, which is likely to progress and lead to eventual destruction of the pavement. ASR is an increasingly important problem for concrete pavements, and is difficult to fix.

ASR is one of two types of alkali-aggregate reactions in concrete, the other being alkali-carbonate reaction. ASR is due to an expansive reaction

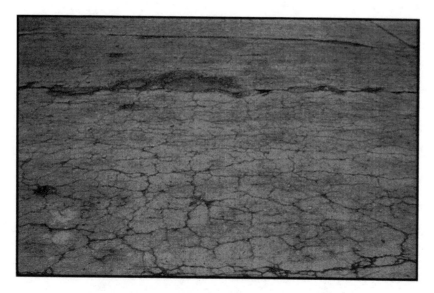

Figure 3.6 Map cracking (Miller and Bellinger 2003: 48).

between alkalis in cement paste and certain reactive forms of silica within aggregate. The reactive aggregate components include opal, silica glass, chalcedony, cristobalite tridymite, and quartz, which can occur in siliceous and volcanic rocks and quartzite. Problems were observed in the United States in structures built between the late 1920s and early 1940s, and problems have been noted primarily in western and southwestern parts of the country. Factors that affect ASR include nature, amount, and particle size of the reactive silica, amount of available alkali, and the presence of moisture. Since concrete pavements cannot be kept free of moisture, the solution is to either use low alkali cements or non-reactive aggregates. In some regions, however, non-reactive aggregates may not be economically available. Fly ash, GGBFS, and silica fume have been found to reduce the effects of ASR. Class F fly ash typically provides adequate protection at 15–20 percent replacement of cement, but class C fly ash is usually less effective (Mindess et al. 2003: 142–152).

Similar reactions include a form of ASR in sand-gravel aggregates from some river systems in the states of Kansas, Nebraska, Iowa, Missouri, and Wyoming, and alkali-carbonate reactions. Alkali-carbonate reactions involve carbonate rocks in some midwestern and eastern states as well as eastern Canada (Mindess et al. 2003: 153–154). ASR and other alkali-aggregate reactions are discussed in detail in *ACI 221.1R-98, Report on Alkali-Aggregate Reactivity* (ACI Committee 221 1998).

Scaling

Scaling is defined as "the deterioration of the upper concrete slab surface, normally 3–13 mm (1/8-1/2 inch), and may occur anywhere over the pavement" (Miller and Bellinger 2003: 47). Scaling may progress from map cracking (Huang 2004: 387). Scaling is shown in Figure 3.7.

Scaling may also occur with repeated application of deicing salts. This type of scaling may be prevented by using an adequately air entrained low permeability concrete with a low water/cement (w/c) ratio. Risk of scaling is higher on concrete surfaces that have not been finished properly (Mindess et al. 2003: 504–505).

Polished aggregates

Polished aggregate problems refer to "surface mortar and texturing worn away to expose coarse aggregate" (Miller and Bellinger 2003: 49). This typically leads to a reduction in surface friction. The reduction in surface friction can make pavements unsafe, particularly in wet weather. A polished aggregate surface is shown in Figure 3.8.

Because cement paste does not have good abrasion resistance, the wear resistance of concrete depends on the hardness of aggregates used. Poor finishing practices may also lead to a weak surface layer and lower abrasion resistance (Mindess et al. 2003: 470–471). Skid resistance may be restored

Figure 3.7 Scaling (Miller and Bellinger 2003: 48).

Figure 3.8 Polished aggregate surface (Miller and Bellinger 2003: 49).

by diamond grinding, but with soft aggregates the treatment may have to be repeated in a few years.

Popouts

Popouts are "small pieces of pavement broken loose from the surface, normally ranging in diameter from 25–100 mm (1–4 inches), and depth from 13–50 mm (1/2–2 inches)" (Miller and Bellinger 2003: 50). Popouts may be caused by "expansive, nondurable, or unsound aggregates or by freeze and thaw action" (Huang 2004: 387). Popouts and D-cracking are caused by similar mechanisms (Mindess et al. 2003: 503).

Miscellaneous distresses

Blowups

Blowups are "localized upward movement of the pavement surface at transverse joints or cracks, often accompanied by shattering of the concrete in that area" (Miller and Bellinger 2003: 52). Blowups are buckling failures due to compressive stresses in the concrete pavement. They may occur in hot weather if a transverse joint or crack does not allow expansion, typically when joints have been allowed to become full of incompressible material. They may also occur at utility cut patches and drainage inlets, or if D-cracking has weakened the concrete at the joint (Huang 2004: 376). A blowup is shown in Figure 3.9.

The most common way of preventing blowups is by maintaining joints so that incompressible materials are kept out, or by installing pressure relief joints (Yoder and Witczak 1975: 640). Expansion joints may be used at bridges, and isolation joints may be used at utility cut patches and drainage inlets.

Under most conditions, properly designed, constructed, and maintained transverse joints will allow enough movement to prevent blowups. This is because the slab contraction due to drying shrinkage will generally be more than enough to overcome any subsequent thermal expansion. Exceptions may occur if pavements are constructed in cold weather, with very high temperatures over the following summers. Blowups may also occur with excessively long joint spacing, as with JRCP, because the joint movement is directly proportional to the slab length.

Faulting of transverse joints and cracks

Faulting is defined as a "difference in elevation across a joint or crack" (Miller and Bellinger 2003: 53). It represents a failure of the load-transfer

Figure 3.9 Blowup (Miller and Bellinger 2003: 52).

Figure 3.10 Faulted transverse joint (photo by author).

system. Faults have a considerable effect on ride and are particularly objectionable to the traveling public. Figure 3.10 shows a faulted transverse crack. Faulting is difficult to capture in a photograph, but is readily noted by the traveling public.

Faulting is most likely for pavements with aggregate interlock joints, or at mid-slab cracks in JPCP. Generally, the approach slab (upstream in traffic) will be higher than the leave slab (Yoder and Witczak 1975: 641). As traffic moves across the joint it imposes impact loads on the leave slab, forcing loose materials from under the leave slab to under the approach slab. This phenomenon is termed pumping (Huang 2004: 380).

Dowels of adequate size are effective for preventing faulting. If dowels are too small, high bearing stresses on the concrete may cause the dowels to wear away the opening and become loose. The light steel used in JRCP may not be enough to prevent faulting of mid-slab cracks. However, there is generally enough steel in CRCP to prevent faulting of the narrow, closely spaced transverse cracks.

Other important factors that affect faulting are drainage, the quality of the subbase or subgrade materials, and the joint spacing. Pumping requires moisture. Certain unstabilized materials under the pavement are also more susceptible to pumping. If the subgrade material is poor, the use of a higher quality subbase material may help prevent pumping. Longer joint spacings lead to more joint movement, and make it more difficult to maintain load transfer.

Traditionally, aggregate interlock joints have been used for JPCP where vehicle weights or speeds are low. If aggregate interlock joints are used, faulting will be reduced if joint spacing is reduced and if larger size and/or harder coarse aggregates are used in the concrete.

Khazanovich et al. (1998: 182–184) found that JPC pavements in wet-freeze climates are more susceptible to faulting, particularly with fine-grained soil subgrades. Placing a thick granular layer beneath the subbase may improve drainage and reduce faulting, particularly for undoweled pavements. Wider slabs also reduce corner deflections and reduce faulting by about 50 percent. Increasing slab thickness alone will not solve joint faulting problems.

Lane-to-shoulder dropoff

Lane-to-shoulder dropoff is a "difference in elevation between the edge of slab and outside shoulder; [and] typically occurs when the outside shoulder settles" (Miller and Bellinger 2003: 54). Dropoff is unlikely to occur if concrete shoulders are used and are connected to the mainline pavement by tie bars. However, concrete pavements may also have asphalt, granular or soil shoulders. These shoulder types can undergo consolidation, settlement, or pumping of underlying soils. They may also have heave, due to frost action or swelling soils, as well as dropoff (Huang 2004: 371). Dropoffs and heaving represent a safety hazard for vehicles that have to go onto the shoulder, and may be corrected by maintaining or replacing the shoulders.

Lane-to-shoulder separation

Lane-to-shoulder separation is "widening of the joint between the edge of the slab and the shoulder" (Miller and Bellinger 2003: 55). It should not be considered a distress if the joint is sealed well enough to keep water out (Huang 2004: 387). Causes are similar to those of lane-to-shoulder dropoff. The most common remedy is to seal the separation as a joint.

Patch deterioration

Patch deterioration refers to "a portion, greater than 0.1 m² [11 ft²], or all of the original concrete slab that has been removed and replaced, or additional material applied to the pavement after original construction" (Miller and Bellinger 2003: 56). Patch deterioration may be caused by traffic, poor materials, or poor construction practices (Huang 2004: 387). Proper patching procedures are discussed in Chapter 16. It is usually easier and cheaper to build the pavement properly in the first place, rather than to resort to premature patching.

Water bleeding and pumping

Water bleeding and pumping is "seeping or ejection of water from beneath the pavement through cracks. In some cases, detectable by deposits of fine material left on the pavement surface, which were eroded (pumped) from the support layers and have stained the surface" (Miller and Bellinger 2003: 58). Water bleeding and pumping may occur at joints, cracks, and pavement edges. Figure 3.11 illustrates water bleeding and pumping.

Yoder and Witczak (1975: 341) provide an excellent discussion of pumping. Active pumping is most pronounced directly after rainfall, and was first noticed when pavements were built directly on plastic clay soils. Compression of the soil under traffic opens up a void space under the pavement, which collects water. This water, combined with a pumpable material such as a high plasticity clay, is then forced out under traffic loading. This process may be gradual and not noticed until years after construction. The muddy water ejected often causes telltale staining on the pavement surface.

Huang (2004: 12) notes that for pumping to take place, the material under the concrete slab must be saturated, there must be frequent heavy wheel loads, and the material under the slab must be erodible. If wheel loads are light, slow moving, or infrequent, pumping may never become a problem even with wet erodible materials. However, for highways and other high volume, high speed pavements, proper drainage should be provided. It may also be necessary to stabilize the soil or subbase under the slab.

Figure 3.11 Water bleeding and pumping (Miller and Bellinger 2003: 58).

Punchouts

Punchouts, which only occur with CRCP, are rectangular chunks of concrete broken loose and punched down below the surface of the adjacent pavement. The LTPP distress manual description for punchout is "the area enclosed by two closely spaced (usually <0.6 m [2 foot]) transverse cracks, a short longitudinal crack, and the edge of the pavement or a longitudinal joint. Also includes "Y" cracks that exhibit spalling, breakup, or faulting" (Miller and Bellinger 2003: 78). Punchouts generally occur at pavement edges. Figure 3.12 shows a high severity punchout.

Punchouts are the key structural distress for CRCP. They start with a loss of aggregate interlock at two of the closely spaced transverse cracks. The portion between the cracks then begins to flex as a cantilever beam until a fatigue crack forms. The cracks break down under repeated loading, the steel ruptures, and pieces of concrete punch down into the subbase and subgrade (Huang 2004: 384–385).

CRCP generally has an excellent performance history where punchouts have been prevented. However, if they occur, punchouts are difficult to repair properly. Full depth repair of punchouts is discussed in Chapter 17. Punchouts may be avoided by designing reinforcement so that transverse cracks are fine enough and far apart enough to stop aggregate interlock break-down. Khazanovich et al. (1998: 189–190) found that punchout

Figure 3.12 High severity punchout (Miller and Bellinger 2003: 79).

distress was reduced with higher steel percentages and cement-treated sub-bases as opposed to asphalt-treated or granular subbases.

Transverse construction joint deterioration

Transverse construction joint deterioration is a "series of closely spaced transverse cracks or a large number of interconnecting cracks occurring near the construction joint" (Miller and Bellinger 2003: 73). It is a "break-down of the concrete or steel at a CRCP construction joint... primary causes... are poorly consolidated concrete and inadequate steel content and placement" (Huang 2004: 388).

Surface characteristics

The traveling public – the pavement users – are most concerned about the pavement surface conditions. The surface conditions are initially "built in" to the pavement during construction, but change over time, generally for the worse. The various distresses discussed above often degrade the pavement surface condition. It is also useful to discuss the contributions of the pavement microtexture and macrotexture to the surface characteristics.

Smoothness

Traditionally, smoothness has been expressed in terms of serviceability, which is "the ability of a specific section of pavement to serve traffic in its existing condition" (Huang 2004: 389). The concept of serviceability was developed by Carey and Irick at the AASHO Road Test. Initially, a present serviceability rating from 0 to 5 (very poor to very good) was used by a panel of raters riding in a vehicle over the pavement. These ratings were correlated to objective measurements of pavement condition to develop a regression equation for present serviceability index (PSI). For rigid pavements, the regression equation used slope variance (a summary statistic for wheelpath roughness) and the sum of cracking and patching. Slope variance was found to be the dominant variable (Yoder and Witczak 1975: 648–653, Huang 2004: 389–398).

The World Bank developed the international roughness index (IRI) as a quantitative measure of smoothness. According to Huang (2004: 399):

> The IRI summarizes the longitudinal surface profile in the wheelpath and is computed from surface elevation data collected by either a topo-graphic survey or a mechanical profilometer. It is defined by the average rectified slope (ARS), which is a ratio of the accumulated suspension motion to the distance traveled obtained from a mathematical model

of a standard quarter car transversing a measured profile at a speed of 50 mph (80 km/h). It is expressed in units of inches per mile (m/km).

In other words, IRI is a not a direct measure of pavement roughness, but a measure of vehicle response to that roughness. General methods for measuring IRI are specified by ASTM E1170 (ASTM E1170 2001).

Agencies have used several types of devices for measuring pavement smoothness. These range from simple straightedges to high-speed, inertial profilers equipped with laser sensors to record elevation measurements (Grogg and Smith 2002: 3). Smoothness measurements are often taken immediately after construction, to determine whether the pavement is acceptable and whether the contractor should be assessed penalties or awarded bonuses. It is also necessary to have devices capable of traveling at highway speeds to monitor network roughness conditions for asset management.

Profilographs consist of a rigid beam or frame with a system of support wheels that serve to establish a datum from which deviations can be measured using a "profile" wheel located at the center of the unit (Woodstrom 1990). The two basic models used are the California and Rainhart profilographs. Profilographs are used exclusively for construction quality control and are light enough to be used to measure pavements 1 day after paving (Grogg and Smith 2002: 3–4).

Response-type road roughness measuring systems (RTRRMS) measure the dynamic response of a mechanical device. Common devices include the Mays ridemeter, the Bureau of Public Roads (BPR) roughometer, and the PCA ridemeter. They are used mainly for network assessment of existing pavements (Grogg and Smith 2002: 4–6).

Inertial road profiling systems (IRPS) are high-speed devices using non-contact sensors (ultrasonic, laser, infrared, or optical) to measure relative displacement between the vehicle frame and the road surface. Due to their accuracy and rapidity, they have largely supplanted RTRRMS for network assessment. Traditionally, these vehicles were too heavy to drive onto freshly constructed pavement. Recently, however, lightweight (golf cart or all terrain vehicle type) profilers have been developed that can drive on day-old concrete pavement with speeds up to 32 kph (20 mph) (Grogg and Smith 2002: 6–7).

The profile index, or PI, has traditionally been used for quality control of new pavements, with the IRI used for network monitoring. The PI is computed from a profilograph trace of the pavement surface, using manual or computer techniques. It is the sum of all individual high and low values exceeding a predetermined value. The predetermined value is termed the blanking band and has historically been 5 mm (0.2 in) for the California profilograph, although some agencies omit the blanking band. Due to the limited span of the profilograph, wavelengths over 15 m (50 ft) cannot be captured, even though those affect ride quality. Because correlation between PI and IRI has been found to be poor, some highway agencies have adopted IRI for construction acceptance testing (Grogg and Smith 2002: 7). The ACPA Technical Bulletin TB–006.0-C (ACPA 1990a) provides an example of a profilograph trace and PI computation.

One of the best ways to provide a smooth pavement over the entire service life is to build it smooth in the first place. Khazanovich et al. (1998: 183–184) found a strong correlation between initial IRI and IRI over time. A good working platform, such as a stabilized subbase, makes it easier to build a smoother pavement. For JPCP, joint faulting is an important contributor to roughness – joints with 38 mm (1½ in) dowels have very little faulting. For CRCP, using higher steel percentages lowered IRI and prevented localized failures.

Skid resistance

Pavement skid resistance is important for safety. Skid resistance is necessary to allow vehicles to stop safely and avoid crashes, particularly in wet weather. For a concrete pavement, the skid resistance depends on the texture provided when the pavement is built as well as subsequent wear due to vehicle tires.

> Texture affects both noise and friction characteristics; these properties must be considered together . . . It is also imperative that texture/friction be specifically addressed to reduce the currently unacceptable level of almost 43,000 fatalities and 3 million injuries annually in highway crashes and to minimize the resultant traffic delays. It has been estimated that poor pavement conditions currently contribute to 13,000 deaths annually. A small but significant portion of the poor pavement condition is due to poor surface texture or friction characteristics.
> (Larson et al. 2005: 506)

One important issue identified by Larson et al. (2005) is the durability of the surface – the ability to retain favorable friction characteristics over time. If the surface becomes polished and loses skid resistance, the texture may be restored through diamond grinding or grooving as discussed in Chapter 17.

On occasion, new concrete pavements are diamond ground to meet smoothness specifications. Because only the high spots are ground, the pavement surface may become a mixture of ground and unground surfaces. "Spot grinding to remove bumps can result in differential friction qualities and can also result in non-uniform appearance" (Larson et al. 2005: 503).

Kohn and Tayabji (2003: 79) note there are two primary functions of surface texture. The first is to provide paths for water to escape from beneath the tires of aircraft or other vehicles. The second is to provide enough sharpness for the tire to break through the residual film remaining after the bulk water has escaped. Huang (2002: 401) lists some factors that affect skid crashes, including polishing of the pavement surface and inadequate cross-slope drainage which allows water to pond on the pavement surface.

Skid resistance is defined using the skid number (abbreviated SK or SN). It is 100 times the kinetic or sliding coefficient of friction measured using a locked-wheel trailer of known weight with standardized tires, riding on wet pavement. Devices for measuring SN include the locked-wheel trailer, Yaw mode trailer, and British Portable Tester. Locked-wheel trailers are the most commonly used devices (Garber and Hoel 2002: 1081).

> Research results indicate that up to 70 percent of wet weather crashes could be prevented with improved texture/friction. Given that wet weather crashes represent about 14 percent of the total fatal crashes, this potential could result in a 10 percent total reduction in fatal and serious injury crashes and also could significantly reduce travel delays... However, a little recognized fact is that 80 to 86 percent of total fatal crashes occur on dry roadways. In the past, most of the emphasis has been on wet weather crashes, with the friction (or macro-texture) on dry roadways often assumed to be adequate... To provide safer roads, the effect of increasing texture/friction on reducing crashes on both wet and dry roadways must be considered.
>
> (Larson et al. 2005: 510)

Locked-wheel trailer testing is performed according to ASTM E274 (ASTM E274 2006). The test is performed at 64 kph (40 mph) with water applied in front of the test wheel. The trailer wheels are then locked and the friction force generated by the trailer is measured. SN decreases with increasing vehicle speed (Huang 2004: 402–403).

Karamihas and Cable (2004: 16) point out that ASTM E274 may be carried out with two types of standard test tires, ribbed or smooth. The two types of tires may rank skid resistance of roads differently, because ribbed tires provide channels for water to escape. Therefore, using ribbed tires may put pavements with good surface drainage at a disadvantage.

Surface texture is attained during construction by dragging various materials or tools across the fresh concrete. This process builds undulations or grooves into the surface before the concrete hardens. The spacing, width and depth of the grooves affect surface friction, skid resistance and tire/road noise. The purpose of a surface texture is to reduce wet-weather accidents caused by skidding and hydroplaning.

The different surface types produced during construction include drags and tining. Drag textures result from dragging brooms, artificial turf, or burlap across the fresh concrete surface. Tined surfaces are produced with metal rakes, and may be either longitudinal or transverse to the pavement centerline. These are illustrated in Chapter 15. Concrete pavement surface textures are discussed in *Concrete Pavement Surface Textures, Special Report SR902P* (ACPA 2000b).

Exposed aggregate is used mainly in Europe, and consists of applying a set retarder to the new concrete surface, and then washing away surface mortar to expose durable chip-size aggregates. This process requires uniformly applying chips to fresh surface and washing away the still-wet mortar. This technique may be used to improve the appearance of city streets.

Hardened concrete may also be textured, either to improve a newly constructed pavement that is not acceptable or will not earn the contractor a bonus, or to rehabilitate an existing pavement with inadequate surface characteristics. The surface may be diamond ground, diamond grooved, or abraded with steel shot (shot blasting). These techniques are discussed in Chapter 17.

Noise

The interaction between pavement texture and vehicle tires produces noise both inside and outside of the vehicle. If the noise is exceptionally loud or high pitched, it will be objectionable to drivers and/or citizens living next to the highway. Noise is becoming an issue of increasing importance in concrete pavement engineering. Hanson and Waller (2005: iv) state that "Noise is the generation of sounds that are unwanted . . . traffic noise can be considered an environmental pollution because it lowers the standard of living."

Two key studies on texturing and noise are NCHRP Synthesis 268 (Wayson 1998) and a comprehensive multistate study by Kuemmel et al. (2000). These and other reports and technical papers may be found on the website of the International Grooving & Grinding Association at http://www.igga.net/downloads/noise.html.

NCHRP Synthesis 268 (Wayson 1998: 1) found that, for concrete pavements,

> Transverse tining causes the greatest sideline (roadside) noise levels and may lead to irritating pure tone noise. Randomized spacing and changing the tine width have been found to reduce the pure tone that is generated and reduce overall noise levels. Texture depth of transverse tining also seems to play an important role in sideline noise levels . . .

Also,

- PCC pavements are in general, more noisy than asphaltic surfaces . . .
- Studies show that sound generation varies with speed. The pavement with the best results for noise may be different with varying

speeds. In addition, the most quiet pavement surface was found to be different for automobiles than for trucks.

- Construction quality is an important consideration in the final overall noise generation no matter which pavement type/texture is selected.
- Texture depth of the transverse tining also seems to play an important role . . .
- The use of porous PCC pavement also results in a noise reduction along the highway. This surface may provide noise attenuation while also being more durable than asphaltic surfaces.

(Wayson 1998: 38–39)

Wayson (1998: 70) concludes that concrete pavements are more durable and provide better skid resistance than asphalt pavements, but generate more noise.

Kuemmel et al. (2000) carried out an extensive study with 57 test sites in six different US states. This included a number of new test sections constructed in Wisconsin in 1997. The report (Kuemmel et al. 2000: 2–3) concluded that:

- Tining depths vary tremendously among the pavements constructed, even within a single test section. This was found to be the case throughout the states analyzed. In many cases, depths specified by the highway agencies were not achieved.
- Uniform tined pavements exhibit discrete frequencies (a whine) and should be avoided.
- For all transverse tined pavements, those with the widest and deepest textures were often among the noisiest.
- Longitudinal tined PCC pavements and an AC pavement exhibited the lowest exterior noise while still providing adequate texture. The performance of longitudinal tining in wet weather has not been documented in any recent accident study.
- One AC pavement, and the longitudinally tined and random skew tined (1:6 skewed) PCC pavements exhibit the lowest interior noise while providing adequate texture. The random skewed can be easily built and eliminates discrete frequencies.
- Random transverse tining can significantly reduce discrete frequencies, but may still exhibit some discrete frequencies unless carefully designed and constructed, and will not substantially reduce overall noise; . . .
- Diamond ground PCC pavements exhibited no discrete frequencies and lower exterior noise levels (compared to random transverse) approximately 3 dB.

These investigators documented the noise reductions of various concrete pavement treatments over uniform transverse tining. They also recommended improving quality control of macrotexture, and research to document the amount of macrotexture necessary for satisfactory wet weather performance of pavements. If noise considerations are paramount, longitudinal tining with 19 mm (3/4 in) spacing may reduce noise while providing adequate safety. Other treatments with potential include 1:6 skew or random transverse tining, or diamond grinding (Kuemmel et al. 2000: 3–4).

Karamihas and Cable (2004: 2) note that "quiet pavement should never be provided at the expense of safety. As such, the challenge is to provide quiet pavement with sufficient drainage and without any reduction in friction." They also found that transverse tining usually causes more noise than random textures. A high-pitched whine may be produced by transverse tining with constant spacing. There remains a problem of lack of standardization of pavement noise measurements (Karamihas and Cable 2004: 7).

Hanson and Waller (2005: 11) reported that, among concrete pavements, diamond ground pavements were the quietest, followed by longitudinally tined, longitudinally grooved, and transverse tined. All were noisier than asphalt pavements. This was based on 300 test sections across the United States Although diamond ground pavements were quieter, they also had the highest variability of noise measurements, probably because grinding does not affect low spots in the pavement. In Minnesota, a turf/broom drag produced low noise levels that were comparable to asphalt (Hanson and Waller 2005: 19–20).

Despite the fact that all of these sources are in agreement that concrete pavements start with better skid resistance and maintain it longer than asphalt pavements, the noise issue will remain a difficult challenge for the concrete paving industry. While virtually all drivers and passengers experience the noise differences while traveling on different pavements, only an unfortunate minority will ever experience the differences in the locked-wheel skid resistance of different types of pavements. Furthermore, some of those who find the skid resistance to be inadequate will not be around to testify on the issue. Therefore, more research into noise reduction for concrete pavements is necessary.

Chapter 4

Subgrades, subbases, and drainage

Pavement performance is strongly influenced by the underlying soil layers, particularly with respect to stability, bearing strength, consolidation over time, and moisture susceptibility. Frequently one or more layers are placed between the soil subgrade and the pavement.

In flexible or asphalt pavements, the layer directly under the pavement is the base and the layer between the base and the subgrade is termed the subbase. Base course materials have to meet tighter quality specifications than subbase materials. Since subbase quality materials may be used directly under a concrete pavement, this layer is typically termed the subbase. This is the terminology used in the 1993 AASHTO Design Guide (AASHTO 1993: I-4).

Concrete pavements distribute loads much more widely than asphalt pavements, and the pressures on the subbase and subgrade are low. As a result, the bearing capacity of the underlying layers is less critical, and there is no need to use stiff base materials except for pavements carrying the heaviest loads.

In the literature, however, both terms, base and subbase, have been used to describe the layer under a concrete pavement. It is rare for concrete highway pavements to have both a base and a subbase layer. In contrast, airfield pavements that handle heavy wide-body aircraft often have both a base and a subbase (Kohn and Tayabji 2003: 41).

These layers may be unstabilized, or stabilized with portland or asphalt cement. They may also be densely graded and relatively impermeable, or deliberately gap graded to provide drainage. The PCA/ACPA publication *Subgrades and Subbases for Concrete Pavements* addresses selection and construction of these layers (PCA/ACPA 1991). Another publication that discusses subbases and bases, specifically for airfields, is *Design and Construction Guide for Stabilized and Drainable Bases* (Hall et al. 2005).

> Careful attention to the design and construction of subgrades and sub-
> bases is essential to ensure the structural capacity and ride quality of
> all types of pavements. For concrete pavements, the requirements may
> vary considerably depending on subgrade soil type, environmental con-
> ditions, and amount of heavy truck traffic. In any case, the objective
> is to obtain a condition of uniform support for the pavement that will
> prevail throughout its service life.
>
> (PCA/ACPA 1991: 1)

As an example, a parking lot carrying only light vehicle traffic would
usually be built directly on the compacted subgrade. A subbase would
be used only to deal with unsuitable soil (ACI Committee 330 2001:
5). For streets and local roads, subbases are generally not used for residential pavements, but would be used to prevent pumping for collector and arterial streets and local roads (ACI Committee 325 2002: 7).
For highways, AASHTO recommends the use of a subbase unless the
subgrade soils are equal in quality to subbase materials, or if the projected traffic is less than one million ESALs (AASHTO 1993: I-21).
ESALs, discussed in Chapter 8, are equivalent single axle loads of 80 kN
(18 kip).

Airfield requirements are similar. For pavements serving only general aviation aircraft, a subbase may not be necessary (ACPA 2002: 2). The FAA
requires a subbase under airport pavements with heavier traffic with the
exception of certain soil type, drainage, and frost susceptibility combinations. For heavier aircraft, stabilized subbases which are essentially bases
are required (FAA 2004: 55).

Subgrades

The subgrade is the native roadbed soil at the site and the engineer must
generally work with what is available. If necessary, better materials may
be brought in or the subgrade may be stabilized, but this represents an
additional cost.

Two common soil classification methods are the ASTM or Unified Soil
Classification System (USCS) (ASTM D2487 2000) and the AASHTO classification system (AASHTO 2005). The AASHTO system is used for highways in the United States, and the ASTM/USCS system is used for other
pavements. Soils are classified by the proportion of the coarse grained
fraction (gravels and sands) and the fine-grained fraction (silts and clays),
as well as the plasticity or response of the fine-grained fraction to moisture. Classification of soil is described by many standard textbooks on
geotechnical or pavement engineering, including Garber and Hoel (2002:
840–850).

For concrete pavement engineering, the important parameters are the quality of the support provided to the pavement, the moisture susceptibility, and the frost susceptibility. The quality of support is characterized by the modulus of subgrade reaction k, in units of MPa/m or psi/inch (sometimes pci or pounds per cubic inch). This modulus represents the spring constant of an imaginary spring or dense liquid foundation supporting the slab. The subgrade reaction increases linearly with the deflection of the slab. It may be measured in the field using a plate bearing test, or predicted from the soil classification or California bearing ratio (CBR).

One commonly used chart for predicting k was developed by the PCA (1966). This chart was used by the PCA 1984 method and is reproduced by Garber and Hoel (2002: 1036) and Huang (2004: 328). It provides approximate relationships for determining k based on ASTM/USCS soil classification, AASHTO classification, FAA classification, resistance value (R), bearing value, and CBR.

Recently Hall et al. (1997: 80) developed a revised k-value chart for the AASHTO 1998 supplement design procedure. The k-values for different soil types are shown in Table 4.1. The AASHTO 1998 guide also provides procedures for correlating k-values to CBR and dynamic cone penetrometer (DCP) data. Charts are provided for fine-grained soils based on moisture content – essentially linear relationships exist between the minimum values shown in Table 4.1 at 100 percent saturation and the maximum values at 50 percent saturation. Procedures are also provided for finding k through back-calculation of FWD test results (AASHTO 1998: 6).

As a practical matter, the design thickness of a concrete pavement is not very sensitive to the k-value. The moisture susceptibility of the soil has a much greater effect on pavement performance. Frost action may also be very damaging.

For heavy duty highway or airport pavements, it is generally necessary to remove any unsuitable material encountered. Unsuitable subgrade materials include peat, organic silt, silt, and soil with high organic content (Kohn and Tayabji 2003: 33).

Subgrades may need to be stabilized to improve low-strength soil, reduce potential for swelling due to moisture, or to improve construction conditions. If the project is on a tight schedule, stabilization reduces the risk of construction delays due to wet weather. Subgrades may be stabilized with lime or cement. Lime stabilization is most suitable for clayey soils with high moisture content. Cement stabilization is used for coarse-grained soils or soils with high silt content. Lime and cement stabilization are discussed in detail by Kohn and Tayabji (2003: 36–38).

Table 4.1 Recommended *k*-value ranges for various soil types (adapted from Hall et al. 1997: 80, AASHTO 1998: 6)

AASHTO Class	Description	ASTM/USCS class	Dry density, kg/m³ (lb/ft³)	CBR (percent)	k-value, MPa/m (psi/in)
		Coarse-grained soils			
A-1-a, well graded	Gravel	GW, GP	2,000–2,240 (125–140)	60–80	81–122 (300–450)
A-1-a, poorly graded			1,920–2,080 (120–130)	35–60	81–108 (300–400)
A-1-b	Coarse sand	SW	1,760–2,080 (110–130)	20–40	54–108 (200–400)
A-3	Fine sand	SP	1,680–1,920 (105–120)	15–25	41–81 (150–300)
		A-2 soils (granular material with high fines)			
A-2-4, gravelly	Silty gravel	GM	2,080–2,320 (130–145)	40–80	81–136 (300–500)
A-2-5, gravelly	Silty sandy gravel				
A-2-4, sandy	Silty sand	SM	1,920–2,160 (120–135)	20–40	81–108 (300–400)
A-2-5, sandy	Silty gravelly sand				
A-2-6, gravelly	Clayey gravel	GC	1,920–2,240 (120–140)	20–40	54–122 (200–450)
A-2-7, gravelly	Clayey sandy gravel				
A-2-6, sandy	Clayey sand	SC	1,680–2,080 (105–130)	10–20	41–95 (150–350)
A-2-7, sandy	Clayey gravelly sand				
		Fine-grained soils*			
A-4	Silt	ML, OL	1,440–1.680 (90–105)	4–8	7–45 (25–165)*
	Silt/sand/ gravel mixture		1,600–2,000 (100–125)	5–15	11–60 (40–220)*
A-5	Poorly graded silt	MH	1,280–1,600 (80–100)	4–8	7–51 (25–190)*
A-6	Plastic clay	CL	1,600–2,000 (100–125)	5–15	7–69 (25–255)*
A-7-5	Moderately plastic elastic clay	CL, OL	1,440–2,000 (90–125)	4–15	7–58 (25–215)*
A-7-6	Highly plastic elastic clay	CH, OH	1,280–1,760 (80–110)	3–5	11–60 (40–220)*

* *k*-value of a fine-grained soil is highly dependent on degree of saturation

Expansive subgrade soils

Excessive differential shrink and swell of expansive soils cause nonuniform subgrade support. As a result, concrete pavements may become distorted enough to impair riding quality. Several conditions can lead to this pavement distortion and warping:

1. If expansive soils are compacted when too dry or are allowed to dry out prior to paving, subsequent expansion may cause high joints and loss of crown.
2. When concrete pavements are placed on expansive soils with widely varying moisture contents, subsequent shrink and swell may cause bumps, depressions, or waves in the pavement.
3. Similar waves may occur where there are abrupt changes in the volume-change capacities of subgrade soils.

(PCA/ACPA 1991: 3)

It is obviously important to identify potentially expansive soils before construction. As a general rule, soils with high colloid content (percent of grains finer than 0.001 mm or 0.0004 in), high plasticity, and low shrinkage limit can be expected to expand the most. These are generally classified as CH, MH, or OH by the ASTM classification method or A-6 or A-7 by the AASHTO method. Concrete pavements can often tolerate clays with medium or low degrees of expansion. The actual amount of expansion depends on the climate, the load conditions from the pavement layers above the subgrade, and the moisture and density conditions at the time of paving (PCA/ACPA 1991: 3–4). Therefore, careful construction is important for control of expansive soils. Construction considerations are discussed in Chapter 13.

Frost-susceptible subgrade soils

"For pavement design purposes, frost action can be evaluated by the effects of (1) frost heave and (2) subgrade softening on spring thawing. Design considerations for controlling frost heave are not necessarily identical to those for controlling subgrade softening" (PCA/ACPA 1991: 7). Subgrade softening is generally very damaging for asphalt pavements, but less damaging for concrete pavements because they can bridge over soft soils during the spring thaw period. The following material has been adapted from PCA/ACPA *Subgrades and Subbases for Concrete Pavements* (1991: 7–8).

Frost heave is caused by the growth of ice lenses in the soil. Water moves by capillary action through soil to ice crystals initially formed, and then freezes. With enough available water, the crystals continue to grow and form thick lenses that lift or heave the pavement. Frost action requires frost-susceptible soil, freezing temperatures that penetrate the subgrade, and a supply of water, usually with a water table within about 3 m (10 ft) of the surface. All three must be present for frost action to occur.

Frost-susceptible soils are often defined as having more than 3 percent of particles smaller than 0.02 mm (0.0008 in). Low plasticity, fine-grained soils with high silt content (0.05–0.005 mm, 0.002–0.0002 in) are particularly susceptible to frost heave. Pore sizes must be small enough to develop capillary action, but large enough to allow water to flow to form the lenses. "Low plasticity silts are most susceptible to frost heave, followed by loams and very fine sands, sandy loams, clay loams, and clays in decreasing order" (PCA/ACPA 1991: 8).

FAA design procedures address the design of pavements for frost and permafrost in detail in Chapters 2 and 3 of FAA document *Pavement Design, Advisory Circular (AC) 150/6320-6D* (FAA 2004). This topic is discussed in more detail in Chapter 10 of this book. Methods of dealing with frost heave are discussed in Chapter 13.

In addition to the dramatic effect of moisture on the k-value of fine-grained soils, moisture in soils also leads to pumping and the formation of ice lenses. This risk can be reduced through the use of subbases.

Subbases and bases

The AASHTO *Guide for Design of Pavement Structures* (AASHTO 1993: I-21) defines a subbase as "one or more compacted layers of granular or stabilized material" between the subgrade and concrete pavement slab. The Guide cites the following reasons for using subbases:

• "to provide uniform, stable, and permanent support,
• to increase the modulus of subgrade reaction k,
• to minimize the damaging effects of frost action,
• to prevent pumping of fine-grained soils at joints, cracks, and edges of the rigid slab, and
• to provide a working platform for construction equipment."

As noted previously, the Guide further states that a subbase is not necessary if the traffic is relatively light (less than one million ESALs) and the roadbed soil quality is equal to that of a subbase. Use of a subbase solely to increase the k is not economical, in the absence of the other reasons (PCA 1984: 6). As a practical matter, the last reason cited in the previous list may be the strongest, because it helps avoid construction delays due to wet weather and helps the contractor build a smoother pavement.

A key consideration is whether the subbase is expected to also function as subdrainage. Conventional subbases are densely graded and compacted granular materials, and materials stabilized with suitable admixtures. AASHTO soil classes A-1, A-2-4, A-2-5, and A-3 granular materials may need to be treated with portland cement to make cement-treated subbases (PCA 1984: 6).

Use of a subbase will increase the k-value over that provided by the subgrade soil. The PCA developed a chart to calculate the increase in k-value due to aggregate (Table 4.2) and cement-treated subbase (Table 4.3) (PCA 1984: 6).

The FAA has developed similar charts. For well-graded aggregate, results are similar to Table 4.2. The k-values for bank run sand and gravel are lower and are shown in Table 4.4. Values for all types of stabilized subbases are provided in Table 4.5 (FAA 2004: 55–56).

The AASHTO *Guide for Design of Pavement Structures* (AASHTO 1993) uses a more elaborate procedure to adjust the k-value upward based on the subbase modulus and thickness, and possibly on a rigid foundation or rock 3 m (10 ft) below the surface, adjust for seasonal effects, and then adjust it downward based on the loss of support factor due to foundation erosion or differential soil movements. That degree of refinement may not be necessary, based on the limited importance of k-value to design and

Table 4.2 Effect of untreated subbases on k-values (PCA 1984: 6)

Subgrade k-value, MPa/m (psi/in)	Subbase k-value, MPa/m (psi/in)			
	100 mm (4 in)	150 mm (6 in)	225 mm (9 in)	300 mm (12 in)
13.5 (50)	17.5 (65)	20 (75)	23 (85)	30 (110)
27 (100)	35 (130)	38 (140)	43 (160)	51 (190)
54 (200)	60 (220)	62 (230)	73 (270)	87 (320)
81 (300)	87 (320)	89 (330)	100 (370)	117 (430)

Table 4.3 Effect of cement-treated subbases on k-values (PCA 1984: 6)

Subgrade k-value, MPa/m (psi/in)	Subbase k-value, MPa/m (psi/in)			
	100 mm (4 in)	150 mm (6 in)	225 mm (9 in)	300 mm (12 in)
13.5 (50)	46 (170)	62 (230)	84 (310)	107 (390)
27 (100)	76 (280)	108 (400)	141 (520)	173 (640)
54 (200)	127 (470)	173 (640)	218 (830)	—

Table 4.4 Effect of untreated subbases on k-values (FAA 2004: 15)

Subgrade k-value, MPa/m (psi/in)	Subbase k-value, MPa/m (psi/in)			
	100 mm (4 in)	150 mm (6 in)	225 mm (9 in)	300 mm (12 in)
13.5 (50)	16.2 (60)	19 (70)	23 (85)	28 (105)
27 (100)	34 (125)	38 (140)	43 (160)	51 (190)
54 (200)	62 (230)	68 (250)	73 (270)	81 (300)
81 (300)	84 (310)	87 (320)	92 (340)	95 (350)

Table 4.5 Effect of stabilized subbases on k-values (FAA 2004: 57)

Subgrade k-value, MPa/m (psi/in)	Subbase k-value, MPa/m (psi/in)			
	100 mm (4 in)	150 mm (6 in)	225 mm (9 in)	300 mm (12 in)
13.5 (50)	23 (85)	30 (110)	46 (170)	60 (220)
27 (100)	46 (170)	62 (230)	73 (270)	87 (320)
54 (200)	76 (280)	84 (310)	100 (370)	107 (390)
81 (300)	95 (350)	103 (380)	113 (420)	116 (430)

given the difficulty of predicting potential erosion. Tables 4.2 and 4.3 are probably sufficient.

There is one further comment worth making on k-values, particularly when stiff cement- or asphalt-treated subbase materials are used. Typically, in engineering, it is thought that more strength and stiffness is better and it is conservative to ignore it. Following that philosophy, it would seem to be conservative to design on the basis of a relatively low k-value regardless of whether or not better materials were used or better compaction were achieved in the field.

That approach would be correct for slab thickness design, but not for joint spacing. As will be shown in Chapter 7, curling stresses increase as k-values increase and thus joint spacings must be reduced to prevent mid-slab cracking. Thus, for determining maximum joint spacing, the k-value must be calculated correctly or overestimated in order to achieve a conservative design.

For airfield pavements, subbases are made of natural or crushed granular materials with CBR from 20 to 100. They are used to provide frost protection or drainage (Kohn and Tayabji 2003: 41). Similar materials are used for highways, streets, and roads (PCA/ACPA 1991: 12–13).

The FAA Advisory Circular *Airport Pavement Design and Evaluation* (FAA 2004) states that "the purpose of a subbase under a rigid pavement is to provide uniform stable support for the pavement slabs. A minimum thickness of 4 inches (100 mm) of subbase is required under all rigid pavements, except as shown" in Table 4.6.

Stabilized bases

New airport pavements, particularly those that carry heavy aircraft with gross aircraft loads of 45,250 kg (100,000 lb) or more, generally require stabilized bases (Hall et al. 2005: 1, FAA 2004: 55). Bases used under airfield pavements may be mechanically or chemically stabilized. Higher quality materials are required for mechanically stabilized bases than for subbases, in terms of crushed aggregate content, deleterious material, and gradation, but they are otherwise similar (Kohn and Tayabji 2003: 42).

Table 4.6 Conditions where no subbase is required (FAA 2004: 55, Table 3.10)

Soil classification	Good drainage		Poor drainage	
	No frost	Frost	No frost	Frost
GW	X	X	X	X
GP	X	X	X	
GM	X			
GC	X			
SW	X			

X = no subbase required

Categories of chemically stabilized bases include soil cement, cement-treated base (CTB), econocrete (lean concrete), and asphalt-treated base (ATB). Soil cement and CTB are similar, but soil cement uses on-site subgrade or fill material while CTB is generally made from processed material and is of higher quality. It is recommended that these materials be specified not to exceed a certain maximum strength, or that other measures be taken, in order to reduce cracking potential.

Mixture proportioning and test methods for soil-cement are discussed in detail in the PCA publication *Soil-Cement Laboratory Handbook, Engineering Bulletin EB052.07S*. This handbook covers determining the correct cement content and testing for freeze-thaw and moisture susceptibility (PCA 1992).

CTB typically has a seven-day compressive strength of 5.3 MPa (750 psi). CTB is listed in the FAA document *Standards for Specifying Construction of Airports, AC 150/5370-10B* (FAA 2005) as Item P-304 – Cement Treated Base Course.

Cement treated base course, FAA Item P-304, is designed to have a minimum seven-day compressive strength of 5,171 kPa (750 psi). In areas subjected to a large number of freeze-thaw cycles, a durability criterion may also apply. A bond breaker is recommended between the CTB and the concrete pavement. The CTB is accepted based on density and smoothness (Hall et al. 2005: 6).

Econocrete is similar to concrete but is made with marginal quality materials. A typical minimum seven-day compressive strength is 5.3 MPa (750 psi) with a 28-day maximum of 8.3 MPa (1,200 psi). The maximum strength requirement is intended to prevent cracking of the econocrete, and reduce the risk that these cracks will cause reflective cracks in the concrete pavement when it is placed. Base materials may also be treated with asphalt. Stiff base materials, particularly econocrete, can increase k-value and curling/warping stresses and therefore decreased joint spacing should be considered (Kohn and Tayabji 2003: 42–46). Econocrete is listed in the FAA document *Standards for Specifying Construction of Airports, AC 150/5370-10B* (FAA 2005) as Item P-306 – Econocrete Subbase Course.

Lean concrete base (LCB), or Econocrete, Item P-306, is weaker than CTB, with seven-and 28-day compressive strength requirements of 3,448 and 5,171 kPa (500 and 750 psi), respectively. To reduce the cracking risk for the concrete pavement, an upper limit on compressive strength of 8,274 kPa (1,200 psi) may be specified. This limit may be waived if a good bond breaker is used between the LCB and the concrete pavement. The minimum cement content is 118 kg/m^3 (200 lb/yd^3), which is considerably lower than conventional concrete. LCB is accepted based on consistency (slump) air content, thickness, strength, and grade (Hall et al. 2005: 6–7).

ATB is similar to conventional hot mixed asphalt, and uses the same FAA specification. ATB may use either Item P-401, which is the same as conventional hot mix asphalt surface layers, or the newer Item P-403. P-403 was developed specifically for ATB and asphalt leveling courses. Mixture design, construction, and acceptance are similar to conventional asphalt pavements (Hall et al. 2005: 7).

Details on materials selection and mix design for CTB, LCB, and ATB are provided by Hall et al. (2005: 35–42). Recycled, crushed, and graded concrete may be used as an aggregate for CTB and LCB if it meets specifications. Similarly, recycled asphalt pavement (RAP) may be used as part of ATB.

Cement-treated subbases and lean concrete subbases are also used under highway pavements. In many areas of the world, aggregates that meet specifications for unstabilized subbases are becoming difficult to find. CTB and lean concrete subbases allow greater use of local materials, substandard aggregates, and recycled materials, generally saving materials and hauling costs. CTB generally contains 4–5 percent cement by weight. Lean concrete subbase has more cement than CTB, but less than conventional concrete (PCA/ACPA 1991: 14–15).

Hall et al. (2005: 5) note that "a uniform, non-erodible base is preferred over a high strength base in a rigid pavement system." The base increases foundation support and reduces stresses and deflections, and improves joint load transfer. Therefore, cracking and faulting potential are reduced.

Selection of stabilized subbase/base type

Since CTB, LCB, and ATB may all be used as stabilized subbases or bases under concrete pavement, it is useful to address the advantages and disadvantages of the materials. Hall et al. (2005: 30–32) discusses the relative merits of these types of layers.

CTB is usually a good choice for a base under rigid airfield pavements. Advantages include resistance to erosion and pumping, good load transfer efficiency, and improved foundation support (increased k-value). CTB can generally be constructed quickly (Hall et al. 2005: 30). However, it is important with CTB to control maximum strength and stiffness and use bond breakers. Joint spacing may also need to be reduced.

LCB subbases and bases offer similar advantages to CTB, but may be more cumbersome because they require slipform pavers and a longer curing period. LCB has a rougher surface than CTB, so bond breakers are important. Thickness and stiffness of the layer should be minimized to reduce cracking risk. LCB may also be notched to control cracking, in a manner similar to joint sawing in concrete pavement (Hall et al. 2005: 31).

ATB generally works as well as CTB or LCB, but has a much lower stiffness. Therefore, friction and restraint are reduced, and a bond breaker is not necessary. The lower stiffness of ATB layer allows it to deflect with the slab and provide more uniform support. ATB layers are easily placed. ATB seems to lead to a lower risk of slab cracking than CTB and LCB layers (Hall et al. 2005: 31).

There are few disadvantages to using ATB layers, but in hot weather it is important to whitewash the surface, to reduce the temperature before placing concrete. Using ATB or asphalt-treated permeable base (ATPB) layers requires mobilization of asphalt paving equipment.

Friction and bond breakers

Concrete slabs expand and contract with temperature and moisture changes. Some high-strength stabilized base layers have a rough finished texture and offer considerable frictional restraint to the concrete pavement movement. This restraint, particularly at early ages when the concrete is weak, can lead to cracking.

The high stiffness and rough finished texture of high-strength stabilized bases, such as CTB and LCB, offer considerable restraint at the PCC slab/base interface. Permeable bases, such as cement-treated permeable base (CTPB), also offer restraint because the concrete penetrates into the open surface. These factors increase the risk of early cracking in concrete pavements built over these base types (Hall et al. 2005: 16).

Furthermore, bond between a concrete and a stiff base can create a "composite slab" and problems with joint saw cutting. A normal saw cut depth of 1/4 to 1/3 of the slab thickness is not deep enough to ensure that the composite slab cracks at the desired joint location.

In order to prevent these friction and bond problems, it is necessary to specify a bond breaker between the base/subbase and the concrete pavement. The most effective bond breaker for CTB, LCB, or CTPB is a thin layer of choke stone spread over the surface just prior to concrete placement. Choke stone is uniformly graded, approximately 13–25 mm (1/2–1 in) in size. This layer breaks bond and also prevents concrete penetration into CTPB and locking. The lower elastic modulus of asphalt generally offers less restraint than cement-treated layers, so no bond breaker is usually necessary unless the surface is milled. Milled surfaces may be treated with a single layer of asphalt emulsion to reduce friction. For cement-based base/subbase layers, waxed base curing compounds provide a bond breaker as well as helping to cure the layer. However, these do not always effectively break bond, so choke stone is preferred (Hall et al. 2005: 26).

For both ATB and ATPB, the black surface can absorb considerable heat and increase the risk of cracking the concrete slab. Therefore, they should be whitewashed with a lime-water solution to lower the surface temperature (Hall et al. 2005: 23).

It is also helpful to limit the strength and stiffness of CTB or LCB layers. A compressive strength of 2.4 MPa (350 psi) is sufficient to support construction traffic. Minimum compressive strengths of 3.4 MPa (500 psi) and maximum of 6.8 MPa (1,000 psi) are desirable. Higher strengths are associated with increased risks of slab cracking. Although higher strengths have often been specified to improve freeze-thaw durability, these are generally not necessary (Hall et al. 2005: 18).

When CTB layers have high cement content and are not adequately cured, they may develop surface plastic shrinkage cracks. These cracks may propagate or reflect upward through freshly placed concrete pavement. If shrinkage cracking is observed, medium to high density geotextile fabric may be placed over the CTB as a crack relief layer (Hall et al. 2005: 23).

Surface drainage

Recognition of the importance of pavement drainage has increased over the years. In 1975, Yoder and Witczak devoted only a few pages to the topic in their classic textbook (Yoder and Witczak 1975: 370–371). Later textbooks devoted entire chapters to the topic (Garber and Hoel 2002: 739–823, Huang 2004: 334–367). Those textbooks cite the same references and describe the same procedures discussed, as adapted from Mallela et al. (2002: 83–94).

Garber and Hoel (2002: 739) cite two reasons for the importance of pavement drainage. The first is that inadequate drainage often results, over time, in severe damage to the pavement structure. The second is that standing water seriously detracts from the safety and the efficiency of the pavement, with increased risk of hydroplaning and loss of visibility. They note that there are two sources of water – surface water from precipitation falling on the pavement in the form of rain or snow, and ground water flowing in the soil beneath the pavement.

The surface drainage system is comprised of those features that carry water away from the pavement and right of way. These include transverse slopes or the crown of the pavement, typically 2 percent for highways.

Longitudinal slopes of at least 0.2 percent are also generally provided. Longitudinal channels or ditches along highways carry away the water that flows from the pavement surface. In urban areas, curbs and gutters control runoff (Garber and Hoel 2002: 739–741).

Subdrainage

Regardless of the efficiency of the surface drainage system, some water will get into the pavement. All pavement surfaces, no matter how impermeable, allow water to enter through joints and cracks, particularly as they age. Unless the water is able to drain away quickly, it will weaken the pavement structure and may cause or accelerate damage.

Some pavements make use of permeable base subdrainage layers. They are generally more costly than conventional subbase layers, and unstabilized granular materials are less stable than densely graded materials. This is because in order to provide the desired permeability the fine portion of the aggregate must be removed (Mallela et al. 2002: 4-2). There is, therefore, a structural penalty paid when an unstabilized granular drainable base is used. A densely graded, well-compacted, crushed limestone base is very stable, but impermeable. In contrast, a uniformly graded, rounded aggregate, such as pea gravel, is very permeable but highly unstable.

Kohn and Tayabji (2003: 46) note that

> A balance between the need for stability and the need for porosity must be considered in the design with stability taking precedence. The thickness of the drainage layer is typically 100–150 mm (4–6 in). The use of unstabilized open graded aggregate drainage layer is not recommended for pavements used by wide bodied aircraft. These layers do not provide the necessary stability and construction related problems (rutting due to construction traffic, etc.) are common. If an unstabilized open graded layer is necessary, it should be placed deeper in the pavement structure to reduce stresses on the layer.

Hall et al. (2005: 5) note that "a well designed and constructed permeable base layer rapidly removes water from within the pavement structure. This leads to a mitigation of PCC durability-related distresses (e.g. D-cracking) and helps increase resistance to joint faulting." However, they are not directly addressed in FAA pavement design procedures. "The structural contribution of permeable base layers is ignored in the design process since they are relatively weak" (Hall et al. 2005: 7).

Aggregate drainage layers can provide a permeability of 300–1,500 m/day (1,000–5,000 ft/day) with a minimum of 300 m/day (1,000 ft/day) typically recommended for high-type highways. This may be provided by using ASTM/AASHTO No. 57 or No. 67 gradations (Mallela et al. 2002: 4–13).

To compensate for the poor stability of open graded aggregate layers, small amounts of asphalt or portland cement may be added. Asphalt- or cement-stabilized bases have only slightly lower permeability than unstabilized bases (Mallela et al. 2002: 4-2), although there is often a higher cost.

Hall et al. (2005: 7–8) discuss the use of stabilized permeable base layers for concrete airport pavements.

> The construction specifications for these layers are typically developed by modifying existing guide specifications, such as Items P-401 or 402 for ATPB and Item P-304 for CTPB, etc. However, the open-graded nature of these materials prevents the application of conventional techniques for performing mix designs and specifying their construction. For example, the ATPB mix designs often are specified on the basis of a gradation and percent binder content. Permeability—an important consideration for this base type—is seldom specified or monitored. Furthermore, field compaction of the mixtures is achieved using method specifications. Acceptance of the mixture is done on the basis of thickness. As can be noted, considerable empiricism is used to specify and construct these mixes, some of which is unavoidable until further research is done.

For airfield pavements, which are difficult to drain because of their width and gentle slopes, permeability values for drainage layers in the range of 150–450 m/day (500–1,500 ft/day) have been found to be adequate in most situations. This can be achieved with a minimum cement content of 148 kg/m^3 (250 lb/yd^3) for CTPB and asphalt content of 2–3.5 percent by mass for ATPB (Hall et al. 2005: 18–19). CTPB is discussed in *Cement-Treated Permeable Base for Heavy-Traffic Concrete Pavements, Information Series IS404.01P* (ACPA 1994c).

CTPB is open graded, and therefore weaker than stabilized bases. These layers must be strong enough to withstand construction traffic without deformation. Due to their inherent weakness, these layers should be used only when there is a need for drainage (e.g. high rainfall, perched water table, low permeability subgrade soils, heavy traffic). If designed and built properly, CTPB layers can improve the long-term performance of pavements by reducing moisture-related distress (Hall et al. 2005: 32). CTPB layers have some of the same problems as CTB and LCB, and concrete paste can penetrate into the open surface structure and create bond. Segregation may also occur during placement.

ATPB layers have similar advantages as CTPB for pavements, but they are generally weaker (less stiff). Due to the lower stiffness, interface restrain is less of an issue and bond breakers are not necessary. "Significant concerns when using ATPB include the overheating of the base in hot-weather paving conditions (similar to ATB) and stripping and stability problems" (Hall et al. 2005: 32).

Details on materials selection and mix design for CTPB and ATPB are provided by Hall et al. (2005: 42–45). Due to the low amount of binder, the long-term stability of the mixture comes from the aggregate structure. Therefore, aggregate quality is an important concern, more so for ATPB than for CTPB.

If the subgrade under the permeable base has fines that are likely to migrate into and clog the permeable base, it may be necessary to provide a separator. This may be a granular layer or a geotextile.

Gharaibeh and Darter (2002: 6) note that permeable ATBs and CTBs have seen limited use with CRCP, because the rough surface texture leads to bonding and a composite pavement layer and reduces the effective steel percentage. A geotextile may be used to prevent bond. Unstabilized drainage layers do not present this problem.

Khazanovich et al. (1998: 184) found that JPCP with better subdrainage had lower joint faulting and overall lower roughness. The effect was stronger with non-doweled JPCP.

Subdrainage design software

The US FHWA has developed a computer program, DRIP 2.0, which may be used to design pavement subdrainage and side drains. DRIP stands for drainage requirements in pavements. This program, along with its user manual, may be downloaded free from the FHWA website: http://www.fhwa.dot.gov/pavement/software.cfm. The program may be used with either SI or US customary units. Chapter 4 of the user manual provides the technical background for the program as well as a useful and thorough review of the topic (Mallela et al. 2002: 4-1–4-29).

The DRIP 2.0 user manual (Mallela et al. 2002: 4-1) outlines the major steps in the design of subdrainage and side drains as:

1 quantifying water inflow;
2 designing the permeable base;
3 designing the separation layer;
4 quantifying flow to edge-drains;
5 computing outlet spacing;
6 checking outlet flow.

The first three items refer to the design of the permeable base, and the latter three to the side drains.

Quantifying water inflow

Water inflow into the subdrainage system is due to surface infiltration, water flow from high ground, groundwater seepage, and meltwater from ice lenses. Generally, infiltration is the single largest source of water flowing into pavements (Mallela et al. 2002: 4–6). Water flowing from high ground may be collected and carried off using intercepting drains to keep it away from the pavement structure (Garber and Hoel 2002: 743–744). A pavement subsurface drainage system is probably not an efficient way to deal with groundwater, and the designer should seek assistance from geotechnical specialists (Mallela et al. 2002: 4–8). Procedures for estimating meltwater were developed by Moulton (1980) and are reviewed by Mallela et al. (2002: 4-6–4-7). Meltwater inflow quantities may be calculated using the DRIP 2.0 computer program.

Infiltration occurs through cracks and joints in the pavement surface. Uncracked concrete is practically impermeable (Mallela et al. 2002: 4–5). Two methods may be used to estimate infiltration – the infiltration ratio method, developed by Cedergren et al. (1973), and the crack infiltration method, developed by Ridgeway (1976). Although the DRIP 2.0 software allows for calculations with both methods, the crack infiltration method is recommended (Garber and Hoel 2002: 807, Mallela et al. 2002: 4-8–4-9).

The design infiltration rate in cubic meters of water per day per square meter of pavement (cubic feet per day per square foot) is

$$q_i = I_c \left(\frac{N_c}{W} + \frac{W_c}{WC_s} \right) + K_p \tag{4.1}$$

where I_c = crack infiltration rate, or the water-carrying capacity of a crack or joint, suggested by Ridgeway (1976) as 0.223 m³/day/m² (2.4 ft³/day/ft²),
N_c = number of contributing longitudinal cracks or joints,
W = width of permeable base in m (ft),
W_c = length of contributing transverse cracks in m (ft), generally equal to the width of the paved surface,
C_s = spacing of transverse cracks or joints in m (ft), and
K_p = rate of infiltration through the uncracked pavement in m³/day/m² (ft³/day/ft²), which is generally very low and may be assumed to be zero or estimated from laboratory tests (Garber and Hoel 2002: 808).

For JPCP, the slabs should be designed to remain uncracked, so that the joint spacing should be used for C_s. JRCP is likely to crack between joints, so a C_s of 4.5–6 m (15–20 ft) is reasonable. For CRCP, it is conservative to use the standard transverse crack spacing of 0.6–2 m (2–6 ft) for C_s, but due to the smaller crack width it might also be reasonable to reduce the crack infiltration rate I_c.

Designing the permeable base

The rapidity of water removal from a permeable base depends on the length of the path that the water must travel and on the permeability of the base.

Mallela et al. (2002: 4-2) provide an equation for computing the resultant slope, S_R, from the longitudinal slope and cross-slope

$$S_R = \sqrt{S^2 + S_x^2} \tag{4.2}$$

where S and S_x are the longitudinal and cross-slope, respectively, in m/m (ft/ft). The resultant slope, S_R, determined from equation 4.2 is then used to compute the resultant length of the flow path through the permeable base resultant slope, L_R, in m (ft):

$$L_R = W \sqrt{1 + \left(\frac{S}{S_x}\right)^2} \tag{4.3}$$

The permeability of the base material, expressed as a coefficient of permeability K, depends on the type of material and the degree to which it is compacted. Generally, a small k is used as a symbol for permeability, but in this text a capital K is used to avoid confusion with the modulus of subgrade reaction. The key material characteristics are the effective grain size D_{10} (corresponding to the 10 percent passing grain size), the porosity n, and the percentage of fines passing the 0.075 mm sieve (No. 200 or 200 spaces per inch) (Mallela et al. 2002: 4-2).

Methods for calculating permeability are reviewed by Mallela et al. (2002: 4-2–4-5). The coefficient K may be computed using DRIP 2.0 software based on the grain size distribution of the material, the in-place density or unit weight in kg/m³ (lb/ft³), the specific gravity of the solids, and the effective porosity. The effective porosity may be determined from a simple test to find the volume of water draining from a known volume of material, or estimated based on the percent and type of fines (mineral filler, silt, or clay). The DRIP 2.0 software provides a material library with ASTM/AASHTO No. 57 or No. 67 gradations as well as granular permeable bases used by a number of US states.

Once the resultant length of the flow path and the permeability of the base material are known, it is necessary to determine the required thickness of the permeable base. Two basic concepts are available for this design – the depth of flow method, and the time to drain method (Mallela et al. 2002: 4-9–4-10). The DRIP 2.0 software may be used to design the layer based on either method.

The depth-of-flow method was developed by Moulton (1980) and may be used to either determine a minimum drainage layer thickness for a given design inflow rate, permeability, and length and slope of flow path, or to find a required coefficient of permeability for a given layer thickness. Moulton's equations and charts are reproduced by Mallela et al. (2002: 4-9–4-11).

The time-to-drain method was developed by Casagrande and Shannon (1952) and is based on a specific time (2 hours to 1 month) and either 50 percent drainage or 85 percent saturation. The DRIP 2.0 software computes time-to-drain by either the Casagrande and Shannon (1952) or Barber and Sawyer (1952) equations. Mallela et al. (2002: 4–13) note that the time for drainage is not sensitive to thickness, so these equations are more useful for determining the required k.

Designing the separation layer

Mallela et al. (2002: 4-15) note that:

> The separator layer may be a granular base material or an appropriate geotextile. The separator layer must (a) prevent fines from pumping up from the subgrade into the permeable base; (b) provide a stable platform to facilitate the construction of the permeable base and other overlying layers; (c) provide a shield to deflect infiltrated water over to its edge-drain, thereby providing protection for the subgrade; and (d) distribute live loads to the subgrade without excessive deflection. Only an aggregate separator layer can satisfactorily accomplish (b) and (d). The granular separator layer is preferred to the fabric since the granular layer will provide the construction platform and distribution of loads to the subgrade. When geotextiles are used as separator layers, they are most often used in connection with stabilized subgrades, which provide the construction platform and load distribution. Both granular and geotextile materials can prevent pumping of fines if they satisfy the filter requirements. The thickness of the granular separator is dictated by construction requirements and can range from 100 to 300 mm (4 to 12 in).

For concrete pavements, distribution of live loads (d) is not important because the rigid concrete slabs significantly reduce pressure on the subgrade. What is important is keeping subgrade fines from migrating up and clogging the permeable base.

Preventing clogging requires that an aggregate layer meets both of the following criteria at the interface:

$$D_{15} \text{ (permeable base)} \leq 5\ D_{85} \text{ (subgrade)} \tag{4.4}$$

$$D_{50} \text{ (permeable base)} \leq 25\ D_{50} \text{ (subgrade)} \tag{4.5}$$

where D_{15}, D_{50}, and D_{85} represent the 15, 50, and 85 percentile particle diameters from the gradation curve. If equations 4.4 and 4.5 are met at the permeable base – subgrade interface, then there is no need for a separator layer. If not, then a separator layer should be used. The same two criteria must be met at the separator layer and permeable base interface. Other requirements for separator layer materials are provided by Mallela et al. (2002: 4-15–4-16).

Geotextiles may also be used to separate the drainable base from the subgrade. They must also meet filter and clogging criteria for the apparent opening size of the geotextile. These are different for woven and non-woven geotextiles and for different soil types (Mallela et al. 2002: 4-16–4-17).

Design example

Mallela et al. (2002: 5-1–5-7) provide a design example for a drainable base under a concrete pavement. The following design parameters are used:

- Pavement width and thickness – two 3.66 m (12 ft) lanes, 225 mm (9 in) thick, with 3.05 m (10 ft) shoulders on each side. Transverse joint spacing is 6.1 m (20 ft).
- Cross-slope – uniform, not crowned, sloped at 2 percent in both longitudinal and transverse directions.
- Permeable base – same width as PCC pavement, AASHTO #57 material with density/unit weight of $2,082$ kg/m^3 (130 pcf), trial thickness 100 mm (4 in) based on construction considerations.
- Subgrade material – Georgia Red Clay, a well-graded clayey-silt, with $K = 0.001$ m/day (0.0033 ft/day). The sieve analysis for the subgrade material is 92 percent passing the 4.75 mm sieve (#4), 67 percent passing the 2 mm sieve (#10), 55 percent passing the 850 μm sieve (#20), 42 percent passing the 300 μm sieve (#50), and 31 percent passing the 75 μm sieve (#200).

Results of the roadway geometry calculations (Figure 4.1) provide $S_R = 0.0283$ and $L_R = 10.35$ m (33.94 ft).

Next, inflow is computed. The inflow computation by the crack infiltration method (equation 4.1) is shown in Figure 4.2 as 1.48×10^{-6} m^3/s/m^2 (0.42 ft^3/day/ft^2). If meltwater is included (not shown), it would add about 10 percent to the inflow.

The permeable base is then designed by the depth-of-flow method. First, the permeability of the base must be determined, using the gradation, specific gravity, and unit weight. This is shown in Figure 4.3. Based on a subgrade permeability of 122 m/day (400 ft/day), the required base thickness is determined to be 194 mm (7.6 in), which is more than the provided thickness of 100 mm (4 in) (Figure 4.4). Therefore, the thickness should be increased to 200 mm (8 in). Alternatively, if the

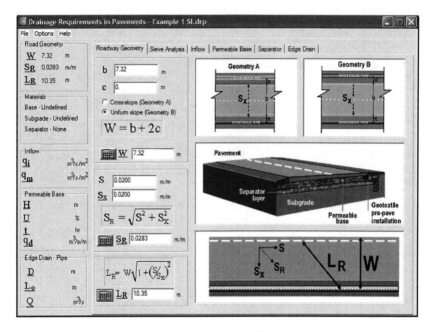

Figure 4.1 Roadway geometry inputs for concrete pavement drainage design.

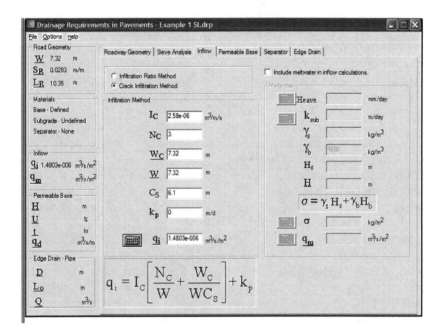

Figure 4.2 Computation of inflow.

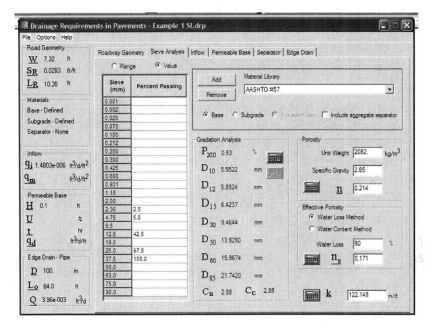

Figure 4.3 Determining the permeability of the base.

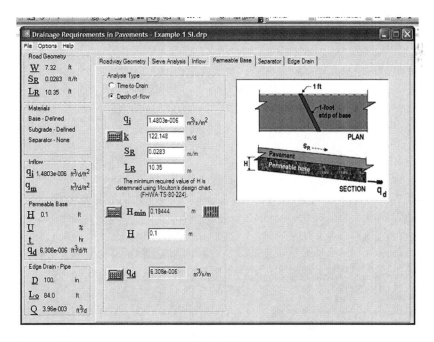

Figure 4.4 Designing the base by the depth-of-flow method.

compaction of the base were reduced so that the in-place density/unit weight were 2,000 kg/m³ (125 pcf), then the 100 mm (4 in) base would be sufficient. This example shows the tradeoff between stability and permeability, as well as the need to avoid overcompacting permeable subbases.

The time-to-drain for the base can also be checked, as shown in Figure 4.5. Even at the lower permeability of 122 m/day (400 ft/day), the base would drain to 50 percent in 4–5 hours, by either the Barber and Sawyer or Casagrande and Shannon equations.

The next step is to determine if a separator is needed, by checking equations 4.4 and 4.5. The gradation curves are plotted for both the permeable base and subgrade materials. Because the drainable base is a standard material (AASHTO #57), the gradation is provided within the DRIP 2.0 software. The subgrade material must be input. The software interpolates the gradation curve to determine D_{15}, D_{50}, and D_{85} for the two materials. Figure 4.6 shows the gradation analysis for the user-defined subgrade material. These values could also be determined by plotting a standard gradation analysis chart.

Next, the separator tab in DRIP 2.0 may be used to determine if the criteria are met. Figure 4.7 indicates that they are for these two materials, so there is no need for an aggregate or geotextile separator.

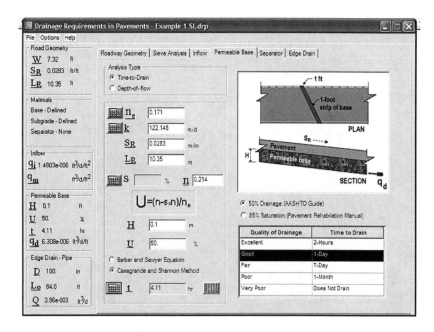

Figure 4.5 Checking the base design by the time-to-drain method.

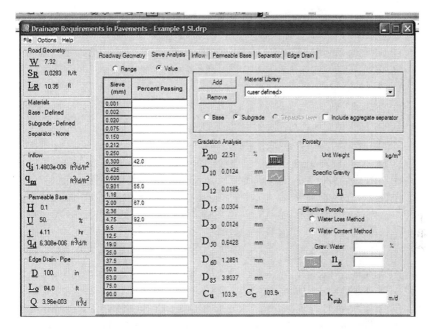

Figure 4.6 User-defined subgrade material.

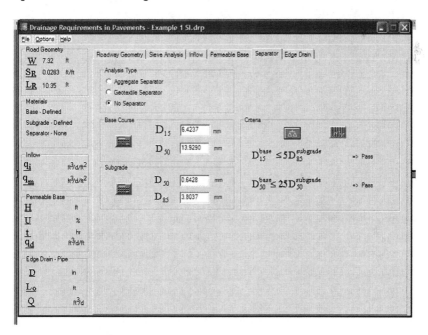

Figure 4.7 Checking need for a separator layer.

Side drains

Water that goes into a permeable base has to come out somewhere. Some water will exit through the subgrade, but if this outflow were sufficient there would be no need for a drainage layer in the first place. Therefore, the drainage layer typically drains to the sides.

The drainage layer may terminate with longitudinal drains, or may extend through the side slope ("daylighting"). Longitudinal drains are more reliable because daylighted drainage layers are susceptible to contamination and clogging. Longitudinal drains may have pipes or may be aggregate French drains without pipes (Huang 2004: 338).

The DRIP 2.0 software includes the design of longitudinal collector pipes. The edge-drains may be designed for pavement infiltration flow rate, peak flow from the permeable base, or the average flow rate during the time to drain the permeable base. It is recommended that the minimum capacity be the peak capacity of the permeable base. It is usually conservatively assumed that all flow is handled by the pipe, with no allowance for outflow through the soil. Pipe flow capacity is computed using the Manning equation, as detailed in many hydraulics and fluid mechanics textbooks (Mallela et al. 2002: 4-17–4-18).

The capacity of the outlet pipes must also be checked, using the Manning equation. The outlet drain slope should be at least 3 percent (Mallela et al. 2002: 4–21).

Generally the only design variables for longitudinal edge-drains are the type of pipe, pipe diameter, and outlet spacing. The type of pipe and outlet spacing are often matters of policy within a transportation organization, and minimums may be specified based on maintenance requirements. Minimum pipe diameters of 75–100 mm (3–4 in) with outlet spacings of 75–100 m (250–300 ft) are common (Mallela et al. 2002: 4-17–4-18).

Separators may also be necessary with edge-drains. "The filter material must be fine enough to prevent the adjacent soil from piping or migrating into the edgedrains but coarse enough to allow the passage of water with no significant resistance." These functions may be performed with aggregates, which may be difficult to place without contamination, or with geotextiles. They must be met with regard to holes or slots in perforated or slotted pipes (Mallela et al. 2002: 4–20). Prefabricated geocomposite edge-drains may also be used (Mallela et al. 2002: 4-21–4-29).

Design example

Continuing the previous design example, corrugated pipes with 100 mm (4 in) diameter are tried, with the "permeable base" peak flow design

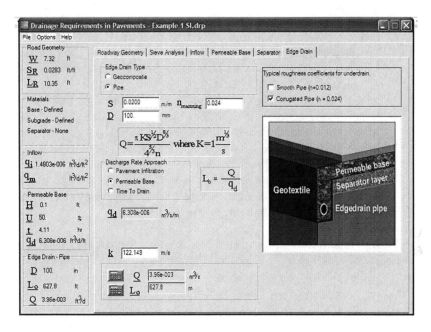

Figure 4.8 Edge-drain design.

method. Results from DRIP 2.0 are shown in Figure 4.8, with a maximum outlet spacing of 628 m (2,060 ft) (Mallela et al. 2002: 5-8–5-9). The time-to-drain criterion may also be checked, but that results in a larger maximum outlet spacing. The software also has a provision for design of geocomposite edge-drains.

Retrofit edge-drains

For an existing pavement, it is obviously impossible to retrofit a permeable base. However, edge-drains can be installed along an existing pavement. These will lower the water table in the vicinity of the pavement and reduce the moisture content of the subgrade and subbase. According to Mallela et al. (2002: 4-20–4-23), a 1989 FHWA study on retrofitting edge-drains for pavement rehabilitation concluded that:

- The edgedrain should be located under the shoulder immediately adjacent to the pavement/shoulder joint.
- By eliminating the filter fabric at the subbase/edgedrain interface, eroded fines cannot clog the filter fabric.

- Trench backfill should be permeable enough to transmit water to the longitudinal edgedrain pipe; asphalt or cement treated backfill increases stability with little or no loss of permeability.
- The most commonly used trench width was 300 mm (12 in); locating the top of the pipe at the bottom of the layer to be drained was recommended.
- Outlet spacing should not exceed 150 m (500 ft); additional outlets should be provided at the bottom of sag vertical curves.
- Because of the tendency of flexible corrugated plastic pipe to curl, use of rigid PVC pipe was recommended for outlet laterals; rigid PVC pipe helps maintain proper outlet pipe grade and provides protection from crushing.
- Headwalls protect the outlet pipe from damage, prevent slope erosion, and ease in the locating of the outlet pipe.

Retrofitting edge-drains as a rehabilitation strategy is discussed in Chapter 17. Maintenance of edge-drains, also important, is discussed in Chapter 16.

Airport drainage

Drainage of airport pavements is addressed in the FAA document *Airport Drainage, Advisory Circular (AC) 150/5320-5B* (FAA 1970). Surface drainage or airport pavements is less effective than for highway pavements, because airport pavements are much wider and must have much gentler transverse and longitudinal slopes for safety reasons. Permeable bases are therefore more useful and important.

Chapter 5

Selection of concrete materials

The successful use of concrete as a construction material extends back two millennia to the days of the Romans – and one of their greatest structures, the Pantheon, is still providing excellent service (Delatte 2001). On the other hand, some concrete structures and facilities built since then have had considerably shorter service lives. Careful selection of concrete materials and careful mixture design and proportioning, combined with good design and construction practices, is a key contributor to success. Durability is addressed in *Guide to Durable Concrete, 201.2R-01* (ACI Committee 201 2001).

Pasko (1998) states that

a good pavement designer should also be a concrete expert. Some facts to be kept in mind are:

- In the United States, there are 118 cement plants, each producing a variety of "unique" products under broad specifications. From personal experience on a research project, five Type I cements from different plants ranged in 28-day strength from 2738 to 4975 psi (19 to 34 MPa).
- There are 420 coal-burning plants in this country, and 28 percent of their fly ash is acceptable for use in concrete in accordance with ASTM C618. These products react differently with various cements, and the result is dependent on the quantities used. This is particularly important with respect to alkali aggregate reaction and sulfate resistance (and, possibly, delayed ettringite formation).
- Thousands of aggregate sources are available for use. Unfortunately, aggregate is not an inert filling. In addition to some aggregates reacting with the cementious materials, there are other characteristics that can cause problems.

Concrete in its most simple form consists of cement, water, and aggregates. In modern practice, chemical admixtures are almost always used, particularly for pavements. The use of supplementary cementitious materials, also called mineral admixtures, is also very common, because these materials reduce the cost and improve the performance and durability of concrete. Three excellent overall references on concrete are Kosmatka et al. (2002), Mindess et al. (2003), and Neville (1997).

Cement

Cement is the hydraulic glue that holds concrete together. In US practice, Portland cements are numbered using the Roman numerals I through V. Type I is a general purpose cement suitable for all uses where no special properties are required. In much of the United States it is by far the most common cement. Type II cement is used where it is important to protect against moderate sulfate attack. Sulfates are found in soils and groundwater, particularly in the western states of the US, and may attack concrete made with type I cement. In regions where sulfates are commonly found, type II cement is the more common type, and not type I. Some cements meet both type I and II and are designated type I/II. Type III cement is ground finer to achieve higher early strength. Type IV cement has slower strength gain for applications where heat of hydration must be minimized – it is now rarely available because the same effect may be achieved, at lower cost, with supplementary cementitious materials. Type V cement provides higher sulfate resistance than type II for more severe environments (Kosmatka et al. 2002: 27–30). Portland cements are manufactured to meet ASTM C150, *Standard Specification for Portland Cement* (ASTM C150 2005).

In Canada, Types I–V are termed types 10–50 (Mindess et al. 2003: 26). Blended hydraulic cements also exist, with several designations depending on whether blast furnace slag or pozzolans are added, along with hydraulic cements and hydraulic slag cements (Kosmatka et al. 2002: 31–33).

Supplementary cementitious materials

Supplementary cementitious materials, or mineral admixtures, include fly ash, slag, and silica fume, as well as other materials. Fly ash and slag are commonly used in pavements. Typically, these materials retard early strength gain of concrete, but improve the ultimate strength and durability. Overall heat of hydration and the rate of heat buildup are both reduced. Workability is improved, and the concrete surface is often easier to finish. Durability is improved because the porosity of the concrete is decreased and susceptibility to sulfate attack and ASR is reduced (Mindess et al. 2003: 107–109).

Supplementary cementitious materials may have pozzolonic properties, cementitious properties, or both. The pozzolonic reaction works with the reaction products of cement hydration to improve strength and decrease porosity of hardened cement paste and concrete (Mindess et al. 2003: 95).

Fly ash

Fly ash, the most widely used supplementary cementitious material in concrete, is a byproduct of the combustion of pulverized coal in electric power generating plants. It produces spherical glassy particles that are finer than Portland cement. Two types of fly ash are available, depending on the type of coal that was burned to make the ash. Class F fly ash has pozzolonic properties, and Class C fly ash has both pozzolonic and cementitious properties. Class F fly ash is typically used at 15–25 percent by mass of the cementitious material, and Class C at 15–40 percent by mass.

(Kosmatka et al. 2002: 58–59)

Requirements for class C and F fly ash are provided in ASTM C618 (ASTM C618 2005).

Kohn and Tayabji (2003: 61) note that "Class C fly ash may be detrimental to the performance of concrete, causing premature stiffening of the fresh concrete, thermal cracking, and/or reduced sulfate resistance . . . (it) is generally not effective in controlling expansions due to ASR." Early stiffening problems are more likely in hot weather with certain water reducing admixtures. The potential for early stiffening may be assessed by making trial batches at the highest temperatures anticipated on the project, and measuring slump loss and setting times.

Slag

Ground granulated blast furnace slag (GGBFS), also called slag or slag cement, is a by-product of metallurgical processes, generally the production of iron from ore. Slags are classified as grade 80, 100, or 120 based on reactivity, which is roughly 80, 100, or 120 percent of the 28-day strength of a reference mortar made with pure cement (Mindess et al. 2003: 102–103). Requirements for slag are provided in ASTM C989 (ASTM C989 2005). GGBFS should be distinguished from slag as an aggregate, which has no cementitious properties.

Ternary blend mixtures, which use Portland cement, fly ash, and slag, can produce very durable, low-permeability concrete. The Department of Aviation for the City of Houston, Texas, had experienced problems of durability of high early strength concrete. For a runway expansion at the George Bush Intercontinental Airport (IAH) in 2002, a concrete with type

I cement (50 percent), class F fly ash (25 percent), and grade 120 slag (25 percent) was developed and extensively tested, and was predicted to have a service life of 120 years (Sarkar and Godiwalla 2003).

Other supplementary cementitious materials

Another material, silica fume or microsilica, is often used in high performance, low-permeability structural concrete but rarely if at all in pavements. It is a by-product obtained during the manufacture of silicon metal and alloys (Mindess et al. 2003: 95). It is more expensive than fly ash or slag and less widely available, and the most likely application in pavements would be for repair materials or thin bonded concrete overlays.

Other materials with potential use in paving concrete include rice husk ash, metakaolin, natural pozzolans, and limestone filler. Rice husk ash is not yet commercially available, but it has the potential to become another important ingredient in concrete. The other three materials are not by-products and have less potential for paving concrete (Malhotra 2006).

Aggregates and water

"Aggregates generally occupy 70 to 80 percent of the volume of concrete and therefore can be expected to have an important influence on its properties" (Mindess et al. 2003: 121). Aggregate is not simply an inert filler in concrete, and its properties deserve careful consideration.

Aggregates are granular materials, usually natural rock (crushed rock or natural gravels) and sands. Aggregates are classified as normal weight, heavyweight, and lightweight, based on specific gravity. For concrete pavements, normal weight aggregates are generally used because they are most available. Aggregates for concrete must meet ASTM C33 *Standard Specification for Concrete Aggregates* (ASTM C33 2003). Aggregate properties that affect concrete pavement performance include shape and texture, size gradation, absorption and surface moisture, specific gravity, unit weight, physical durability, chemical durability, and strength. Strength of the aggregate particles rarely governs concrete strength, because the lower strength of the paste or the aggregate-paste bond is likely to govern instead.

Aggregate shapes may be rounded or angular, with rounded materials occurring naturally and angular materials produced by crushing and processing. Within those two main divisions, aggregate shapes may be further classified as spherical, irregular, highly irregular, flat or oblate/flaky, and elongated. Textures may be glassy, smooth, granular, rough, crystalline, or honeycombed. Rounded aggregates are more workable, but angular particles may develop higher flexural strength, which is important for pavements (Mindess et al. 2003: 122–125).

Grading of an aggregate is determined by a sieve analysis, where the mass of an aggregate sample retained on each of a number of standard sieves is recorded. Two key parameters are the maximum aggregate size and the shape of the gradation curve.

Absorption and surface moisture are of significance for calculating water that aggregate will add to or subtract from paste, and are used in mixture proportioning. Specific gravity is used to establish weight–volume relationships, also for mixture proportioning. Unit weight differs from specific gravity in that it includes not only the volume of the particles but the volume of the space between them when they are densely packed (Mindess et al. 2003: 133–140). These properties are all necessary when proportioning concrete mixtures by either the ACI 211.1 or 211.3 procedures (ACI Committee 211 1991, 2002).

Physical durability of aggregates refers to soundness and wear resistance. Aggregates are unsound if they deteriorate due to volume changes caused by repeated cycles of freezing and thawing or of wetting and drying. Unsound aggregates lead to surface popouts and D-cracking, discussed in Chapter 3. Wear resistance of aggregate plays some part in wear resistance of concrete under traffic, particularly in areas where studded tires are allowed (Mindess et al. 2003: 140–142).

For concrete pavements, the wear resistance of fine aggregates is more important than that of coarse aggregates for retaining skid resistance over time (Huang 2004: 408). Aggregates with poor wear resistance lead to polished pavement surfaces with poor skid resistance. It also seems logical that the wear resistance of coarse aggregates would also be an important factor in the performance of aggregate interlock joints over time.

"Most chemical-durability problems result from a reaction between reactive silica in aggregates and alkalis contained in the cement" (Mindess et al. 2003: 142). These encompass ASR and alkali-carbonate reaction, and lead to map cracking as discussed in Chapter 3. Fly ash, class F in particular, and slag have been used to control ASR. Other measures include the use of low-alkali cements and, when possible, the avoidance of susceptible aggregates (Mindess et al. 2003: 149–151).

It has been recognized that the concrete linear coefficient of thermal expansion (CTE) is an important performance parameter. This coefficient determines how much joints and cracks open and close, and how much concrete slabs curl due to temperature gradients. The concrete thermal coefficient may range from 7.4–13 $\mu\varepsilon/°C$ (4.1–7.2 $\mu\varepsilon/°F$). Since aggregates make up such a large portion of the concrete, they effectively determine the thermal coefficient for the concrete. Limestone has a low thermal coefficient of 6 $\mu\varepsilon/°C$ (3.3 $\mu\varepsilon/°F$), while sandstone has a higher thermal coefficient of 11–12 $\mu\varepsilon/°C$ (6.1–6.7 $\mu\varepsilon/°F$) (Mindess et al. 2003: 460).

Therefore, for a given temperature differential, concrete made with sandstone will attempt to move nearly twice as much as concrete made with limestone, and will develop twice as much stress if the movement is restrained. Joints must be designed to accommodate the larger displacements. In addition, the temperature-induced curling of the slab will be twice as much.

It may not be economically feasible to change coarse aggregate sources on a paving project to reduce the concrete CTE. However, reducing the joint spacing will decrease curling stresses and the risk of cracking (Hall et al. 2005: 22).

Coarse aggregate

For pavements, it is generally preferable to use the largest available coarse aggregate size, limited to 1/3 of the thickness of the pavement. This is because using the largest available aggregate reduces the proportion of paste, thus reducing shrinkage. Most ready-mix concrete equipment can handle aggregates up to 50 mm (2 in) in size. One caveat is that if the aggregate is susceptible to freeze-thaw damage, or D-cracking, a reduction in aggregate size will improve durability (Mindess et al. 2003: 125–127). For aggregate interlock pavement joints, use of larger maximum size aggregate should improve load transfer.

Use of a continuously or densely graded aggregate will also reduce paste requirements, since the smaller aggregate fills gaps in the larger aggregate. Uniformly graded or gap graded aggregates require more paste. Pervious concrete, unlike conventional concrete, uses a uniformly graded coarse aggregate (Mindess et al. 2003: 126–131).

Fine aggregate

ASTM C33 sets limits for fine and coarse aggregates separately. For fine aggregates, the particle distribution is represented by a fineness modulus FM, which is the sum of the cumulative percentage retained on seven standard sieves (150 μm–9.5 mm or No. 100-3/8 in), divided by 100. The FM is typically between 2.3 and 3.1, with a smaller number indicating finer sand. FM determines the effect of the fine aggregate on workability, which is important for mixture proportioning (Mindess et al. 2003: 126–131). Traditionally, natural sands have been used for concrete, but where these deposits are being depleted, manufactured sands made from crushed rock are used instead.

Hall et al. (2005: 20–21) suggest instead using a well-graded coarse sand with an FM in the range of 3.1–3.4, particularly with the high cement contents often used in airfield pavement construction. Coarser sands reduce volumetric shrinkage.

Optimized combined aggregate grading

Aggregate grading research for soils, base, asphalt, and other applications has proven that the best performance is derived from that blend of equi-dimensional particles that are well-graded from coarsest to finest. Optimum combined aggregate grading is important for portland cement concrete because it minimizes the need for the all-important second mix component—the paste—and has a significant effect on the air-void structure in the paste. The paste volume should be no more than is necessary to provide lubrication during placement and bind the inert aggregate particles together to resist the forces that will affect the mass during its service life... Gap grading (especially at the No. 4 and 8 sieves) and excessive fine sand and cementitious materials content were found to cause problems. Corrections to fill gaps in the aggregate grading led to significant reductions in water, improvements in mobility and finishability, and increases in strength.

(Shilstone and Shilstone 2002: 81)

The nos 4 and 8 sieves are 2.36 and 4.75 mm, respectively. The 1.16 mm and 600 μm (nos 16 and 30) sieves are also important.

Standard coarse and fine aggregate grading specifications such as those published by ASTM and public agencies seldom facilitate development of well-graded combined aggregate. ASTM C-33-03 (ASTM C33 2003), paragraph 1.3 makes provision for the specifier to cite that standard for quality of materials, the nominal maximum aggregate size, and other grading requirements. This allows the specifier to cite the coarseness factor chart (Figure 5.1) and limits within the parallelogram in zone II for combined aggregate grading. The one proportioning the mixture can use locally available non-standard or asphalt standard aggregate to fill voids.

Shilstone and Shilstone (2002) discuss aggregate gradation in detail. More detailed information is provided by Shilstone (1990). A key feature of the Shilstone approach is the use of the combined aggregate relationship nomograph, shown in Figure 5.1.

The x-axis of the chart represents the coarseness factor of the combined aggregates. This is the percentage of aggregate retained on the 2.36 mm (no. 8) sieve that is also retained on the 9.75 mm (3/8 in) sieve. The y-axis of the chart is the combined aggregate workability factor. This is the percentage of aggregate passing the no. 8 sieve, adjusted for cementitious content. To adjust for the cementitious material content, add or subtract 2.5 percentage points on the y-axis for every 43 kg/m^3 (94 lb/yd^3) of cementitious material more or less than 335 kg/m^3 (564 lb/yd^3) (Shilstone and Shilstone 2002: 82).

The chart provides a trend bar and five zones. "The diagonal bar is the Trend Bar that divides sandy from rocky mixtures. Zone I mixtures segregate during placement. Zone II is the desirable zone. Zone III is an extension of Zone II for 0.5 in (13 mm) and finer aggregate. Zone IV has too

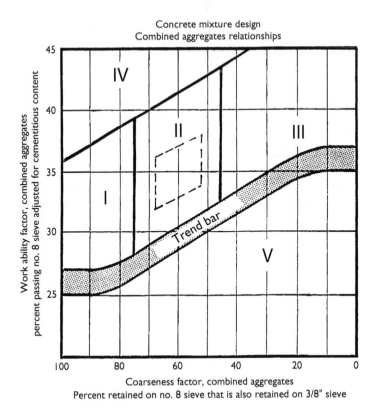

Figure 5.1 Combined aggregate relationship (coarseness factor) chart (courtesy: Shilstone and Shilstone 2002).

much fine mortar and can be expected to crack, produce low strength, and segregate during vibration. Zone V is too rocky" (Shilstone and Shilstone 2002: 82). Paving concrete should fall into zone II since the coarse aggregate is almost always larger than 13 mm (1/2 in).

The Shilstone Companies (undated) suggest the following guide specification for highway pavement concrete:

This guide is meant to provide a basis for development of concrete pavement mixtures using locally available aggregate. It is based upon the objective of optimizing the combined aggregate grading to minimize the need for paste (water, cementitious materials, and entrained air).

A. The minimum concrete (compressive) (flexural) strength shall be _____ [MPa] psi.

B. Cementitious materials shall include portland cement and may include Fly Ash or Ground-Granulated Blast Furnace Slag as appropriate for the local conditions.
C. Admixtures, if used, shall meet the requirements of ASTM C-494 and ASTM C260 and be compatible with the cementitious materials. Develop an air content of _____ % ± 1½ %
D. The aggregate shall comply with the Agency or ASTM C-33-04 specifications and meet the following criteria:

- The aggregate shall meet quality requirements of _____ (Agency or ASTM C33).
- The nominal maximum aggregate size shall be [38 mm] 1–1/2 inch.
- The grading of the combined aggregate shall fall within Zone II when plotted on the Coarseness Factor Chart (Figure 5.1). The supplier shall establish his mixture design within a parallelogram indicating limits of allowable production variations. When aggregate tests lead to results outside of the parallelogram, adjust mixture proportions to fall within the design limits.
- The grading of the combined aggregate shall be such that the sum of the percent retained on 2 adjacent sieves shall not be less than 13%.
- When aggregate tests lead to results outside of the limits of the mixture design parallelogram, adjust mixture proportions to fall within the design limits.
- When tested in accordance with ASTM C642, the permeable pores shall be 13% or less.

Kohn and Tayabji (2003: 63) note that concrete produced with well-graded aggregate combinations will have less water, provide and maintain adequate workability, require less finishing, consolidate without segregation, and improve strength and long-term performance. In contrast, gap-graded aggregate combinations tend to segregate, contain more fines, require more water, shrink more, and impair long-term performance.

Shilstone and Shilstone pointed out that the coarseness factor chart is a guide. Two other graphics, percent retained on each sieve and the 0.45 power chart, define details not reflected in the three size groupings in the nomograph. Gaps in grading can occur in other sieve sizes such as the 1.18 mm and 600 μm (nos 16 and 30). Optimized mixtures have produced excellent results for building construction, highways, and airfields.

The maximum density grading or 0.45 power chart is more widely used in asphalt paving than in concrete paving. This relationship may be represented as:

$$P = 100(d/D)^{0.45} \tag{5.1}$$

where d = sieve size in question, P = percent finer than (passing) sieve d, and D = maximum size of aggregate. This may need to be adjusted for aggregate angularity, shape, surface roughness, size, and compaction method (Barksdale 1991: 3-22–3-23). A gradation that follows equation 5.1 closely would use the least paste, but would probably need to be modified to allow for workability.

Lightweight aggregate

Lightweight aggregates have been used in bridges and buildings to reduce dead load. Since dead load is not significant for pavements, there has not been perceived to be a benefit to the use of lightweight aggregate. There are, however, potential advantages that have not been traditionally considered.

Some key references on lightweight aggregate concrete are provided by the ACI, including ACI 211.2-98 *Standard Practice for Selecting Proportions for Structural Lightweight Concrete (Re-Approved 2004)* (ACI Committee 211 1998) and ACI 213R-03 *Guide for Structural Lightweight-Aggregate Concrete* (ACI Committee 213 2003). These documents focus on structural concrete rather than pavement, but they identify some of the desirable characteristics of lightweight aggregate (LWA) concrete that also apply to pavements:

- lower modulus of elasticity, which in pavements translates to less tensile stress (cracking risk) for the same deformation or strain;
- flexural and splitting tension strengths comparable with conventional concrete;
- freeze-thaw durability may be equal to or better than that of conventional concrete;
- reduced risk of alkali-aggregate reactions;
- some types of LWA provide superior abrasion resistance.

In 1963 and 1964, a CRCP LWA concrete test section was built on an interstate highway frontage road in Houston, Texas, along with a standard aggregate CRCP test section for comparison. The sections were evaluated in 1974, 1984, and 1988. A 24-year performance survey in Texas found that the CRCP pavements built with lightweight aggregate concrete had relatively less surface distress than standard aggregate sections (Won et al. 1989, Sarkar 1999). At 34 years, another detailed investigation was carried out. The LWA section showed high durability, low permeability, and little cracking and spalling (Sarkar 1999).

LWA has also been investigated as a very thin non-skid surface overlay for concrete pavements. Natural sand, expanded shale LWA, and slag were compared (Gomez-Dominguez 1978).

Research has recently focused on the use of a replacement of a portion of fine normal weight aggregate with saturated lightweight aggregate, which has much higher absorption, in order to improve concrete curing (Bentz et al. 2005, Lam 2005, Mack 2006). Since pavements have high surface to volume ratios and are exposed to the environment, and are thus hard to cure, this concept seems to hold considerable promise for concrete pavement engineering.

Typical LWAs are expanded shales, slates, clays, and similar materials, with dry unit weights of 550–1, 050 kg/m^3 (35–65 lb/ft^3) and 5–15 percent absorption of water by weight. They are heated in rotary kilns and bloat up similar to popcorn (Mindess et al. 2003: 158–159).

Waste materials as aggregates

As supplies of natural aggregates deplete and landfills fill up, interest in recycling waste materials as aggregates increases. There is a need to proceed with caution, however, because there would be little benefit to turning our concrete pavements into "linear landfills" with all of the maintenance and performance problems that the term implies.

Solid wastes that have been considered as aggregate for concrete include mineral wastes from mining and mineral processing, blast furnace slags, metallurgical slags, bottom ash, fly ash not meeting class C or F specifications, municipal wastes (including commercial or household), incinerator residues, and building rubble (including demolished concrete). Factors that must be considered when deciding whether to use these materials include economy, compatibility with other materials, and concrete properties. The latter two factors often rule out the use of waste materials (Mindess et al. 2003: 156–158). However, hard mineral wastes and slags offer some potential for improving surface friction if used as fine aggregate.

The RMC Research Foundation (2005) publication *Ready Mixed Concrete Industry LEED Reference Guide* discusses the environmental benefits of incorporating post-consumer recycled content into concrete construction. The importance of this has been reviewed in Chapter 1. The most commonly used aggregate substitute is crushed recycled concrete. Air-cooled blast furnace slag may be used as an aggregate, but has no cementitious or pozzolanic properties and should not be confused with GGBFS or "slag" used as a cement replacement, as discussed above (RMC Research Foundation 2005: 38). For commercial or industrial projects that are attempting to achieve a specific level of LEED certification, it may be useful to consider recycled materials as aggregate.

"Crushed recycled concrete aggregate generally has lower specific gravity and higher water absorption than natural stone aggregate. New concrete made with such aggregate typically has good workability, durability, and resistance to saturated freeze-thaw action. Fine recycled aggregate should only be used in very limited quantities... strength using only coarse aggregate replacement is similar to that of concretes using natural aggregates" (RMC Research Foundation 2005: 38). Other cautions are that the crushed concrete should come from a known source and be clean, for example, free of chemicals, clay coating, and other fine materials.

Air-cooled blast furnace slag may be a dense and hard aggregate, and therefore desirable from the standpoint of wear and abrasion resistance. Crushed glass is generally not recommended as an aggregate because of the risk of ASR.

On occasion, use of recycled materials has led to problems.

> Heaving of pavements and a building foundation became progressively worse on a project at Holloman Air Force Base (AFB) N.M. The cause of heaving was identified as sulfate attack on recycled concrete used as fill and base course below the buildings and pavements. This recycled concrete came from sulfate-resistant airfield Portland concrete pavement that had existed for decades at Holloman AFB without distress. However, severe sulfate exposure conditions, ready availability of water, the more permeable nature of the crushed recycled concrete, less common thaumasite attack, possible soil contamination as a secondary source of alumina, or some combination of these factors allowed sulfate attack to develop in the recycled material even though it had not in the original concrete pavement.
>
> (Rollings et al. 2006: 54)

Water

The traditional rule of thumb is that potable water is good for making concrete. It does not, however, follow that non-potable water cannot be used, although there are limits on dissolved solids and organic material. Seawater should be avoided for any concrete containing reinforcement (Mindess et al. 2003: 115–120).

Admixtures

Chemical admixtures are used in concrete to affect either the fresh or hardened concrete properties. While a wide variety of admixtures are available to the industry, concrete pavement almost always uses air-entraining admixtures, with set-controlling admixtures (accelerators and retarders) and water reducers or plasticizers used under specific circumstances. Admixture technology is advancing rapidly, and new admixtures are always being developed which may be useful for paving applications.

Air-entraining admixtures

Air-entraining admixtures are used to protect concrete from damage due to freezing and thawing. Their primary purpose is to develop an entrained air void system of tiny spherical bubbles, 0.05–1.25 mm (0.002–0.05 in) in diameter, throughout the concrete, with an average spacing of no more than 0.2 mm (0.008 in). The air should be 9 percent of the mortar fraction, or 7.5 percent of the total concrete volume for 9.5 mm (3/8 in) maximum size coarse aggregate to 4 percent for 64.5 mm (2½ in) aggregate. This is because the air void system protects the cement paste, and with larger coarse aggregate there is less paste. Air entrainment also makes concrete more workable (Mindess et al. 2003: 168–176).

Concrete pavements are exposed to the environment, and unless the climate makes freezing and thawing very unlikely, air-entraining admixtures should be used. Some admixture manufacturers provide products specifically tailored for paving concrete.

Accelerating admixtures

Set-accelerating admixtures hasten the normal processes of setting and strength development of concrete. Calcium chloride is a popular accelerator because of its low cost, but it has the major disadvantage of increasing the rate of corrosion of reinforcement, tie bars, and dowels. Non-chloride accelerators are available for reinforced concrete applications (Mindess et al. 2003: 185–187). For paving applications, accelerating admixtures are likely to be useful in cold weather or when a pavement repair or overlay must be opened to traffic quickly.

Set-retarding admixtures

Set-retarding admixtures delay set, allowing more time for placement, consolidation, and finishing. Subsequent strength development is not significantly affected (Mindess et al. 2003: 182–183). For paving applications, retarding admixtures are likely to be useful in hot weather or when haul distances between concrete production and placement are exceptionally long.

Water-reducing admixtures

Three types of water reducing admixtures are currently available – low range (regular), mid-range, and high range (superplasticers). Low-range water reducers allow 5–10 percent water reduction, mid-range 10–15 percent, and high range 15–30 percent. These materials, particularly the latter, may be used to produce flowing concrete with very high slump (Mindess et al. 2003: 177–181).

Concrete pavements are typically placed at low slump, particularly when slipformed, so flowing concrete is not needed. However, water reducing admixtures may be used to reduce water content, and thus the amount of cement required to achieve a specific water/cement ratio. A reduction in cement and paste reduces the amount of shrinkage and thermal deformation of the pavement during curing. Some water-reducing admixtures also act as set retarders. Water-reducing admixtures may be useful for fixed form paving or for small areas of hand work.

Compatibility of materials

As more different materials are put into concrete, the compatibility of those materials becomes more of an issue. Mindess et al. (2003: 167) note that "effects that the admixtures may have on other concrete properties should be taken into account." Other admixtures may alter the effectiveness of air-entraining admixtures, as may finely divided mineral admixtures such as fly ash (Mindess et al. 2003: 172). Generally, admixtures from the same manufacturer will have been tested for compatibility with each other and are therefore less likely to present problems.

Kohn and Tayabji (2003: 62) note that "some concretes exhibit undesirable characteristics because of incompatibility among different concrete materials. Undesirable characteristics include:

1. Early loss of workability (early stiffening)
2. Delayed set (retardation)
3. Early-age cracking due to excessive autogenous and drying shrinkage of concrete
4. Lack of proper air-void system."

Early stiffening problems may be attributed to individual cementitious materials, interactions between cementitious materials, admixtures, and temperature effects. The factors that result in incompatibility are poorly understood and difficult to determine through testing, and many problems are triggered by higher or lower temperatures (Kohn and Tayabji 2003: 62–63).

To minimize incompatibility problems, Kohn and Tayabji (2003: 63) recommend using admixtures only from a single manufacturer and keeping dosages under the manufacturer's recommended maximums, and using only cementitious materials that meet project specifications and/or ASTM standards. In areas with significant seasonal temperature differences, it is advisable to have separate mixture designs for hot and cold weather. As stated earlier, Kohn and Tayabji (2003: 61) have noted that class C fly ash may cause compatibility problems with certain water-reducing admixtures in hot weather.

The FHWA Techbrief *Protocol to Identify Incompatible Combinations of Concrete Materials* defines incompatibility as "interactions between acceptable materials that result in unexpected or unacceptable performance" (FHWA Techbrief 2006: 1). Some of the findings on incompatibility include:

- A number of relatively simple field tests are available to warn of potential incompatibility.
- Use of a fly ash containing tricalcium aluminate may result in flash set because of insufficient sulfate to control hydration.
- Some type A water-reducing admixtures accelerate tricalcium aluminate hydration.
- Increasing temperatures increase the rate of chemical reactions and may make marginally compatible material combinations incompatible.
- Use of non-agitating transporters for paving concrete may exacerbate false set problems.
- Conversely, delayed setting increases the risk of plastic shrinkage cracking and makes timing joint sawing more difficult.
- Material chemistry can provide clues – fine cementitious materials with high tricalcium aluminate or low sulfate contents or fly ashes with high calcium oxide contents may cause problems.
- Possible field adjustments to correct problems include adjusting supplementary cementitious material type, source, or quantity; adjusting chemical admixture type or dosage; or changing the batching sequence or mix temperature (FHWA Techbrief 2006).

Two FHWA reports provide more information on this topic (Taylor et al. 2006a,b).

Van Dam et al. (2005) investigated durability of early-opening-to-traffic paving concrete. Because these concretes must typically achieve flexural strength in 6–8 or 20–24 hours, based on the construction window available, they use high cement contents and a wider variety and greater quantity of admixtures. Therefore, there is a greater potential for cement/admixture interactions that can lead to later durability problems (Van Dam et al. 2005: 6).

Fiber reinforcement

Fiber-reinforced concrete has not been widely used for pavements. The exception is bonded concrete overlays and UTW, which have made use of both steel and synthetic fibers, although synthetic fibers have been much more widely used. Fibers are intended to improve the flexural toughness and fatigue performance of the concrete (ACI Committee 325 2006: 9). The benefits, if any, of using fibers in concrete overlays and pavements have proven difficult to quantify. Use of fibers in concrete is discussed in an ACI Committee 544 *Report on Fiber Reinforced Concrete (Reapproved 2002) ACI 544.1R-96* (ACI Committee 544 1996).

Steel fiber-reinforced concrete has been used to design thinner pavements with longer joint spacings, but this has led to difficulties. "During the 1980s several steel fiber reinforced concrete airfield pavements were built at civil and Navy airfields using the new methodology. These tended to be relatively thin and large (sometimes only 100- to 150-mm (4- to 6-in) thick and 15 or even 30 m (50 and even 100 ft) between contraction joints). There were soon reports of widespread corner breaks at these airfields . . . the large plan dimension of the slabs relative to their thin cross section required very little differential shrinkage between the top and bottom of the slab to get curling in the field . . . Once opened to traffic, these slabs developed widespread corner breaks" (Rollings 2005: 170–171).

Chapter 6

Mixture design and proportioning

Once quality materials have been selected, as discussed in Chapter 5, it is necessary to design a concrete mixture for the project that will meet the requirements for fresh and hardened concrete properties, while keeping economy in mind. The concrete must meet project requirements in both the fresh and hardened state. In the fresh state, the chief characteristic is workability. In the hardened state, requirements for strength and durability must be met. If the fresh concrete properties are not satisfactory, the concrete will not be placed, consolidated, and finished properly, and the hardened properties will not matter.

The most commonly used method for proportioning concrete mixtures is the ACI Committee 211 *Standard Practice for Selecting Proportions for Normal, Heavyweight, and Mass Concrete, 211.1-91* absolute volume method (ACI 211 1991). This method is discussed in Chapter 10 of Mindess et al. (2003: 221–242) and Chapter 9 of Kosmatka et al. (2002: 149–177). This procedure uses the following steps (Mindess et al. 2003: 227–236):

- Assemble the required information. This includes:
  - sieve analysis (gradation distribution) of coarse and fine aggregates;
  - dry rodded unit weight of coarse aggregate;
  - bulk specific gravities and absorption capacities of coarse and fine aggregates;
  - slab thickness (since the maximum coarse aggregate size must be less than or equal to one-third of the least dimension); w
  - required strength;
  - exposure conditions.

- Choose the slump – typically 25–75 mm (1–3 in) for pavements, higher if fixed forms are used, possibly lower if the pavement is slipformed.
- Choose maximum aggregate size – one-third of pavement thickness, although availability or D-cracking concerns may dictate a smaller aggregate. In many areas, 19 or 25 mm (3/4 or 1 in) may be the largest

size available. If larger sizes are available, the pavement performance may be improved.

- Estimate the mixing water and air content. Less water is typically used for paving concrete than for structural concrete because the desired slump is lower. The amount may be further reduced with water reducing admixtures. Use of an optimized aggregate gradation, as discussed in Chapter 5, will also reduce the mixing water requirement. Air content is based on the maximum size aggregate (and thus the paste proportion) and whether the exposure is mild, moderate, or severe. ACI 211.1 provides tables for maximum coarse aggregate size up to 75 mm (3 in). Pavements are often subjected to severe exposure conditions.

- Determine the w/c or water/cementitious materials (w/cm) ratio. This is based on either strength or durability, whichever is lower.

- Calculate cement or cementitious materials content. This is determined by dividing the water by w/c (or w/cm). This gives either the total cement content, or total cementitious materials if fly ash and/or slag are used. If fly ash or slag are used, they replace cement on either a weight or volume basis – these are not equal because the specific gravities of these materials are not equal to that of cement. Reducing water content by using water-reducing admixtures also reduces the cement content required for the same w/c ratio.

- Estimate the coarse aggregate content. This depends on the maximum size of the coarse aggregate and the FM (or lubricating potential) of the sand. With larger coarse aggregates, a larger volume of coarse aggregate is used, as the larger aggregate particles replace smaller coarse aggregate, fine aggregate, and cement paste. With lower FM fine aggregate, more coarse aggregate may be used because less fine aggregate is required to lubricate the mixture and provide workability.

- Estimate the fine aggregate content. At this point the masses of water, cement, other cementituous materials, and coarse aggregate are known. Using the respective specific gravities, the volumes of these materials per cubic meter or cubic yard are known, plus the percent air. Everything else in the concrete volume has to be fine aggregate. The volume of fine aggregate and the specific gravity of the fine aggregate are used to determine the mass.

- Adjust for aggregate moisture. The mixing water estimates developed earlier are on the basis of SSD aggregate. If the aggregate has surface moisture, it will add water to the mixture, and a larger quantity of wet aggregate will be needed. If the aggregate is dry, it will subtract mixing water.

- Prepare a trial batch. The batch weights determined up to this point are estimates that need to be verified in the laboratory. Water-and air-entraining admixtures will need to be adjusted to achieve the correct slump and air content. Unit weight should be verified to determine the yield of the mixture. Trial batches are important – the steps outlined up to this point represent the starting point and not the end of determining a satisfactory mixture.

Recently, ACI Committee 211 has published the *Guide for Selecting Proportions for No-Slump Concrete, 211.3R-02* (ACI Committee 211 2002). This document covers mixtures with slump less than 25 mm (1 in), including roller compacted (appendix 3) and pervious concrete (appendix 6). The charts provided in this document are limited to a maximum aggregate size of 38 mm ($1\frac{1}{2}$ in). This document does not appear to address conventional paving concrete, which is generally proportioned by the methods previously outlined.

For paving concrete, the water and paste content should be minimized to reduce shrinkage. A total water content of less than 145 kg/m^3 (250 lb/yd^3) and a total paste volume of less than 60 percent are preferred (Hall et al. 2005: 22). This may require the use of water-reducing admixtures. As noted earlier, an optimized aggregate gradation also reduces the water requirement.

Workability

Workability refers to ease of placing, consolidating, and finishing freshly placed concrete without segregation. Although it is often addressed by slump test measurements, this rapid, cheap, and simple test does not measure all of the factors that are necessary for satisfactory workability. For slipformed concrete pavements, factors affecting workability include:

- segregation during transportation and placement;
- ease of consolidation;
- well-formed slipformed edges with little or no edge-slump;
- minimal or no hand finishing required (Kohn and Tayabji 2003: 57).

The desired slump is used to select the water content for trial batches, following the procedure outlined above. Slipformed concrete needs to be stiff enough to stand up on its own without the edges slumping as it emerges from the paver, so typical slumps are 25–50 mm (1–2 in) or less. Edge-slump often indicates too much water in the mixture. The slump of concrete for fixed form paving may be higher.

Although workability is thought of as being governed principally by water content, other factors also affect it:

- Aggregate – Size, grading, particle shape, water demand, variability.
- Cement – Cement content, water demand.

- Fly ash (if used) – Effect on initial set, water demand, effect on finishing.
- Slag cements and . . . GGBFS – Effect on finishing and saw cutting.
- Water – Total water demand.
- Admixtures – Air entrained concrete exhibits better workability; water reducers reduce water demand while improving workability.

(Kohn and Tayabji 2003: 57)

Strength

Both ACI 211.1 and 211.3 use required compressive strength to select a w/c or w/cm ratio. ACI 211.3 (ACI Committee 211 2002: 5–6) states that "If flexural strength is a requirement rather than compressive strength, the relationship between w/cm and flexural strength should be determined by laboratory tests using the job materials."

As a general rule of thumb, flexural strength is on the order of 10 percent of compressive strength of concrete. It is relatively higher with crushed aggregates, which provide better bond, and lower with rounded aggregates such as gravel.

The historic emphasis on strength, however, will not necessary lead to good pavement performance. "As Hardy Cross once wrote: Strength is essential but otherwise not important" (Murray 2000: 30). Although Professor Cross was most likely speaking of structural concrete, his statement is even more true of concrete pavement. It can probably be stated with confidence that no concrete pavement has ever failed in compression, although it is possible that thin pavements subjected to overloads have failed in flexure or in punching shear. In fact, many problems have been caused by the addition of too much cement to paving concrete in a misguided attempt to enhance strength.

Concrete pavements are designed for fatigue. "Fatigue is the degradation of a material's strength caused by a cyclically applied tensile load that is usually below the yield strength of the material. Fatigue is a concern because a material designed to withstand a safe load one time may fail when the same load is applied cyclically one time too many. The cyclically applied load causes a crack to initiate and propagate from the areas of highest stress concentrations. The material finally fails when the crack grows to a sufficient length so that applied load causes a stress that exceeds the material's ultimate strength" (Titus-Glover et al. 2005: 30).

Design for fatigue is based on a stress ratio (SR) between traffic-induced stress and the flexural strength of the concrete.

$$SR = (\sigma/MOR) \tag{6.1}$$

where σ = load-induced tensile stress in the concrete, and MOR = modulus of rupture or flexural strength of the concrete. At a stress ratio at or near one, the concrete will fail after very few cycles. With a low SR, many load applications may be made.

Flexural strength of concrete, or MOR, is determined using a third point (ASTM C78 2002) or center point (ASTM C293 2002) bending test. The beams tested are $150 \times 150 \times 500$ mm ($6 \times 6 \times 20$ in). Third point testing (ASTM C78 2002) is preferred because the center third of the beam is in pure bending, with no shear. Some agencies use center point testing instead (ASTM C293 2002). Third point testing gives lower but more consistent flexural strengths than center point testing (Mindess et al. 2003: 379). Other tests that can be used include splitting tensile strength (ASTM C496 2004) and compressive strength (ASTM C39 2005).

The test method used for concrete acceptance varies by agency. According to the ACPA database of state practices as of 1999, 12 states used flexural strength for acceptance of concrete, with nine of those using third point and three using center point loading. One state used both third point flexural and compressive strength. Of the others, 21 states used compressive strength, and one state used splitting tensile strength. For US states using flexural strength testing, the required MOR ranges between 3,100 and 4,500 kPa (450 and 650 psi), measured at 7, 14, or 21 days (ACPA 1999c).

FAA requires a minimum allowable flexural strength of 4,136 kPa (600 psi) at 28 days for airport pavements, although individual project specifications may be set higher. For pavements carrying only lighter aircraft, gross weight limit 13,500 kg (30,000 pounds), 28-day compressive strength may be used for acceptance with a minimum of 30.7 MPa (4,400 psi). Requirements are provided by item P-501, Portland Cement Concrete Pavement, of *Standards for Specifying Construction of Airports, AC 150/5370-10B* (FAA 2005: P-501-8).

It should be noted that the MOR used for acceptance and the MOR used for design are almost certainly not the same. Because any concrete below the required MOR may be rejected, the contractor will furnish concrete with a higher strength. Also, testing may be done as early as 7 days to provide early feedback to the contractor and the engineer, but the fatigue life of the pavement extends over decades. Therefore, a higher MOR should be used for pavement design than for concrete acceptance.

For a given SR, a certain number of load repetitions are predicted before the pavement fails in flexural fatigue. The following fatigue equations were developed by Packard and Tayabji (1985) for the PCA design procedure (PCA 1984):

$$\text{For SR} \geq 0.55, \quad \log N_f = 11.737 - 12.077(\text{SR}) \tag{6.2}$$

$$\text{For } 0.45 < \text{SR} < 0.55, \quad N_f = \left(\frac{4.2577}{\text{SR} - 0.4325} \right)^{3.268} \tag{6.3}$$

And for SR ≤ 0.45, the number of repetitions is unlimited.

To achieve a longer fatigue life for a pavement, the SR should be reduced. This is achieved either by reducing the flexural stress in the pavement for a given load (by making it thicker) or by increasing the MOR.

To combine the fatigue effects of different loading magnitudes and configurations, the SR from each vehicle is determined. Next, fatigue equations such as 6.2 are used to calculate the allowable number of repetitions N_f for each SR. Finally, Miner's fatigue hypothesis is applied to sum up the cumulative damage function (CDF) for fatigue:

$$\text{CDF} = \sum_i \frac{n_i}{N_{fi}} \tag{6.4}$$

Where $i =$ the number of load groups or configurations, $n_i =$ the actual or projected number of repetitions of load group i, and $N_{fi} =$ the allowable number of repetitions for load group i. The ratio n_i/N_{fi} represents the fraction of the pavement fatigue life consumed by load group i. The CDF may be expressed as a ratio, in which case it must be ≤ 1.0, or it may be multiplied by 100 and expressed as a percent.

A new concrete pavement fatigue model was developed for the ACPA StreetPave concrete pavement design computer program. The main enhancements were the inclusion of additional fatigue data that has become available since equation 6.2 was developed, and the inclusion of a reliability parameter. The stress ratio and cumulative damage remain as defined in equations 6.1 and 6.4. The revised fatigue expression is:

$$\log N_f = \left[\frac{-\text{SR}^{-10.24} \log(1-P)}{0.0112} \right]^{0.217} \tag{6.5}$$

where $P =$ probability of failure and therefore $(1 - P) =$ probability of survival, and other variables are as previously defined. The fatigue relationships represented by equations 6.2 and 6.5 are compared in Figure 6.1.

It was found that the PCA fatigue equation 6.2 is equivalent to approximately 90 percent reliability, which indicates that only 10 percent of pavements should fail before the end of design life (Titus-Glover et al. 2005). The StreetPave program is discussed in detail in Chapter 9.

Durability

For pavements, durability is almost certainly a more important issue than strength. Pavements are exposed outside, often in harsh environments. In areas such as the author's home in Northeast Ohio, pavements, bridges, and other infrastructure are routinely bathed in harsh deicing materials during the winter. There are very few civil engineering materials that have their properties improved by repeated treatments of brine, particularly during

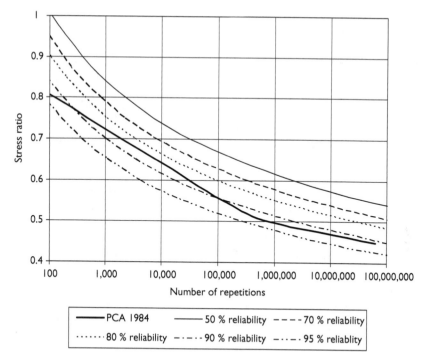

Figure 6.1 Comparison of PCA 1984 and StreetPave fatigue relationships.

repeated cycles of freezing and thawing. Freeze-thaw cycles and chlorides present significant challenges for pavement durability. In addition to using durable aggregates, a low w/c or w/cm ratio should be used. This ratio may be lower than that required for strength.

Freeze-thaw durability

Repeated cycles of freezing and thawing can damage concrete over time unless it has a well-distributed system of finely divided air voids. The air is entrained using the air-entraining admixtures discussed in Chapter 5. Although the air void system in the hardened concrete is important, generally the total air content of fresh concrete is measured. The proper air content may be determined using the tables in ACI 211.1 (ACI Committee 211 1991). It is possible to evaluate the air void system of hardened concrete using petrographic analysis, but this is expensive and rare except for large projects or as part of forensic investigations.

Kohn and Tayabji (2003: 60) note:

- Trial batches are necessary to determine the proper amount of air entraining admixture.
- The concrete for the trial batch should be allowed to sit for an amount of time representative of the haul to the project site, because 1 or 2 percent of the air may be lost during the haul.
- Increasing air content reduces concrete strength if other factors are equal.
- Slipform paving can reduce the air content 1 or 2 percent during consolidation.

The freeze-thaw durability of roller compacted and pervious concrete remains an issue of concern. ACI 211.3 (ACI Committee 211 2002: 13) states that

> Although the resistance of RCC to deterioration due to cycles of freezing and thawing has been good in some pavements and other structures, RCC should not be considered resistant to freezing and thawing unless it is air-entrained or some other protection against critical saturation is provided. If the RCC does not contain a sufficiently fluid paste, proper air entrainment will be difficult, if not impossible, to achieve. In addition, a test method for measuring the air content of fresh RCC has not been standardized.

This contrasts with the observations of Piggott (1999) on the favorable long-term performance of RCC in cold regions.

On pervious concrete, ACI 211.3 (ACI Committee 211 2002: 25) states that

> Freezing-and-thawing tests of pervious concrete indicate poor durability if the void system is filled with water. Tests have indicated that durability is improved when the void structure is permitted to drain and the cement paste is air-entrained. No research has been conducted on resistance of pervious concrete to the aggressive attack by sulfate-bearing or acidic water that can percolate through the concrete. Therefore, caution should be used in applications where aggressive water may exist.

Therefore, the key is to ensure that the pervious concrete is not saturated during freezing weather. The National Ready Mixed Concrete Association (NRMCA) report *Freeze Thaw Resistance of Pervious Concrete* defines three different exposure conditions – dry freeze and dry hard freeze, wet freeze, and hard wet freeze. In dry freeze and dry hard freeze areas, such as many high areas of the Western US, a 100–200 mm (4–8 in) layer of clean aggregate base should be provided under the pervious concrete. Wet freeze areas, such as the middle part of the Eastern US, should use the same precaution. Possible safeguards in the wet hard freeze areas (subjected to long continuous periods below freezing) include a 100–600 mm (4–24 in) layer of base, air entrained paste, and/or perforated PVC drainage pipes. A pervious concrete pavement at Pennsylvania State University, in the hard wet freeze area, has performed well over five winters (NRMCA 2004, ACI Committee 522 2006).

Scaling

Mindess et al. (2003: 505) note that "the use of an adequately air-entrained, low w/c ratio, low permeability concrete is the best protection against salt scaling." These are the same mixture considerations needed to provide freeze-thaw durability. As noted earlier, construction practices also have an important effect on scaling.

> Deicer-scaling resistance of concrete is greatly improved when these fundamentals are followed:
>
> 1. Minimum cement content – 564 lb per cubic yard (335 kg per cubic meter)
> 2. Maximum water-cementitious materials ratio – 0.45
> 3. Low slump – not more than 4 in (100 mm) (unless a water reducer is used)
> 4. Sound, clean, durable, well-graded aggregate
> 5. Adequate air void system
> 6. Proper proportioning, mixing, placing, and finishing.
>
> (ACPA 1992a: 1)

Two other important elements are adequate curing and a period of air drying. Curing is discussed in Chapter 15. A period of air drying greatly improves scaling resistance – the pavement should be allowed to dry 30 days before de-icers are used, although earlier exposure is possible for concrete with w/c of 0.40 or less (ACPA 1992a).

Sulfate attack

Kohn and Tayabji (2003: 59) point out that "if the soils or groundwater contain sulfates, the cementitious material(s) need to be appropriately resistant to sulfate attack *and* the water/cementitious materials ratio needs to be reduced appropriately." Appropriate cementitious materials include pozzolans, slags, and cement with low C_3A content. Type V cement has the lowest C_3A limit, followed by type II.

Corrosion of reinforcement and dowel bars

Traditionally reinforcement and dowel bars for concrete pavements have been made from mild steel, which is subject to corrosion. In areas subject to freezing and thawing, epoxy coated dowels and reinforcement are now often used. Epoxy coated dowels and tie bars are shown in Figures 2.4 and 2.6.

Snyder (2005) investigated the use of passive cathodic protection to protect dowel bars and reported extensive test results. This system uses a 1.2 mm (50 mil) zinc sleeve mechanically bonded to a carbon steel dowel. The zinc sleeve acts as both a barrier to corrosion and a sacrificial anode for cathodic protection. These dowels were also evaluated through four million cycles in an accelerated loading facility and provided load transfer efficiency comparable to conventional dowels, indicating that the zinc coating did not inhibit performance. Corrosion of unprotected carbon steel dowels seems to be an important contributor to poor joint performance. This may be manifested in joint lockup and failure of adjacent joints, and mid-panel cracks. Other protection techniques, such as barrier techniques, corrosion resistance, and non-corroding dowels, were found to be either ineffective or prohibitively expensive. Although conventional dowels are probably satisfactory for pavements with projected service lives on the order of 20 years, zinc coated dowels provide an alternative worthy of consideration for long-life (e.g. 50 year) pavements.

Shrinkage and thermal deformation

The contraction and expansion of concrete pavement slabs due to shrinkage and thermal effects are important for performance. Both the absolute deformation and the relative deformation between the top and the bottom of the slab are important. The absolute deformation, due to uniform strain through the thickness of the slab, determines how much joints open and close, and restrained contraction due to friction between the slab and subbase or subgrade causes tensile stress in the slab.

Strain gradients due to temperature are called curling, and those due to moisture are called warping. These cause cupping upward or downward of the slab. Where the slab lifts off the subbase, it is poorly supported. This can lead to corner breaks as shown in Figure 3.1, particularly for undoweled joints.

During the daytime, the top of the slab is warmer than the bottom, and the slab curls downward. The slab curls upward at night, and if the corners lift off the subbase there is an increased risk of corner breaks. The slab goes through one complete cycle in 24 hours.

Differential moisture changes in the concrete slab cause warping. During curing of the concrete, it is generally easier for moisture to escape from the top of the slab than the bottom, and differential drying shrinkage leads to a permanent upward cupped distortion of the slab, sometimes referred to as "built-in curl" although "built-in warp" is a more accurate term.

Drying shrinkage and thermal deformations should both be minimized. Since drying shrinkage occurs in the cement paste, the amount of paste should be reduced by using larger maximum size coarse aggregate (if available and not subject to D-cracking) and by blending the aggregate to achieve a dense gradation. The drying shrinkage of the paste itself may be minimized by using the lowest possible water content, possibly through use of water reducers. Thermal deformations are reduced by using aggregate with a low coefficient of thermal expansion.

Some factors that influence shrinkage and premature cracking include high water demand, gradation of combined aggregates, and the type of coarse aggregate. High cement contents (more than 295 kg/m^3 or 500 lb/yd^3) and fine sands (FM lower than 3.1) may increase water demand. Aggregate gradation affects workability and amount of paste. Different coarse aggregates have different thermal coefficients, affecting thermal deformations and temperature sensitivity. Also, some admixtures such as calcium chloride as an accelerator and water reducers with accelerators may increase drying shrinkage (Hall et al. 2005: 19–20).

Mixture proportioning example

Two sample mixture proportions are shown below. The first uses 38 mm (1 ½ in) maximum size coarse aggregate, and the second uses 19 mm (3/4 in) coarse aggregate. If available, the larger size coarse aggregate will reduce shrinkage.

Large size coarse aggregate

1 Required information: type I cement with a specific gravity of 3.15; fine aggregate bulk specific gravity (SSD) = 2.63, fineness modulus 2.70; coarse aggregate maximum size 38 mm (1½ in), bulk specific gravity (SSD) 2.68, dry-rodded unit weight 1,600 kg/m^3 (100 lb/ft^3). Pavement is 250 mm (10 in) thick. Required 28-day flexural strength is 4.1 MPa (600 psi) roughly corresponding to a compressive strength of 31 MPa (4,500 psi) based on experience with local materials. The pavement will be subject to freezing and thawing.

2 Choose slump: for slipform paving, slump is 25–50 mm (1–2 in).

3 Choose maximum aggregate size: 38 mm (1½ in) is less than one-third of 250 mm (10 in).

4 Estimate mixing water and air content: from ACI Committee 211 (1991) tables, mixing water is 145 kg/m^3 (250 lb/yd^3). Mixing water

may be reduced if water-reducing admixtures are used. Air content for severe exposure conditions is 5.5 percent.

5 w/c ratio: from ACI Committee 211 (1991) tables, a w/c ratio of 0.44 is necessary to achieve the desired strength. The maximum permissible w/c ratio based on durability is 0.50, so 0.44 should be used.

6 Calculate cement content: divide the water by the w/c ratio to get the cement content of 330 kg/m³ (556 lb/yd³).

7 Estimate coarse aggregate content: from ACI Committee 211 (1991) tables, cross-index the maximum coarse aggregate size with the fineness modulus of sand to find a volume of dry-rodded coarse aggregate per unit volume of concrete of 0.73. Multiply this by the dry-rodded unit weight of the coarse aggregate to obtain $1,168$ kg/m³ ($1,971$ lb/yd³).

8 Estimate the fine aggregate content: determine the volume of each constituent material by dividing it by its unit weight. The constituent materials so far add up to 0.74 cubic meter (20.1 cubic feet), so the remainder, or 0.26 cubic meter (6.9 cubic feet) must be fine aggregate. Multiply by the unit weight of fine aggregate to obtain 682 kg/m³ ($1,131$ lb/yd³).

9 Adjust for aggregate moisture: the water, coarse aggregate, and fine aggregate weights must be adjusted if the aggregates are not in SSD condition.

10 Prepare and test trial batches.

Normal size coarse aggregate

Design for 19 mm (3/4 in) aggregate is adapted from Mindess et al. (2003: 236–239).

1 Required information is the same except for the maximum aggregate size.

2 Slump is the same.

3 Maximum aggregate size is less than one-third of the pavement thickness.

4 Estimate mixing water and air content: from ACI Committee 211 (1991) tables, mixing water is 165 kg/m³ (280 lb/yd³). Mixing water may be reduced if water-reducing admixtures are used. Air content for severe exposure conditions is 6 percent. More water is required with smaller size coarse aggregate.

5 w/c ratio is the same.

6 Calculate cement content: with more water and the same w/c ratio, more cement will be necessary. The cement content 375 kg/m³ (636 lb/yd³), for an increase of 45 kg/m³ (80 lb/yd³).

7 Estimate coarse aggregate content: from ACI Committee 211 (1991) tables, cross-index the maximum coarse aggregate size with the fineness

modulus of sand to find a volume of dry-rodded coarse aggregate per unit volume of concrete of 0.63. Multiply this by the dry-rodded unit weight of the coarse aggregate to obtain 1,008 kg/m³ (1,701 lb/yd³).

8 Estimate the fine aggregate content: the constituent materials so far add up to 0.72 cubic meter (19 cubic feet), so the remainder, or 0.28 cubic meter (8 cubic feet) must be fine aggregate to make one cubic meter (one cubic yard, 27 cubic feet) of concrete. Multiply by the unit weight of fine aggregate to obtain 736 kg/m³ (1,307 lb/yd³).

9 Adjust for aggregate moisture: the water, coarse aggregate, and fine aggregate weights must be adjusted if the aggregates are not in SSD condition.

10 Prepare and test trial batches.

Comparison with the previous example shows that with the smaller coarse aggregate, more water and cement are needed. This will add to the cost of the mixture, and increase shrinkage.

Batch adjustments for aggregate moisture condition

Continuing the previous example, assume that the coarse aggregate has a surface moisture of 0.5 percent and the fine aggregate has a surface moisture of 4.2 percent. Unless the quantities are adjusted, the concrete mixture will have too much moisture, the slump will be too high, and the strength and durability will be reduced. Therefore, the coarse aggregate weight must be multiplied by 1 + surface moisture/100 percent, or 1.005, to get 1,013 kg/m³ (1,710 lb/yd³) of the wet coarse aggregate. Similarly, the fine aggregate must be multiplied by 1.042 to get 766 kg/m³ (1,263 lb/yd³) of the wet fine aggregate. Now, since 5 kg/m³ (9 lb/yd³) of the wet coarse aggregate and 31 kg/m³ (55 lb/yd³) of the wet fine aggregate batched is actually water, this must be subtracted from the mixing water added. Therefore, the batch water becomes 129 kg/m³ (217 lb/yd³). The moisture condition of the aggregate must be monitored, and if it changes the proportions must be adjusted again.

Batch adjustments for supplementary cementitious materials

If a portion of the cement is replaced by supplementary cementitious materials, the replacement may be in terms of either mass or volume, although mass is commonly used. Since supplementary cementitious materials have a different (usually lower) specific gravity from cement, the fine aggregate content will also be adjusted (Mindess et al. 2003: 230–231).

As an example, assume that the 19 mm (3/4 in) aggregate mixture discussed above is to be modified for a 30 percent replacement of cement with fly ash with a specific gravity of 2.4. The new cement content is 262 kg/m^3 (445 lb/yd^3). The mass of fly ash is 30 percent of the total cementitious material, or 113 kg/m^3 (191 lb/yd^3).

However, with a lower specific gravity the fly ash occupies a larger volume than the cement it replaces. The new cement volume is 0.083 cubic meter (2.97 cubic feet) and the fly ash volume is 0.047 cubic meter (0.97 cubic foot).

The coarse aggregate content is unchanged. This increases the volume of all of the constituents except the fine aggregate to 0.73 cubic meter (19.7 cubic foot) so the volume of sand is reduced to 0.27 cubic meter (7.3 cubic foot). Therefore, the revised weight of the sand will be 706 kg/m^3 (1, 191 lb/yd^3). With a fly ash replacement, trial batches become important because the fly ash increases workability and it will probably be possible to remove some water from the mixture.

Batch adjustments for water reducing admixtures

Adjustments for water-reducing admixtures are a little simpler. For example, consider the adjustment of the original 19 mm (3/4 in) aggregate mixture (with cement only) with a 10 percent reduction in water due to the use of a low or mid-range water-reducing admixture. The batch water is reduced to 149 kg/m^3 (252 lb/yd^3). The cement content may now be reduced to 338 kg/m^3 (572 lb/yd^3).

As with the adjustment for supplementary cementitious materials, the coarse aggregate content is unchanged. This reduces the volume of all of the constituents except the fine aggregate to 0.69 cubic meter (18.7 cubic foot) so the volume of sand is reduced to 0.31 cubic meter (8.3 cubic foot). Therefore, the revised weight of the sand will be 808 kg/m^3 (1, 360 lb/yd^3). Once again, trial batches are necessary to evaluate the properties of the adjusted mixture.

Accelerated construction

Often it is necessary to repair or reconstruct concrete pavements in service within very narrow time windows. In many cases, closure windows may be overnight or over a weekend. Van Dam et al. (2005) refer to concrete pavement mixtures that may be opened to traffic after short closures as early-opening-to-traffic (EOT) concrete. One important reference on accelerated concrete pavement construction is ACI 325.11R *Accelerated Techniques for Concrete Paving* (ACI Committee 325 2001). Another reference on the topic is *Fast-Track Concrete Pavements, Technical Bulletin TB004.02* (ACPA 1994a).

For airports under the FAA jurisdiction, item P-501-4.18, Portland Cement Concrete Pavement, of *Standards for Specifying Construction of*

Airports, AC 150/5370-10B (FAA 2005: P-501-25) requires a flexural strength of 3,792 kPa (550 psi), or a wait of 14 days before opening the pavement to traffic. For lighter load airfields using compressive strength, the requirement is 24.13 MPa (3,500 psi). In airport construction, high early strength concrete may be necessary for bridging areas where taxiways cross or high traffic volume apron areas (Kohn and Tayabji 2003: 58).

If the facility can be closed down for a few days, even three or four, then it will often be found that conventional concrete will achieve sufficient strength and no special mixtures or materials will be necessary. In one case, the construction of a new runway at Cleveland Hopkins International Airport in Cleveland, Ohio, two concrete mixtures were developed. A high early strength concrete was proposed for critical tie-ins to existing runways and taxiways. Early in the project, problems were encountered with the use of high early strength concrete, and it was found that with careful scheduling the conventional concrete could be used for all construction, including the tie-ins. Use of the conventional mixture simplified the construction (Peshkin et al. 2006a,b).

Overnight construction

For an overnight closure window, where the facility is closed in early evening and reopened in early morning, the concrete must achieve adequate flexural strength to carry traffic in 6–8 hours. The narrow closure window must incorporate removal of existing pavement as well as placement and curing of the new pavement. Typical specifications require either a compressive strength of 8.3–24 MPa (1,200–3,500 psi) or a flexural strength of 1.8–2.8 MPa (260–400 psi) (Van Dam et al. 2005: 9). Airports may require 5–7 MPa (750–1,000 psi) compressive strength at 4–6 hours for overnight construction (Kohn and Tayabji 2003: 58).

Overnight closures are complicated by the fact that these are typically pavements carrying heavy traffic, with heavy user costs and delays due to construction. For freeways and toll roads, often only overnight closures are permitted.

Occasionally airport pavement reconstruction is limited to overnight closures. The airports in Charleston, South Carolina, and Savannah, Georgia, each have two runways that cross. When the runway intersections needed to be repaired, there were no alternate runways to handle aircraft. The Charleston project in 1990 and the Savannah project in 1996 used similar techniques. During short overnight closures, individual slabs were removed and replaced with very fast setting concrete. If the concrete could not be placed and cured in time, precast panels were installed as a temporary pavement and removed during a subsequent closure. All of the pavement slabs in the intersection were replaced during multiple overnight closures, allowing daytime aircraft traffic to continue uninterrupted. Complete case study details are provided by Peshkin et al. (2006a,b).

Van Dam et al. (2005: A-1–A-2) provide examples of three mixture designs that provided good performance for 6–8 hour EOT concrete as well as three for 20–24 hour EOT concrete. These are shown in Table 6.1.

Many 6–8 hour EOT mixtures in the past have relied on proprietary cements, but satisfactory mixtures may be made with conventional cement. Degussa admixtures has developed a patented system (4×4^{TM}) to achieve 2.8 MPa (400 psi) at 4 hours. The system uses either type III cement or a very reactive type I or I/II cement, plus extended set-controlling admixtures to allow the mixture to be placed and consolidated and a non-chloride accelerating admixture. The system was first employed by the California Department of Transportation (CALTRANS) and has since been employed for Interstate Highways in California, Alabama, and West Virginia, as well as the Philadelphia International Airport (Bury and Nmai 2005).

Weekend construction

For weekend closures, concretes that achieve the required flexural strength in 20–24 hours are satisfactory. These are typically larger projects and involve more construction and more demolition. Examples of mixture designs that provide satisfactory performance under these conditions are shown in Table 6.1. Opening strength requirements are typically the same as those for overnight closures, but more time is available to achieve them. Airports may require 14–21 MPa (2,000–3,000 psi) compressive strength at 24 hours for overnight construction (Kohn and Tayabji 2003: 58).

Durability of EOT concrete

The durability issue presents a serious problem, because those pavements that are so heavily trafficked that only short closure windows are available are precisely those pavements that need to have durable concrete. Van Dam et al. (2005: 19) note that "In comparison with 20- to 24-hour EOT concrete, 6- to 8-hour EOT concrete will have higher cement contents and lower w/c ratios. Type III cement, accelerators, and water reducers are more often used in 6- to 8-hour EOT concrete than in 20- to 24-hour EOT concrete. It was clearly observed from the laboratory test results that the 6- to 8-hour mixtures had less desirable durability characteristics than the 20- to 24-hour mixtures. This observation was reflected in the overall poorer performance in freeze-thaw and scaling tests, increased shrinkage, increased difficulties in achieving desirable air-void system characteristics, increased amounts of paste microcracking, decreased paste homogeneity, and increased absorption. This is not to suggest that durable 6- to 8-hour EOT concrete mixtures cannot be made; it simply points out the difficulty

Table 6.1 Sample concrete mixture designs for early opening to traffic (Van Dam et al. 2005: A-1–A-2)

Mixtures 1, 2, and 3, all of which were made using a vinsol resin air-entraining agent, provided good performance for six- to eight-hour EOT concrete

Constituent/property	Mixture 1	Mixture 2	Mixture 3
Type I cement	525 kg/m^3 (885 lb/yd^3)	525 kg/m^3 (885 lb/yd^3)	425 kg/m^3 (716 lb/yd^3)
w/c ratio	0.40	0.36	0.40
Accelerator type	Non-chloride	Non-chloride	Calcium chloride
Water reducer	None	None	None
Coarse aggregate (crushed limestone)	1,030 kg/m^3 (1,736 lb/yd^3)	1,030 kg/m^3 (1,736 lb/yd^3)	1,030 kg/m^3 (1,736 lb/yd^3)
Fine aggregate (natural sand)	427 kg/m^3 (720 lb/yd^3)	482 kg/m^3 (812 lb/yd^3)	425 kg/m^3 (716 lb/yd^3)
Average slump	140 mm (5.5 in)	70 mm (2.75 in)	65 mm (2.5 in)
Average air content	5%	5%	5.6%
Eight-hour compressive strength	16.4 MPa (2,375 psi)	20.4 MPa (3,000 psi)	17.0 MPa (2,465 psi)
28-day compressive strength	44.0 MPa (6,400 psi)	56.3 MPa (8,150 psi)	53.8 MPa (7,800 psi)
Eight-hour flexural strength	2.4 MPa (350 psi)	3.0 MPa (435 psi)	2.4 MPa (350 psi)

Mixtures 4, 5, and 6, all of which were made using a vinsol resin air-entraining agent, provided good performance for 20- to 24-hour EOT concrete

Constituent/property	Mixture 4	Mixture 5	Mixture 6
Type I cement	400 kg/m^3 (678 lb/yd^3)	400 kg/m^3 (678 lb/yd^3)	475 kg/m^3 (805 lb/yd^3)
w/c ratio	0.43	0.40	0.43
Accelerator type	Calcium chloride	Non-chloride	None
Water reducer	None	None	None
Coarse aggregate (crushed limestone)	1,030 kg/m^3 (1,736 lb/yd^3)	1,030 kg/m^3 (1,736 lb/yd^3)	1,030 kg/m^3 (1,736 lb/yd^3)
Fine aggregate (natural sand)	628 kg/m^3 (1,060 lb/yd^3)	659 kg/m^3 (1,010 lb/yd^3)	659 kg/m^3 (1,110 lb/yd^3)
Average slump	85 mm (3.35 in)	50 mm (2 in)	150 mm (6 in)
Average air content	6.6%	5.7%	5.9%
20-hour Compressive strength	24.5 MPa (3,550 psi)	19.9 MPa (2,890 psi)	17.8 MPa (2,580 psi)
28-day compressive strength	46.0 MPa (6,670 psi)	40.6 MPa (5,890 psi)	39.3 MPa (5,700 psi)
20-hour flexural strength	3.4 MPa (490 psi)	3.8 MPa (550 psi)	3.6 MPa (520 psi)

in achieving the desired characteristics of a durable mixture in these higher early strength mixtures. Thus, there is a higher level of risk associated with using a 6- to 8-hour EOT concrete than a 20- to 24-hour EOT concrete that must be considered when selecting a specific mixture to reduce lane closure time." This suggests that if six- to eight-hour EOT concrete is to be used, there should be extensive preliminary testing to ensure durability.

One example of poor durability of EOT concrete occurred at Houston Hobby Airport in Texas. The intersection of runways 4-22 and 12R-30L at Houston Hobby airport is heavily used and closures for repairs of this area must be limited because of the impact on commercial operations: the intersection carries 95 percent of the total aircraft traffic at the airport. When the intersection is closed, Hobby has only one visual flight rules (VFR) runway that can handle air carrier operations, and even short periods of rain or fog shut down the airport.

The intersection was repaired in 16 days in 1994 using very high early strength concrete. The original 150 mm (6 in) thick runway pavement at the time of this repair was more than 50-years old and had been over-laid several times to increase the section by 450 mm (18 in). The project concrete consisted of significant quantities of type III cement and fly ash, with a superplasticizer and an accelerating admixture. The specified flexural strengths were 5.2 MPa (750 psi) at 24 hours and 5.9 MPa (850 psi) at 28 days. The 24-hour strength was much higher than the typical requirement at 28 days. Because the original amount of accelerator used was too great and the concrete lost workability before finishing could be completed, the amount of accelerator was reduced, leading to decreased 24-hour strengths. By the late 1990s, this repaired area had experienced significant deterio-ration, resulting in operational warnings to pilots and emergency closures for repairs. The problem was initially identified through pilot complaints about rideability problems on runway 12R-30L. Although the center line profile was smooth, the wheel track profiles for the main gear had an abrupt transition to 1 percent slopes. Observed distresses included surface cracking, differential joint movement, concrete expansion and shoving, and full depth and lateral cracking. A forensic investigation found that the con-crete used in the project had produced delayed ettringite crystal formations, which expanded the volume of the concrete pavement. The movement was so great that several light bases in the intersection were damaged by shear-ing forces as the pavement expanded and slid relative to the base. Asphalt adjacent to the intersection was cracked and heaved due to the pressure of the expanding concrete (Sarkar et al. 2001).

Chapter 7

Design fundamentals

Concrete pavement design is often thought of solely in terms of thickness design, but there are other important elements. These include the type and spacing of joints (if any) – both transverse joints for JPCP and longitudinal joints for all pavement types. The design of the drainage system has been discussed previously in Chapter 4.

> Theoretical analysis of a concrete pavement is a surprisingly daunting task: geometric and joint issues, material shrinkage factors, environmental effects on the structure, uncertain in-situ material properties and strengths, complex nonlinear material response, perplexing fatigue behavior, moving loads, etc. Consequently, pavement design has tended to evolve with relatively simple analytical and material models, and a number of empirical and experience-based guidelines on joints, pacing, etc. that tend to work and keep us out of trouble.
>
> (Rollings 2005: 170)

"Any of the major designs – plain, reinforced, continuously reinforced, or prestressed – will give good performance if the subgrade, subbase, jointing, and use of steel, if any, are properly considered and provided for in the design, and if traffic is reasonably well forecast" (Ray 1981: 8).

> The determination of pavement thickness requirements is a complex engineering problem. Pavements are subject to a wide variety of loading and climatic effects. The design process involves a large number of interacting variables, which are often difficult to quantify. Despite considerable research on this subject, it has been impossible to arrive at a direct mathematical solution for thickness requirements. For this reason, pavement engineers must base pavement thickness on a theoretical analysis of load distribution through pavements and soils, the analysis of experimental pavement data, and a study of the performance of pavements under actual service conditions.
>
> (FAA 2005: 23)

Any transitions between concrete pavement and adjacent asphalt pavement sections or bridges should also be carefully designed. For CRCP pavements, the design also includes the selection of the steel percentage, by volume. CRCP reinforcement, transitions, and other special details are discussed in Chapter 12.

Pavement support

Pavement support is generally reduced to a k-value, based on the soil type. Table 4.1 may be used to select a k-value for the subgrade soil, and Tables 4.2 through 4.5 may be used to adjust the value based on an untreated aggregate or stabilized subbase.

While it is conservative to use a low k-value for thickness design, it is not conservative to use a low value to determine joint spacing. Therefore, if a stiff stabilized layer is used under the pavement – CTB, ATB, etc. – a higher k-value must be used for joint spacing determination.

For most concrete pavement design procedures, the thickness of the pavement is not very sensitive to the k-value. This is in contrast to flexible or asphalt pavements, where thickness is highly dependent on the soil stiffness. Therefore, the subgrade and subbase materials are not particularly important for determination of pavement thickness, but are very important for constructability, pumping, and frost susceptibility. These issues are addressed in Chapters 4 and 13.

Traffic

There are three approaches to dealing with traffic for pavement design. The simplest, by far, is if only one vehicle type needs to be considered in the design. If a mixed traffic stream must be considered, then the effects of different axle groups may be considered separately, or all axle groups may be converted to equivalent 80 kN (18 kip) single axles.

Single vehicle

For some pavements, primarily airfield or industrial pavements, it may be possible to use a single vehicle approach. This method is valid if a single heavy vehicle or aircraft type dominates the traffic stream, and other loads on the pavement are so light that they cause minimal damage. This is the case if the lighter loads are not more than about 80 percent of the design vehicle weight.

To design for a single vehicle, the pavement flexural stress for a trial pavement thickness is determined. Next, the flexural stress is divided by the modulus of rupture to obtain the stress ratio SR, and fatigue models such as equation 6.2 or 6.5 are used to find the allowable number of repetitions. The trial pavement thickness is then adjusted until the design will carry the projected number of load repetitions.

Mixed traffic stream

Most highways and airports carry mixed traffic streams, and the single vehicle assumption cannot be used. If this is the case, the design approach adopted by the 1984 PCA method may be used. This design procedure calculates the pavement damage caused by each vehicle type in the traffic stream, and then adds up the damage over the projected life of the pavement to ensure that it does not exceed some allowable value.

First, a trial pavement thickness is assumed. Next, the damage caused by each type of vehicle in the mixed traffic stream is calculated, based on the expected number of load repetitions for that vehicle type over the projected life of the pavement. For highways, vehicles are typically grouped by type (single, tandem, or tridem/triple axle) into 8.9 or 17.8 kN (2,000 or 4,000 pound) increments.

Damage due to each vehicle increment is then added up and compared to allowable maximums. For the 1984 PCA method, the limits are 100 percent fatigue damage and 100 percent erosion damage, computed using equation 6.4. Either fatigue or erosion may determine the design thickness (PCA 1984).

The NCHRP/AASHTO M-EPDG applies a similar but significantly more sophisticated approach. Damage is calculated for each load group over each season. Total damage over the life of the pavement is then compared to allowable maximums for transverse cracking, joint faulting, etc., and then the design is adjusted if necessary (Applied Research Associates 2006).

For airport pavements, only a limited number of types of aircraft need to be considered. The number of repetitions or coverages per type may be predicted fairly accurately. The predicted air traffic is not likely to change significantly, unless there is a change of airline tenants.

The FAA Advisory Circular *Airport Pavement Design, AC 150/5320-6D* states that "the pavement design method is based on the gross weight of the aircraft. The pavement should be designed for the maximum anticipated takeoff weight of the aircraft. The design procedure assumes 95 percent of the gross weight is carried by the main landing gears and 5 percent is carried by the nose gear. The FAA Advisory Circular AC 150/5300-13, *Airport Design*, lists the weight of many civil aircraft. The FAA recommends using the maximum anticipated takeoff weight, which provides some degree of conservatism in the design. This will allow for changes in operational use and forecast traffic, which is approximate at best. The conservatism will be offset somewhat by ignoring arriving traffic" (FAA 2004: 24).

Axle equivalency (ESAL) concept

The various editions of the AASHTO Design Guide for rigid and flexible pavements, culminating in the 1993 Guide (AASHTO 1993) and 1998 Supplement (AASHTO 1998), have relied on the axle equivalency concept to design pavements for mixed traffic. Axle equivalency factors have been developed for various magnitudes of single, tandem, or tridem/triple axles by determining the relative damaging power of each axle, in loss of pavement serviceability, to that of an equivalent 80 kN (18 kip) single axle.

These equivalency factors are then multiplied by the number of axles in each category to determine a number of 80 kN (18 kip) equivalent single axle loads, or ESALs. The pavement is then designed to carry a specific number of ESALs.

Equivalency factors are roughly proportional to the fourth power of the axle load magnitude (Yoder and Witczak 1975: 154). Therefore, doubling the load on an axle type causes 16 times as much pavement damage, as a rule of thumb. This relationship is not exact, and varies by pavement type, but is useful as an estimate.

There are two significant problems with using ESALs for pavement design. The first is that AASHTO has developed different equivalency factor tables for flexible and rigid pavements, and the numbers are different for different pavement thicknesses. This makes it difficult to compare nominally equivalent designs for asphalt and concrete pavements, because even if they are designed for the same traffic they will not be designed for the same number of ESALs.

The second is that some lighter axles do essentially no damage to the pavement, and thus should have an equivalency factor of zero. Using the PCA fatigue equations, loads that produce a SR $\leq$0.45 do no damage and therefore unlimited repetitions are allowed. However, the AASHTO ESAL tables assign equivalency factors to those loads.

Curling and warping stresses

Stresses in rigid pavements are due to environmental effects and traffic loads. Stresses due to traffic loads determine the fatigue life of the pavement, while environmental effects due to curling and warping determine the maximum joint spacing for the pavement. As noted in Chapter 2, stresses are resisted by the flexural strength of the concrete, and not by any included reinforcing steel. Reinforcing steel for JRCP and CRCP is simply used to hold cracks together.

Stresses due to curling and warping depend on the ratio between the length of the slab L and the radius of relative stiffness ℓ. The radius of relative stiffness ℓ is provided by equation 7.1:

$$\ell = \sqrt[4]{\frac{ED^3}{12(1 - v^2)k}} \qquad (7.1)$$

where $E =$ the modulus of the elasticity of the concrete, D is the pavement thickness, k is the modulus of subgrade reaction, and v is the Poisson's ratio of concrete, typically taken as 0.15.

Because ℓ is in units of length, the ratio L/ℓ is dimensionless. Huang (2004: 151–152) provides an example of calculating the radius of relative stiffness for a slab 200 mm (8 in) thick, with $E = 27.6$ GPa (4,000,000 psi) and $k = 54.2$ MPa/m (200 psi/in). From equation 7.1, $\ell = 776$ mm (30.6 in).

The slab edge-stress due to temperature curling is provided by:

$$\sigma = \frac{CE\alpha_t \Delta t}{2(1 - v^2)} \tag{7.2}$$

where $\alpha_t =$ concrete thermal coefficient of expansion, discussed in Chapter 5, and $\Delta t =$ temperature difference between the top and bottom of the slab. C is a correction factor for a finite slab, determined using L/ℓ, with a chart developed by Bradbury (1938) (Figure 7.1).

Note that a stiffer base or subbase material (higher k) leads to a smaller ℓ, and thus higher L/ℓ ratios for the same joint spacing, higher C coefficients, and higher curling stresses. Also, with a stiffer k, the required pavement thickness D will often decrease, which will reduce ℓ further still. Use of improved, stiffer base and subbase layers and thinner pavements, without correspondingly reducing joint spacing, will lead to a greater incidence of mid-slab cracking due to curling and warping.

Curling stresses may be calculated for a slab 7.6 m (25 ft) long and 3.66 m (12 ft) wide using the example from Huang (2004: 151–152) above.

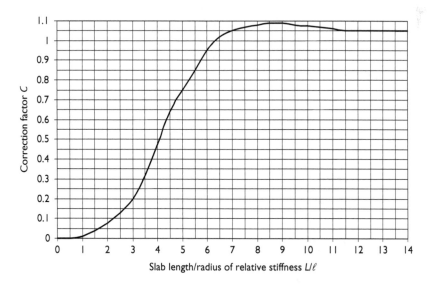

Figure 7.1 Curling stress correction factors for a finite slab (after Bradbury 1938).

$L_x/\ell = 9.81$ and $L_y/\ell = 4.71$, so $C_x = 1.07$ and $C_y = 0.63$. If the temperature differential between the top and bottom of the slab is $11.1\,°C$ ($20\,°F$) and the thermal coefficient of the concrete $\alpha_t = 9 \times 10^{-6}/°C$ ($5 \times 10^{-6}/°F$), then the edge-stress due to curling is 1.48 MPa (214 psi) in the longitudinal direction. If the slab length is reduced to 4.6 m (15 ft), then $L_x/\ell = 5.89$ and $C_x = 0.93$. The shorter slab length reduces the edge-stress to 1.28 MPa (186 psi).

Because thermal strain $\varepsilon_t = \alpha_t \Delta_t$, equation may be modified for differential shrinkage $\Delta\varepsilon_{sh}$ between the top and bottom of the slab:

$$\sigma = \frac{CE\Delta\varepsilon_{sh}}{2(1 - v^2)} \tag{7.3}$$

Curling occurs in a daily cycle, with heating during the day causing curling downward (positive temperature gradient) and cooling at night causing curling upward (negative temperature gradient). The bottom of the slab, against the earth, remains at a more constant temperature. Tensile stresses are induced in the top of the slab during the day and in the bottom of the slab at night.

In contrast, warping occurs one time as the pavement is curing and drying. Generally it is easier for moisture to escape from the top of the slab than from the bottom, particularly if curing practices are not adequate or the subbase drains poorly. There will, therefore, be differential shrinkage between the top and bottom of the slab. This builds a permanent upward warp in the slab.

LTPP data has been used to investigate locked-in curvatures for more than 1,100 JPCP 152 m (500 ft) test site profiles from 117 different sites. The worst sections were found to have differential elevations between the mid-slab and the joints of as much as 10 mm (0.4 in). The most extreme curvatures occurred in pavements over low plasticity sandy-gravely clay or clayey sand-gravel. Slabs with severe upward curvature are at risk for excessive joint deflections, premature faulting and spalling, pumping, and mid-slab cracking. Curvatures were much higher for pavements with undoweled joints (Byrum 2000, 2001).

Clearly, differential elevations of this magnitude can affect pavement ride quality. Pavement construction practices, particularly curing, are likely to have an important effect on the magnitude of locked-in curvature, with well-cured doweled pavements likely to have the least curvature.

Traffic loading stresses

The stresses caused by traffic loading on a concrete pavement slab depend on the location of the load, among other factors. If the load is carried in the interior of the slab, away from any corners, the stress level is generally

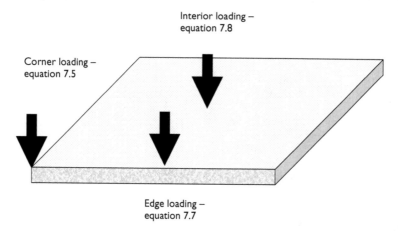

Interior loading –
equation 7.8

Corner loading –
equation 7.5

Edge loading –
equation 7.7

Figure 7.2 Corner, edge, and interior loading.

low. Therefore, analytical solutions have been developed for loads applied at corners, edges, and at the interior of concrete pavement slabs, as shown in Figure 7.2.

Stresses from corner loading

Corner stresses are associated with corner breaks, while edge and interior stresses are associated with mid-slab transverse cracking. Stresses depend on the load applied and load configuration, slab thickness, modulus of subgrade reaction k, and radius of relative stiffness ℓ.

The oldest and simplest corner stress equation can be developed directly from equilibrium. This formula was developed by Goldbeck (1919) and Older (1924) and is discussed by Huang (2004: 154–155). It is based on some very conservative assumptions – the load is placed at the corner of the slab, and there is no subgrade support ($k = 0$). It approximates the situation that may occur with pumping under pavement corners, or curling and warping that lifts the corner from the subgrade.

The formula is developed from the standard bending stress equation $\sigma = Mc/I$. The failure plane is assumed to occur a distance x from the corner, and forms a crack of length $2x$. Therefore, if D is the slab thickness, $c = D/2$ and $I = bh^3/12 = (2x)D^3/12$. The bending moment for the cantilever is Px. Thus, the corner stress becomes:

$$\sigma_c = \frac{Mc}{I} = \frac{Px(D/2)}{(2x)\,D^3/_{12}} = \frac{3P}{D^2} \tag{7.4}$$

This formula is conservative, because the load is distributed over the contact area of the tire and thus the moment must be less. Furthermore, it is unlikely that there would be no subgrade support under the corner at all. It is, however, useful to note that stress decreases with the square of the pavement thickness D.

This equation was updated by Westergaard (1926) and subsequently by Ioannides et al. (1985), and is discussed by Huang (2004: 154–155).

$$\sigma_c = \frac{3P}{D^2}\left[1-\left(\frac{c}{\ell}\right)^{0.72}\right] = \frac{3P}{D^2}\left[1-\left(\frac{1.772a}{\ell}\right)^{0.72}\right] \tag{7.5}$$

Where $c =$ the side length of a square contact area and $a =$ radius of a circle with the same contact area. Westergaard and Ioannides et al. use h for slab thickness instead of D, but D is used in equation 7.5 for consistency with the rest of this book.

Because of the term in brackets, the corner stress calculated using equation 7.5 will always be lower than that calculated using equation 7.4. In pavement stress calculations, the load may be considered to be evenly distributed as a uniform tire pressure p over a circle of radius a, so:

$$P = p\,\pi\,a^2 \text{ or } a = \sqrt{\frac{P}{p\pi}} \tag{7.6}$$

For a slab 200 mm (8 in) thick, and a single axle load equal to half an ESAL or 40 kN (9,000 lb), equation 7.4 gives a corner bending stress of 3 MPa (435 psi). If the concrete has $E = 27.6$ GPa (4,000,000 psi) and the subgrade $k = 54.2$ MPa/m (200 psi/in), then $\ell = 776$ mm (30.6 in) as calculated previously in this chapter. If the tire pressure is 620 kPa (90 psi), then the radius a is 143 mm (5.6 in). The corner stress then is multiplied by 0.556, which is the term in brackets in equation 7.5, so the stress becomes 1.67 MPa (242 psi).

Stresses from edge-loading

For mid-slab cracking, the edge-loading stress is critical because it is much higher than the stress due to interior loading. Westergaard developed several stress equations. The final equation for a circular loading patch is provided by Ioannides et al. (1985) and is discussed by Huang (2004: 157–158):

$$\sigma_e = \frac{0.803P}{D^2}\left[4\log\left(\frac{\ell}{a}\right)+0.666\left(\frac{a}{\ell}\right)-0.034\right] \tag{7.7}$$

This assumes a Poisson's ratio of concrete $v = 0.15$. Continuing with the previous example slab and loading case [$D = 200$ mm (8 in), $P = 40$ kN

(9,000 lb), $E = 27.6$ GPa (4,000,000 psi), the subgrade $k = 54.2$ MPa/m (200 psi/in), $\ell = 776$ mm (30.6 in), and a is 143 mm (5.6 in)] we calculate an edge-loading stress $\sigma_e = 2.36$ MPa (343 psi). This stress is 42 percent higher than the corner loading stress calculated above using equation 7.5. This suggests that, as a general rule, edge-loading stress is higher than corner loading stress, and thus mid-slab cracking is more likely than corner breaks unless the pavement corners are heavily loaded or unsupported. Also, dowels near corners help prevent corner breaks.

Stresses from interior loading

The Westergaard (1926) interior loading stress equation may be written as:

$$\sigma_i = \frac{0.316P}{D^2} \left[4 \log \left(\frac{\ell}{b} \right) + 1.069 \right]$$ (7.8)

Where $b = a$ if $a \geq 1.724D$, and otherwise

$$b = \sqrt{1.6a^2 + D^2} - 0.675D$$ (7.9)

This equation is discussed by Huang (2004: 156–157). As with equation 7.7, this assumes a Poisson's ratio of concrete $v = 0.15$. For the slab and loading parameters discussed in the examples above, equation 7.9 applies because a is 143 mm (5.6 in) and is not $\geq 1.724D$, or 345 mm (9.65 in). Therefore, $b = 134$ mm (5.29 in) and the interior-loading stress $\sigma_i = 1.26$ MPa (183 psi). This is much lower than the edge-loading stress, and so the edge-loading stress is used to analyze the risk of mid-slab cracking.

Combining stresses from curling, warping, and traffic loading

Obviously stresses due to curling, warping, and loading are present simultaneously and affect the performance of the pavement. Since the service life of the slab is predicated on fatigue, the number of cycles or reversals of each type of stress is important.

Over the life of a pavement, the stress cycles due to curling and warping are much fewer than those due to loading. Temperature effects occur once per day, with the top of the slab heating and cooling while the bottom of the slab remains at a more constant temperature. Even over 20 years, therefore, there are only about 7,300 fatigue cycles due to temperature. Moisture warping occurs only once, as the pavement cures and dries out, unless there are significant seasonal moisture changes that are not uniform through the slab, which would not be a common occurrence.

In contrast, traffic loads produce fatigue stresses in pavements for any-where from tens or hundreds of thousands of repetitions, for airports, up to tens or hundreds of millions of repetitions for busy highways. Therefore, in fatigue analysis, curling and warping stresses are generally ignored.

In Yoder and Witczak's classic textbook, they state "It is important to note that warping stresses are not considered when determining thickness of pavement. The philosophy that governs the design, simply stated, is 'Joints and steel are used to relieve and/or take care of warping stresses, and the design, then, is based upon load alone when considering thickness.' This principle is so important that it must be clearly understood by the designer. Recall that a joint is nothing more than a 'designed crack'" (Yoder and Witczak 1975: 125). In this passage, "warping" encompasses both curling and warping as they are defined today.

At the time the paragraph above was written, lightly reinforced slabs with only temperature steel (JRCP) were often used. Today, the philosophy for JPCP is to ensure that the joint spacing is short enough to reduce curling stresses. Therefore, with proper joint spacing, there is no explicit need to consider curling and warping stresses in design. With CRCP, the closely spaced cracks divide the pavement into miniature slabs short enough that curling is not an issue.

Slab deflections

Westergaard (1926) and Ioannides et al. (1985) also developed equations for slab deflections due to corner, edge, and interior loading (Δ_c, Δ_e, and Δ_i, respectively). These equations are discussed by Huang (2004: 155–158).

$$\Delta_c = \frac{P}{k\ell^2}\left[1.205 - 0.69\left(\frac{c}{\ell}\right)\right] = \frac{P}{k\ell^2}\left[1.205 - 0.69\left(\frac{1.772a}{\ell}\right)\right] \quad (7.10)$$

$$\Delta_e = \frac{0.431P}{k\ell^2}\left[1 - 0.82\left(\frac{a}{\ell}\right)\right] \quad (7.11)$$

$$\Delta_i = \frac{P}{8k\ell^2}\left\{1 + \frac{1}{2\pi}\left[\ln\left(\frac{a}{2\ell}\right) - 0.673\right]\left(\frac{a}{\ell}\right)^2\right\} \quad (7.12)$$

Deflections, particularly at slab corners, are important from the standpoint of pumping. Slabs that deflect more apply more pressure and energy to the subgrade, with a higher probability of ejecting pumpable water and materials. This is the basis of the PCA 1984 erosion design.

On the other hand, it is not an efficient design practice to attempt to solve pumping problems by increasing slab thickness. It is better to provide dowels at joints and to provide a subbase. Doweled joints support slab edges and corners, transferring loads to adjacent slabs through dowel shear action. This is a more certain and probably more economical approach to pumping.

General stress and deflection calculation from finite element programs

The analytical solutions developed by Westergaard and refined by Ioannides and his colleagues represent idealized situations that rarely occur in practice. The corner and edge solutions are very conservative. Even with the obsolete practice of using narrow lanes only 2.7–3 m (9–10 ft) wide, it would not be realistic to expect traffic to drive right along the edge of the pavement, nor would it be safe.

Typically modern highways are designed with 3.7 m (12 ft) traffic lanes. Therefore, for highway pavements neither edge-loading nor interior loading really applies, and the actual stress condition is in between. It is always conservative to design on the basis of edge-loading stress. Some highway pavements use widened slabs for the outside lane that carries the most truck traffic – typically widened lanes are 4.3 m (14 ft) wide. The definition of a widened slab is "concrete pavement slab that is paved wider (usually at least 18 in. [450 mm] wider) than a conventional 12 ft (3.7 m) traffic lane to increase the distance between truck tires and slab edge, thereby reducing edge stresses due to loading. Concrete shoulders that are tied to the travel lane with bars also reduce edge loading stresses" (ACI Committee 325 2006: 3). Doweled transverse and longitudinal joints also reduce edge-stresses.

Due to the difficulty of computing stresses for various pavement and loading geometries, finite element stress analysis programs have been developed. Early versions were used to develop the design tables and charts for the PCA 1984 procedure. Other subsequent finite element programs include JSLAB, ILLI-SLAB, KENSLABS, and EVERFE.

The program KENSLABS, which is provided with Huang's pavement design textbook, has many features in common with the other programs. Key features are summarized in Chapter 5 of the textbook (Huang 2004) and are listed below:

- Each pavement slab is divided into rectangular plate elements. The number of elements per slab is determined from the grid geometry input by the program user. A larger number of elements improves the accuracy of the solution but requires more computation time, which is generally not a problem with modern computers.
- The slab may have one or two layers. If there are two layers, they may be bonded or unbonded. This may be used to model bonded and unbonded overlays, or concrete pavement bonded to a base of a different material such as asphalt or CTB. Each of the two layers may be assigned its own modulus of elasticity and Poisson's ratio.
- Three different foundation models may be used. The simplest is the liquid, or spring foundation using k-values. A more realistic, and complex, model is the solid foundation. The subgrade and subbase are modeled as a single elastic halfspace, or Boussinesq model. The most complex

foundation is the layer model, also called a Burmister foundation. With this foundation, the modulus of elasticity and Poisson's ratio of each layer (subgrade, subbase, base if any) are used directly. This behaves as a multilayer flexible pavement under the slab. For most problems, the liquid k-value foundation is sufficiently accurate. Gaps between the slab and the foundation, representing pumping or subsidence, may also be considered.

- Dowel bars are modeled as transfer of shear and bending forces between slabs. A shear spring constant may be specified, or the diameter and spacing of the dowels as well as their material properties may be provided. It is also possible to model loose dowels, by specifying a gap between the dowel and the concrete.
- Aggregate interlock joints are modeled as shear transfer across the joint between slabs.
- Multiple slabs may be modeled – thus, load transfer across doweled or aggregate interlock transverse joints or across tied longitudinal joints may be included in the problem. A maximum of six slabs may be analyzed.
- Traffic loads are represented as rectangular patches with uniform tire pressure over each patch. Different pressures may be used for different patches, and point loads may be applied at nodes.
- Temperature curling may be included in the problem. This feature may also be used indirectly to model moisture warping.
- The output of the program includes stresses, strains, and deflections. These may be reported at all nodes, or only at the nodes with the highest values. Dowel bearing stresses may also be calculated.

Finite element programs appear to have satisfactory accuracy, and have been shown to agree with analytical solutions. There are also limitations to these tools, and the limitations of KENSLABS are described by Huang (2004: 208–209).

Another rigid pavement finite element program is EVERFE. This is available at no cost on the web at http://www.civil.umaine.edu/EverFE/.

Thickness design

Thicker pavements can carry more loads, because flexural stresses are reduced and the fatigue life of the pavement is increased. There are two basic approaches to determining the proper thickness for a concrete pavement. These are empirical and mechanistic designs. Empirical design is based on the observed performance of pavements in the field, under normal or accelerated traffic loading. Mechanistic design relies on calculation of stresses and strains. In reality, all design procedures combine both approaches to some extent.

The AASHTO design procedure up through the 1993 guide (AASHTO 1993) is empirically based. The design equations are based on regression of the results of a two-year road test carried out in Ottawa, Illinois, in the late 1950s, by AASHO (before the T for transportation was added). The AASHO Road Test is discussed in Chapter 1. Loops with different pavement designs were loaded by trucks with different axle weights and configurations, and the distresses and serviceability of the pavements were measured over time. Because approximately 20 years of traffic were applied in 2 years, this represents an accelerated test. The basic equations were modified in succeeding editions of the AASHTO Design Guide, often adding adjustment factors based on mechanistic principles (Huang 2004: 568–570).

The problem with relying on empirical designs becomes evident when it is necessary to extend them to new conditions – either new environments, new pavement designs, or new vehicle types. The AASHTO design procedure has, of course, been used for environments other than Ottawa, Illinois, and triple axles and CRCP have been added to the design procedure, although they were not part of the original road test. Furthermore, the pavements at the AASHO Road Test were much more vulnerable to pumping than pavements built with more recent details and construction techniques. Details of the AASHTO 1993 and 1998 supplement design procedures are provided in Chapter 8.

The PCA 1984 procedure (PCA 1984) is mechanistic. Two failure mechanisms are identified, fatigue cracking and erosion (pumping), and the pavement thickness is determined so that neither failure mode occurs before the end of the required pavement life. This approach forms the basis for the PCA and ACPA methods for airport and industrial pavement design also. FAA airport pavement design procedures are also mechanistic.

The latest proposed AASHTO M-EPDG procedure combines both approaches, as the name indicates. Results from finite element modeling are used to determine stresses.

It should be noted that thicker pavements cannot compensate for defective details, poor materials, or poor construction practices. Adequate thickness is a necessary but not a sufficient condition for long pavement life.

Transverse joint design and spacing

Distortion of pavement slabs due to curling and warping has been discussed previously. With greater joint spacing and longer slabs, the slab corners lift and flexural stresses at the midpoint of the slab increase.

The joint design problem breaks down into several elements:

- Selection of an appropriate joint spacing. Some agencies have used variable spacing to reduce the likelihood of rhythmic excitation of vehicles.

- Determination of whether aggregate interlock joints or doweled joints will be used.
- If doweled joints are used, what diameter of dowels to use.
- Determination whether the joint is transverse to the centerline of the pavement or skewed.

In the past, keyed joints have been used. These are described in Chapter 2. Problems occur when either the key or the top and bottom of the slab above the slot break off. Yoder and Witczak (1975: 582) reported that they should not be used for slabs less than 225 mm (9 in) thick, or for airfields with heavy traffic. In current practice, keyed joints are avoided and tied or doweled joints are used instead.

Joint spacing

With JPCP, the joint spacing should be short enough to prevent high curling stress buildup. This was discussed earlier in the section on calculation of curling and warping stresses. Smith et al. (1990) found that joint faulting and transverse cracking increase with longer joint spacing. Cracking, in particular, increases when joint spacing exceeds 5.5 m (18 ft) for pavements with random joint spacing. This trend would be expected to hold true for pavements with uniform joint spacing. "Nussbaum and Lokken (1978) recommended maximum joint spacings of 20 ft. (6.1 m) for doweled joints and 15 ft. (4.6 m) for undoweled joints" (Huang 2004: 15).

The AASHTO 1993 Pavement Design Guide states

> In general, the spacing of both transverse and longitudinal contraction joints depends on local conditions of materials and environment... the spacing to prevent intermediate cracking decreases as the thermal coefficient, temperature change, or subbase frictional resistance increases; and the spacing increases as the concrete tensile strength increases. The spacing also is related to the slab thickness and the joint sealant capabilities... As a rough guide, the joint spacing (in feet) for plain concrete pavements should not greatly exceed twice the slab thickness in inches. For example, the maximum spacing for an 8-inch [200 mm] slab is 16 feet [4.8 m].
>
> (AASHTO 1993: II-49)

This relationship may be expressed as:

$$S = 0.024 \, D \tag{7.13}$$

where S = joint spacing in m and D = slab thickness in mm, or

$$S = 2D \tag{7.14}$$

where $S =$ joint spacing in feet and $D =$ slab thickness in inches.

ACI Committee 325 (2002: 14) provides a chart of maximum joint spacing as a function of pavement thickness and k-value. This is based on a maximum L/ℓ ratio of 4.44. For example, a 280 mm (11 in) slab with a k-value of 12 MPa/m (44 psi/in) should have a maximum joint spacing of 6.2 m (20 ft). In contrast, a 130 mm (5.1 in) slab on stiff support of 110 MPa/m (400 psi/in) should have a joint spacing of no more than 2 m (6.6 ft). This maximum joint recommendation is clearly conservative. Smith et al. (2002: 4–9) notes that the maximum recommended L/ℓ for unbonded overlays is 4.5–5.5 (ACI 325.13R 2006, ERES 1999). As Figure 7.1 shows, the stress coefficient C is 0.64 for $L/\ell = 4.5$ and 0.85 for $L/\ell = 5.5$.

For highways, the joint spacing

> varies throughout the country because of considerations of initial costs, type of slab (reinforced or plain), type of load transfer, and local conditions. Design considerations should include: the effect of longitudinal slab movement on sealant and load transfer performance; the maximum slab length which will not develop transverse cracks in a plain concrete pavement; the amount of cracking which can be tolerated in a jointed reinforced concrete pavement; and the use of random joint spacings.
>
> (FHWA 1990a)

However,

> For plain concrete slabs, a maximum joint spacing of [4.6 m] 15 feet is recommended. Longer slabs frequently develop transverse cracks. It is recognized that in certain areas, joint spacings greater than [4.6 m] 15 feet have performed satisfactorily. The importance of taking local experience into account when selecting joint spacing (and designing pavements in general) cannot be overstated. Studies have shown that pavement thickness, base stiffness, and climate also affect the maximum anticipated joint spacing beyond which transverse cracking can be expected. Research indicates that . . . there is an increase in transverse cracking when the ratio L/ℓ exceeds 5.0.
>
> (FHWA 1990a)

As Figure 7.1 shows, the stress coefficient C is 0.75 for $L/\ell = 5.0$. Figure 7.3 shows maximum slab length as a function of pavement thickness D and modulus of subgrade reaction k if $L/\ell \leq 5.0$. A concrete modulus of elasticity of 27.6 GPa (4 million psi) is assumed.

As an example, a 500 mm (20 in) thick pavement with slab support of 54 MPa/m (200 psi/in) could have a joint spacing up to 7.6 m (25 ft). If the k is increased to 135 MPa/m (500 psi/in), then the joint spacing should be reduced to 6.1 m (20 ft).

Often thicker airfield pavements [greater than 300 mm (12 in)] are built with 7.6 by 7.6 m (25 by 25 ft) squares. Another option is to pave using 11.4 m (37.5 ft) paving passes and sawcut a longitudinal joint, which reduces the number of paving passes required by a third and produces 5.7 by 6.1 m (18.75 by 20 ft) panels.

For airports, the trend has been to use stiffer base materials such as Econocrete or bituminous base course. With stiffer base materials, there is a greater risk of mid-slab cracking unless the joint spacing is also reduced. Therefore, the 7.6 m by 7.6 m (25 by 25 ft) joint spacing may be excessive with stabilized bases.

The FAA *Airport Pavement Design, Advisory Circular (AC) 150/5320-6D* provides a chart for maximum joint spacing for unstabilized bases,

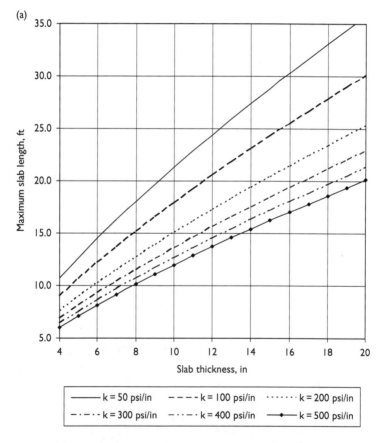

Figure 7.3 Maximum joint spacing based on slab thickness D and modulus of subgrade reaction k for $L/\ell \leq 5.0$ in (a) U.S. customary units and (b) metric units.

(b)

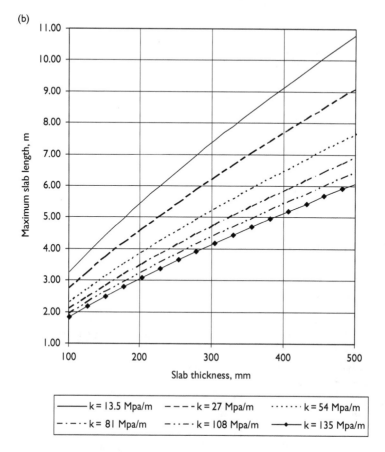

Figure 7.3 (Continued)

which is provided in Chapter 10 as Table 10.3. For stabilized bases, the joint spacing is limited to five times the radius of relative stiffness, equation 7.1 (FAA 2004: 87). Furthermore, although the FAA places a maximum value of 135 MPa/m (500 psi/in) on the subgrade modulus for pavement thickness design, the actual value should be used for calculating the radius of relative stiffness. Hall et al. (2005: 15) suggest a maximum joint spacing of 6.1 m (20 ft) for pavements more than 305 mm (12 in) thick. Shorter spacing should be used for thinner pavements.

If maximum joint spacing is exceeded, the pavement slabs will often crack between the joints. "On a runway in Texas 200-mm (8-in) thick concrete pavement was placed in 4.6 by 5.7m (15 by 18.75 ft) slabs. Slabs of this thickness are limited to a maximum joint spacing of 4.6 m (15 ft) by Air

Force practice. Almost every slab in this project quickly cracked to bisect the 5.7 m (18.75 ft) length" (Rollings 2005: 168).

For crowned pavements, a joint should always be provided at the crown. "Slabs that are crowned in the middle without a joint tend to crack down the crowned axis of the slab. During the 1990s three Air Force runways and taxi lanes were designed with unjointed crowned slabs by a government design agency. In every case, the slabs cracked. One of the examples is a concrete runway that developed a continuous 3,048-m (10,000-ft) long centerline crack much to the disgust of the Air Force owner" (Rollings 2005: 168–169). It could be argued that a concrete slab "knows" what its proper joint spacing should be, and if the designer doesn't provide for the joint the slab will crack.

Traditionally, with JRCP, longer joint spacings were used and it was assumed that the slabs would crack and that the temperature steel would hold the cracks together. In practice, these pavements have often performed poorly, particularly at the longer joint spacings.

Variable joint spacing

Evenly spaced joints, if faulted, can induce harmonic motion in vehicle suspensions at certain speeds. For this reason, some agencies have used variable joint spacing. The ACPA (2006) notes that this type of motion was primarily a problem with large cars on California freeways with 4.6 m (15 ft) spacing of undoweled joints. In California in the 1960s and 1970s, variable joint spacing of 4, 5.8, 5.5, and 3.7 m (13, 19, 18, and 12 ft) became the standard. Because the 5.8 and 5.5 m (19 and 18 ft) panels were found to be too long and prone to mid-slab cracking, the spacing was changed to 3.7, 4.6, 4, and 4.3 m (12, 15, 13, and 14 ft). The AASHTO Design Guide also discusses randomized or irregular spacing patterns and suggests avoiding multiples of 2.3 m (7.5 ft) (AASHTO 1993: II-49).

> Random joint spacings have been successfully used in plain undoweled pavements to minimize resonant vehicle responses. When using random joint spacings, the longest slab should be no greater than [4.6 m] 15 feet, to reduce the potential for transverse cracking. Some States are successfully using a spacing of [3.7-4.6-4.0-4.3 m] 12'-15'-13'-14'. Large differences in slab lengths should be avoided.
>
> (FHWA 1990a)

Pavement construction standards have improved, and there are few 1959 Buicks still on the road. If pavements are built smoothly with doweled joints to prevent faulting, then there is probably little benefit to the use of variable joint spacing.

Aggregate interlock joints

> Loads applied by traffic must be effectively transferred from one slab to the next in order to minimize vertical deflections at the joint. Reduced deflections decrease the potential for pumping of the base/subbase material and faulting. The two principal methods used to develop load transfer across a joint are: aggregate interlock; and load transfer devices, such as dowel bars ... Aggregate interlock is achieved through shearing friction at the irregular faces of the crack that forms beneath the saw cut. Climate, and aggregate hardness have an impact on load transfer efficiency. It can be improved by using aggregate that is large, angular, and durable. Stabilized bases have also been shown to improve load transfer efficiency. However, the efficiency of aggregate interlock decreases rapidly with increased crack width and the frequent application of heavy loads to the point that pavement performance may be affected. Therefore, it is recommended that aggregate interlock for load transfer be considered only on local roads and streets which carry a low volume of heavy trucks.
>
> (FHWA 1990a)

Early concrete pavements were all built with aggregate interlock joints. As pumping and faulting of joints began to cause problems on heavily trafficked pavements, dowels were introduced.

Nevertheless, there remains a role for aggregate interlock joints in concrete pavements. Aggregate interlock joints work well with low-speed traffic, relatively light loads, and shorter joint spacing. This corresponds to thinner pavements – ACI Committee 325 (2002: 15) suggests that dowels are not necessary for pavements less than 200 mm (8 in) thick. This is supported by ACPA (1998: 26–27) for plain concrete pavements, although it is suggested that thinner reinforced pavements with longer joint spacing require dowels. There is also an issue of constructability – it is more difficult to place dowels accurately in thinner pavements.

Aggregate interlock joints transfer load through shear of the irregular interlocking faces of adjacent slabs. Generally, joints are sawn 1/4–1/3 of the way through the slab, and then the slab is allowed to crack naturally. The load transfer is aided by the stiffness of supporting layers, such as a stabilized subgrade or subbase.

Efficiency of aggregate interlock joints depends on the following factors:

- The width of the joint or crack opening. This is directly related to slab length. ACI Committee 325 (2002: 14) suggests that 5.5 m (18 ft) is a practical maximum for undoweled joints.
- The opening width also depends on drying shrinkage and the thermal coefficient of the concrete. Lower shrinkage and thermal coefficients work best.

- Maximum size of the coarse aggregate, because when the crack forms and goes around the particles it will be rougher. Larger coarse aggregate also means less paste and less shrinkage.
- Hardness of the coarse aggregate, because harder stones will not break down under traffic.

One advantage of aggregate interlock joints in regions where deicing salts are used to remove snow and ice is that a pavement with no reinforcement or dowels is not susceptible to corrosion of embedded steel. This makes the pavement both cheaper to build and more durable.

Doweled joints

Dowels are used in transverse joints to prevent pumping and faulting. Dowels differ from tie bars in that they are lubricated to allow the joints to open and close without longitudinal stress buildup, while preventing differential vertical movement across the joint. Dowel bars are generally used for all pavements 200 mm (8 in) or more in thickness (ACI Committee 325 2002: 15). When dowels are used, they are typically 460 mm (18 in) long and spaced 300 mm (12 in) on center across the slab, for 11 dowels in a typical 3.66 m (12 ft) highway pavement lane. In contrast, when dowels are installed in an existing pavement with dowel bar retrofit, as discussed in Chapter 17, generally only three are put in each wheel path for a total of six across the lane.

The FHWA recommends the use of dowel bars on all highway pavements.

> Dowel bars should be used on all routes carrying more than a low volume of heavy trucks. The purpose of dowels is to transfer loads across a joint without restricting joint movement due to thermal contraction and expansion of the concrete. Studies have shown that larger dowels are more effective in transferring loads and in reducing faulting.
>
> (FHWA 1990a)

Dowel bar diameter

Proper minimum dowel size is based on having sufficient diameter to resist shear and bending forces transmitted from one slab to another, and to reduce the bearing stress of the steel dowel against the concrete to an acceptable value. Generally, the concrete bearing stress is the critical design parameter. If it is too high, the dowel will wear away the concrete and become loose.

In theory, the allowable dowel-bearing stress should be a function of the concrete compressive strength, and it would be possible to use smaller diameter dowels with higher strength concrete. In practice, recommendations have traditionally been based on the thickness of the pavement, because thicker pavements are assumed to carry larger numbers and heavier vehicle loads. In the past 19 and 25 mm (3/4 and 1 in) dowels were recommended for thinner pavements, but in current practice the minimum recommended dowel diameter is 32 mm (1.25 in) for 200 mm (8 in) pavements.

For highway pavements

> It is recommended that the minimum dowel diameter be D/8, where D is the thickness of the pavement. However, the dowel diameter should not be less than [32 mm] 1 1/4 inches. It is also recommended that [450 mm] 18-inch long dowels be used at [300 mm] 12-inch spacings. Dowels should be placed mid-depth in the slab. Dowels should be corrosion-resistant to prevent dowel seizure, which causes the joint to lock up. Epoxy-coated and stainless steel dowels have been shown to adequately prevent corrosion.
>
> (FHWA 1990a)

Table 7.1 shows recommended dowel diameters, lengths, and spacings for different pavement thicknesses, from 200 mm (8 in) pavements to 50 mm (2 in) dowels for 400 mm (16 in) thick pavements.

Smith and Hall (2001) recommend a different approach to dowel bar diameter selection for highways based on the traffic level. If the pavement is designed to carry less than 30 million ESALs, 30 mm (1.25 in) dowel bars are recommended. Dowels 38 mm (1.5 in) in diameter should be used for pavements carrying between 30 and 90 million ESALs. For traffic over 90 million ESALs, the bar diameter should be 41 mm (1.625 in). Since thicker pavements are needed to carry heavier traffic, the two recommendations produce similar dowel diameters.

Table 7.1 Dowel bar diameter recommendations (ACPA 1998: 27, ACI Committee 2002 325: 15)

Slab thickness, mm (in)	Dowel diameter, mm (in)	Dowel length, mm (in)	Dowel spacing, mm (in)
<200 (<8)		Dowels not required	
200 (8)	32 (1.25)	450 (18)	300 (12)
250 (10)	32 (1.25)	450 (18)	300 (12)
280 (11)	38 (1.5)	450 (18)	300 (12)
300 (12)	38 (1.5)	450 (18)	300 (12)
350 (14)	44 (1.75)	500 (20)	300 (12)
400 (16) and up	50 (2)	600 (24)	450 (18)

Skewed joints

Skewed joints represent another approach to addressing the problem of joint faulting. The idea is that by skewing joints, only one wheel crosses the joint at a time, making faulting less objectionable (Yoder and Witczak 1975: 609).

Skewed joints and variable spacing may be combined.

> Skewed joints with randomized spacings, say 13-19-18-12 ft. (4.0-5.8-5.5-3.7 m), have also been used. The obtuse angle at the outside pavement edge should be ahead of the joint in the direction of traffic, since that corner receives the greatest impact from the sudden impact of wheel loads. The advantage of skewed joints is that the right and left wheels do not arrive at the joint simultaneously, thus minimizing the annoyance of faulted joints. The use of randomized spacings can further reduce the resonance and improve the riding comfort.
>
> (Huang 2004: 179)

The AASHTO Design Guide also discusses skewed joints and notes that they may reduce deflections and stresses at joints, and provide a smoother ride for vehicles (AASHTO 1993: II-49).

> While they do not affect joint spacing, skewed joints have been used in plain pavements to provide a smoother ride. A skew of [0.6 m] 2 feet in [3.7 m] 12 feet is recommended, with the skew placed so that the inside wheel crosses the joint ahead of the outside wheel. Only one wheel crosses the joint at a time, which minimizes vehicle response and decreases stresses within the slab. Skewed joints are most commonly used when load transfer devices are not present. While skewed joints may be used in conjunction with load transfer devices, studies have not substantiated that skewing doweled joints improves pavement performance and are not recommended. Dowels in skewed joints must be placed parallel to the roadway and not perpendicular to the joints.
>
> (FHWA 1990a)

> On new JPCP, skewed joints can be effective in reducing faulting on nondoweled pavements, but have no effect when used on properly doweled pavements (Yu et al. 1998; Khazanovich et al. 1998). Furthermore, JPCP designs with skewed joints constructed on a stiff base (treated cement or lean concrete) are prone to corner breaks.
>
> (ACI Committee 325 2006: 4)

It is notable that skewed joints do not prevent joint faulting, but only mitigate its effects. It is more reliable to use conventional perpendicular transverse joints with dowels.

Longitudinal joints and tie bar design

As noted in Chapter 2, longitudinal joints are often used between highway traffic lanes. Tie bars may be placed across longitudinal joints between traffic lanes and between traffic lanes and a concrete shoulder. They differ from dowel bars in that they do not have to allow movement. Also, traffic does not cross these joints except when changing lanes, so load transfer is not a major concern. One common tie bar design is 13 mm (#4) conventional reinforcing bars 910 mm (36 in) long, spaced at 760–1020 mm (30 to 40 in) intervals. Tie bar length is governed by bond stress (Huang 2004: 171).

Most steel available today for tie bars would be classified as Grade 60, with 413 MPa (60,000 psi) yield stress. "When using Grade 60 steel, [16 mm] 5/8-inch by [1 m] 40-inch or [13 mm] 1/2-inch by [0.8 m] 32-inch tiebars should be used. These lengths are necessary to develop the allowable working strength of the tie bar. Tie bar spacing will vary with the thickness of the pavement and the distance from the joint to the nearest free edge" (FHWA 1990a).

Huang (2004: 170) provides an equation for calculating the required area of steel per unit length (A_s) of the pavement slab:

$$A_s = \frac{\gamma_c DL' f_a}{f_s} = 0.13 DL' \text{ mm}^2/\text{m} = 469 \times 10^{-6} DL' \text{ in}^2/\text{ft} \qquad (7.15)$$

Where γ_c = unit weight of concrete (23.6×10^{-6} N/mm^3, or 0.0868 lb/in^3), D = slab thickness (mm, in), L' = distance from the longitudinal joint to a free edge (m, ft), f_a = average coefficient of friction between the slab and subgrade, usually taken as 1.5, and f_s = the allowable stress in the steel (2/3 of the yield stress, conservatively 276 MPa or 40,000 psi with modern steel). To calculate the bar spacing, divide the cross-sectional area of a single bar by the required area of steel per unit length of the slab.

As an example, determine the spacing for 13 mm (#4) bars for a 250-mm thick pavement with two lanes of total 7.32 m (24 ft) to the outside edge. The required $A_s = 238$ mm^2/m (0.113 in^2/ft). Each bar has a cross-sectional area of 127 mm^2 (0.20 in^2), so the required spacing is 533 mm (21 in).

Figure 3.13 of the AASHTO Design Guide provides two charts for estimating tie bar spacing based on the slab thickness and the distance to the closest free edge of the pavement. Charts are provided for 13 mm (#4) and 16 mm (#5) bars. Tie bar spacing of greater than 1.2 m (48 in) is not recommended. For the example given above, the recommended tie bar spacing is 400 mm (16 in) (AASHTO 1993: II-62–II-64). Table 7.2 may also be used to select tie bar spacing.

Recommended tie bar lengths are 800 mm (32 in) for 13M/#4 bars, 1 m (39 in) for 16M/#5 bars, 1.18 m (46 in) for 19M/#6 bars, and 1.35 m (53 in) for 22M/#7 bars (ACPA 1998: 29–30).

Table 7.2 Tie bar spacing recommendations in mm (in)

Slab thickness, mm (in)	Distance to free edge or untied joint, m (ft)							
	1.2 (4)	3.0 (10)	3.6 (12)	4.3 (14)	5.0 (16)	6.0 (20)	7.0 (23)	8.0 (26)
	13M/#4 bars (ACPA 1998: 29–30)							
100 (4)	1,200 (48)							NA – too far to free edge
125 (5)	1,200 (48)							NA – too far to free edge
150 (6)	1,200 (48)							NA – too far to free edge
175 (7)	1,200 (48)				1,130 (45)			NA
200 (8)	1,200 (48)			1,150 (45)	990 (39)	820 (32)		NA
225 (9)	1,200 (48)			1,020 (40)	880 (35)	730 (29)	620 (24)	490 (19)
250 (10)	1,200 (48)		1,100 (43)	920 (36)	790 (31)	660 (26)	560 (22)	Larger
275 (11)			1,000 (39)	830 (32)	710 (28)	600 (24)	510 (20)	Larger
300 (12)	1,200 (48)	1,100 (43)	910 (36)	760 (30)	650 (26)	550 (22)	470 (18)	Larger
350 (14)		940 (37)	780 (31)	650 (26)	560 (22)	470 (18)		Larger
400 (16)		820 (32)	680 (27)	570 (22)	490 (19)			Larger
450 (18)		730 (29)	610 (24)	510 (20)				Larger
	16M/#5 bars (ACPA 1998: 29–30)							
225 (9)	Use smaller bars					1,140 (45)	980 (39)	NA
250 (10)						1,030 (40)	880 (35)	770 (30)
275 (11)						930 (36)	800 (31)	700 (27)
300 (12)	Use smaller bars			1,200 (48)	1,030 (40)	850 (33)	730 (29)	640 (25)
350 (14)				1,020 (40)	880 (35)	730 (29)	630 (24)	550 (22)
400 (16)	Use smaller bars		1,070 (42)	900 (35)	770 (31)	640 (25)	550 (22)	480 (18)
450 (18)	Smaller bars	1,140 (45)	950 (37)	800 (31)	680 (26)	570 (22)	490 (19)	Larger

Table 7.2 (Continued)

Slab thickness, mm (in)	Distance to free edge or untied joint, m (ft)				
	4.3 (14)	5.0 (16)	6.0 (20)	7.0 (23)	8.0 (26)
19M/#6 bars (ACPA 1998: 25–30)					
275 (11)	Use smaller bars				1,010 (39)
300 (12)					920 (24)
350 (14)	Use smaller bars			910 (35)	790 (31)
400 (16)	Use smaller bars		920 (36)	790 (31)	690 (26)
450 (18)	Use smaller bars	990 (39)	820 (32)	700 (27)	19M/#6 @ 610 (24) or 22M/#7 @ 840 (33)

Tiebars should not be placed within [380 mm] 15 inches of transverse joints. When using tiebars longer than [810 mm] 32 inches with skewed joints, tiebars should not be placed within [450 mm] 18 inches of the transverse joints. The use of corrosion-resistant tiebars is recommended, as corrosion can reduce the structural adequacy of tiebars. It is recommended that longitudinal joints be sawed and sealed to deter the infiltration of surface water into the pavement structure. A [9.5 mm wide by 25 mm deep] 3/8-inch wide by 1-inch deep sealant reservoir should be sufficient.

(FHWA 1990a)

Chapter 8

Highway pavement design

High speed, heavy duty highways are an important concrete pavement application. Concrete pavement has been an essential component of the interstate highway system in the United States as well as similar systems around the world.

For many years in the United States, the AASHTO 1993 Design Guide and its predecessors have been used to design both asphalt and concrete pavements. Other available concrete pavement design methods include the PCA 1984 procedure and the AASHTO 1998 Supplement procedure. These are expected to be replaced by the M-EPDG, but full implementation of the new procedure will be dependent on local calibration studies and may take many years.

AASHTO 1993 design guide

Based on the results of the AASHO Road Test, AASHTO published design guides in 1972, 1986, and 1993. These were all similar in overall concept but incorporated various modifications to account for different environmental conditions and new technologies. The development of the design equations is discussed by Huang (2004: 568–570).

The basic AASHTO Design Guide rigid pavement design equation, in US customary units, is (AASHTO 1993: II-45):

$$\log_{10} W_{18} = Z_R S_0 + 7.35 \, \log_{10}(D+1) - 0.06 + \frac{\log_{10}\left[\dfrac{\Delta \mathrm{PSI}}{4.5 - 1.5}\right]}{1 + \dfrac{1.624 \times 10^7}{(D+1)^{8.46}}}$$

$$+ (4.22 - 0.32 p_t) \left[\frac{S'_c C_d \left[D^{0.75} - 1.132 \right]}{215.63 J \left[D^{0.75} - \dfrac{18.42}{\left(\dfrac{E_c}{k} \right)^{0.25}} \right]} \right] \tag{8.1}$$

where W_{18} = traffic carried in ESALs, Z_R = the standard normal deviate for the desired reliability, S_0 = overall standard deviation, D = slab thickness in inches, ΔPSI = design change in serviceability index = $p_o - p_t$, p_o = initial serviceability index, p_t = terminal (final) serviceability index, S'_c = flexural strength of the concrete (psi), C_d = drainage coefficient, J = load transfer coefficient, E_c = modulus of elasticity of concrete, in psi, and k = modulus of subgrade reaction in psi/in.

This equation must be solved for D, the design slab thickness in inches. The following sections discuss the input values for the design equation, and methods for solving the equation.

The easiest way to use this equation in SI units is to convert S'_c and E_c/k, solve the equation, and then multiply D in inches by 25.4 to get mm. To convert S'_c in kPa to psi, multiply by 0.145. To convert the ratio E_c/k in [MPa/(MPa/m] to [psi/(psi/in)], multiply by 39.3.

Design inputs

The AASHTO Design Guide (1993) provides detailed instructions for determining the inputs for equation 8.1.

Traffic in ESALs W_{18}

Axle equivalency factors that may be used to convert single, tandem, and tridem axles to ESALs are provided in appendix D of the Design Guide, in tables D.10 through D.18 (AASHTO 1993: D-12–D-20). The equivalency factors depend on the type of axle (single, tandem, and tridem), the axle load magnitude (in 8.9 kN or 2 kip increments), the slab thickness D in inches (6 through 14 in in 1 in increments), and the pavement terminal serviceability index p_t (2.0, 2.5, or 3.0).

In theory, this must be an iterative process, because the design thickness D depends on W_{18}, which depends on the axle equivalency factors, which in turn depend on D. In practice, it is generally possible to estimate the pavement thickness in advance to determine which table to use, based on how heavy the projected traffic is.

Approximate equivalency factors can be calculated using the 4th power rule. Examination of the equivalency factor tables shows that 129 kN (29 kip) tandem axles and 173 kN (39 kip) tridem axles are each approximately equivalent to 80 kN (18 kip) single axles. Therefore, an approximate equivalency factor EF in SI units may be calculated:

$$EF = (P_1/80)^4 \text{ or } (P_2/129)^4 \text{ or } (P_3/173)^4 \tag{8.2}$$

where P_1, P_2, and P_3 = load in kN on a single, tandem, or tridem axle, respectively. In US customary units, the relationship is:

$$EF = (P_1/18)^4 \text{ or } (P_2/29)^4 \text{ or } (P_3/39)^4 \tag{8.3}$$

where P_1, P_2, and P_3 = loads in kip.

If the projected traffic is not known accurately, it is possible to simply estimate the number of ESALs. Because equation 8.1 is logarithmic, small errors in calculation of ESALs are not likely to make a great difference in the required pavement thickness D. After all, how easy is it to predict how many tandem axles of a given weight will be using a given highway over the next 20 years?

Reliability $Z_R S_0$

Pavement performance is variable. If it were possible to perfectly predict the average life of pavements, and design on that basis, then 50 percent of pavements would fail before the end of their expected design lives. Although this might be satisfactory for city streets and local roads, busy high speed highways carrying heavy traffic variability demand a higher level of reliability.

Reliability is an important topic and is therefore covered in full chapters in pavement design textbooks by Yoder and Witczak (1975: Chapter 13) and Huang (2004: Chapter 10). There are two approaches to reliability. One approach, used by the PCA 1984 method and traditional structural engineering procedures, is to use a factor of safety. This is discussed further below. The other approach is to assign a specific reliability – for example, an 85 percent probability that the pavement will not fail before 20 years. Failure is defined as falling below the terminal serviceability index p_t.

The reliability term in equation 8.1 is the product $Z_R S_0$. The overall standard deviation S_0 should be selected based on local conditions, and may be estimated as 0.35 based on the AASHO Road Test as compared to 0.45 for flexible pavements, and should be between 0.30 and 0.40 for rigid pavements. If variability of traffic is considered, 0.39 should be used, and 0.34 if traffic is not considered (AASHTO 1993: I-62, II-9–II-10). Therefore, it is probably best to use 0.39 for design.

Table 8.1 Reliability and standard normal deviate Z_R (AASHTO 1993, I-62 and II-9)

Functional classification	Recommended level of reliability			
	Urban		Rural	
	Percent	Z_R	Percent	Z_R
Interstate and other freeways	85–99.9	−1.037 −3.75	80–99.9	−0.841 −3.75
Principal arterials	80–99	−0.841 −2.327	75–95	−0.674 −1.645
Collectors	80–95	−0.841 −1.645	75–95	−0.674 −1.645
Local	50–80	0 −0.841	50–80	0 −0.841

The desired reliability and corresponding standard normal deviate Z_R range from 50 to 80 percent and $Z_R = 0$ to -0.841 for local roads to 85–99.9 percent and $Z_R = -1.037$ to -3.750 for interstate and other freeways. Suggested values of reliability and corresponding Z_R are shown in Table 8.1.

Initial and terminal serviceability p_o, p_t, ΔPSI

The initial serviceability p_o depends on how smoothly the pavement can be built, and the terminal serviceability p_t depends on how rough the agency is willing to let the pavement get before undertaking major rehabilitation. Based on the results of the AASHTO Road Test, rigid pavements may be constructed to an initial serviceability p_o of 4.5, which is 0.3 points higher than flexible pavements. At terminal serviceability p_t of 2.0, 2.5, and 3.0, respectively, AASHTO estimates that 85, 55, and 12 percent of people would find the pavement objectionable (AASHTO 1993: II-10).

Flexural strength of concrete S'_c

The flexural strength of concrete S'_c used in equation 8.1 is the average 28-day third point bending strength. Many agencies have minimum flexural strength specifications for acceptance at 7, 14, or 28 days. Because of the incorporation of reliability, discussed above, the minimum flexural strength for acceptance should not be used for design, because the actual pavement strength will always be greater. Actual flexural strength from testing and construction records may be used or the design S'_c may be estimated as the acceptance strength plus an additional margin (AASHTO 1993: II-16–II-17).

Drainage coefficient C_d

As discussed earlier, drainage of concrete pavements is important. The drainage coefficient C_d ranges from 0.70 to 1.25 based on the quality of drainage (very poor to excellent) and the percent of time that the pavement moisture levels approach saturation, from less than 1 percent to greater than 25 percent. Drainage coefficients are provided in Table 8.2 (AASHTO 1993: II-26).

The effect of the drainage coefficient is to increase the required pavement thickness to compensate for poor drainage. Obviously, this will not always be an effective approach. If drainage is really a problem, it is better to fix that problem rather than to try to build the pavement thicker to compensate.

Load transfer coefficient J

The load transfer coefficient J actually encompasses three different design factors – whether or not load transfer devices (dowels) are used, whether the shoulder is asphalt or tied concrete, and whether the pavement type is JPCP/JRCP or CRCP. Within each category ranges are given. The total range is from 2.3 (low end for CRCP with tied concrete shoulders) to 4.4

Table 8.2 Drainage coefficients C_d (AASHTO 1993, II-26)

Quality of drainage	Percent of time exposed to moisture levels approaching saturation			
	<1%	1–5%	5–25%	>25%
Excellent	1.25–1.20	1.20–1.15	1.15–1.10	1.10
Good	1.20–1.15	1.15–1.10	1.10–1.00	1.00
Fair	1.15–1.10	1.10–1.00	1.00–0.90	0.90
Poor	1.10–1.00	1.00–0.90	0.90–0.80	0.80
Very poor	1.00–0.90	0.90–0.80	0.80–0.70	0.70

Table 8.3 Load transfer coefficients J by pavement type (AASHTO 1993, II-26)

Pavement type	Shoulder			
	Asphalt		Tied concrete	
	Load transfer devices			
	Yes	No	Yes	No
JPCP, JRCP	3.2	3.8–4.4	2.5–3.1	3.6–4.2
CRCP	2.9–3.2	N/A	2.3–2.9	N/A

(high end for JPCP/JRCP, no dowels, and asphalt shoulders). Load transfer coefficients are provided in Table 8.3 (AASHTO 1993: II-26).

The load transfer coefficient J has a strong effect on the design thickness. The most common case is doweled JPCP with asphalt shoulders, with $J = 3.2$. For highway pavements that carry a substantial number of heavy truck loads, particularly at high speeds, it is more logical to provide load transfer devices than to increase the pavement thickness.

Concrete modulus of elasticity E$_c$

AASHTO recommends using the ACI relationship between compressive strength of concrete and modulus of elasticity E_c (AASHTO 1993: II-16):

$$E_c = 4,730\sqrt{f'_c}\ MPa = 57,000\sqrt{f'_c}\ psi \tag{8.4}$$

Modulus of elasticity E_c does not vary substantially across the narrow strength range of paving concrete, and thus does not have a strong effect on pavement thickness.

Modulus of Subgrade Reaction k

The AASHTO Design Guide procedure for determining the effective modulus of subgrade reaction k is complex, and follows these steps (AASHTO 1993: II-37–II-44):

- First, the subgrade k-value is estimated by dividing the resilient modulus M_R by 19.4, in US customary units.
- If the subgrade modulus varies substantially from season to season due to freezing and/or moisture effects, a seasonally weighted k should be determined.
- The k-value is then adjusted upward for the subbase type and thickness.
- The k-value is then adjusted downward for potential loss of support due to erosion of the subbase.
- Finally, if bedrock lies within 3.05 m (10 ft) of the surface the k-value is adjusted upward yet again.

The effect of all of these steps is a k-value usually fairly close to the original estimate. Loss of support should be avoided by using non-pumping materials under the slab, not by making the pavement thicker to compensate for weak support. Furthermore, the slab thickness D is not very sensitive to the k-value. The simpler methods for determining k outlined in Tables 4.1–4.5 are sufficient.

Solution methods

Equation 8.1 is difficult to solve directly for D. The AASHTO Design Guide provides a nomograph for determining the solution – with a sharp pencil and a lot of care, a reasonable degree of accuracy may be achieved. The equation may also be put into a spreadsheet or other equation solver, or a programmable calculator.

Two software packages are available, DarWIN from AASHTO and Win-PAS from the ACPA. Both packages provide design solutions for both asphalt and concrete pavements. WinPAS is less expensive and is available for US $495 from the ACPA bookstore (www.pavement.com). WinPAS may be used to solve for the pavement thickness D, or for any other variable. For example, it is possible to determine the reliability of a given pavement design for a certain projected traffic.

Design example

The design example provided here is the example used in the AASHTO Design Guide to illustrate the use of the nomograph (AASHTO 1993: II-45), with one exception. The example uses a standard deviation $S_0 = 0.29$, which is lower than the recommended range, and WinPAS will not accept this value. Therefore, $S_0 = 0.30$ is used instead. The input values for the design example are provided in Table 8.4. The design solution is shown in Figure 8.1. WinPAS provides a design thickness of 9.77 in or 248 mm, which rounded up to 10 in or 254 mm matches the value from the AASHTO Design Guide.

A few comments may be made. The subgrade value is low, and probably not compatible with the high level of reliability assumed, which is

Table 8.4 Input values for design example (AASHTO 1993, II-45)

Input	Value		
	SI units		US customary
Modulus of subgrade reaction k	19 MPa/m		72 pci (psi/in)
Modulus of elasticity of concrete E_c	34.5 GPa		5,000,000 psi
Concrete flexural strength S'_c	4.5 MPa		650 psi
Load transfer coefficient J		3.2	
Drainage coefficient C_d		1.0	
Standard deviation S_0		0.29 (increase to 0.30)	
Reliability and standard normal deviate Z_R		95%, − 1.645	
Initial serviceability		4.2	
Terminal serviceability		2.5	
Traffic in ESALs W_{18}		5,100,000	

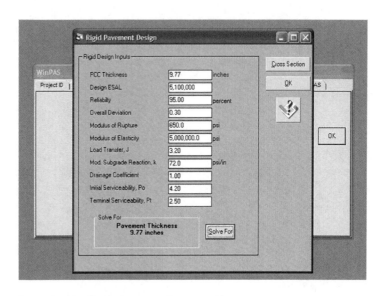

Figure 8.1 WinPAS solution to design problem.

Table 8.5 Sensitivity analysis – effect of design parameters on thickness D

Input value	Thickness D inches	mm	Percent change
Baseline case (Table 8.4)	9.77	248	N/A
Traffic 510,000 ESALs	6.69	170	−31.5
Traffic 4,080,000 ESALs (−20%)	9.45	240	−3.2
Traffic 6,120,000 ESALs (+20%)	10.05	255	+2.9
Traffic 51,000,000 ESALs	13.77	350	+40.9
Reliability 80%	8.97	228	−8.2
Reliability 99.9%	12	305	+22.8
Overall deviation 0.49	10.9	277	+11.6
Low strength concrete – $S = 550$ psi, $E = 3,700,000$ psi	10.56	268	+8.1
High strength concrete – $S = 750$ psi, $E = 5,062,500$ psi	9.07	230	−7.1
Load transfer $J = 2.6$, CRCP with tied shoulders	8.76	222	−10.3
Load transfer $J = 4.1$, no dowels	11.11	282	+13.7
Subgrade $k = 50$ psi/in	9.89	251	+1.2
Subgrade $k = 200$ psi/in	9.37	238	−4.1
Saturated >25% with very poor drainage, $C_d = 0.70$	11.74	289	+20.1
Saturated <1% with excellent drainage, $C_d = 1.25$	8.98	220	−8.1
High terminal serviceability 3.0	10.28	261	+5.2
Low terminal serviceability 2.0	9.42	239	−3.6

at the upper range even for an interstate highway. For this type of pavement a subbase would be required, leading to a higher k-value. The load transfer coefficient J corresponds to a doweled pavement (JPCP or JRCP) with asphalt shoulders. Table 8.5 provides a sensitivity analysis showing the effect of changing various design variables on the required design thickness.

It may be seen that traffic has a significant effect when increased or decreased by a factor of 10, but less of an effect for smaller variations, say plus or minus 20 percent. The required overall reliability, load transfer coefficient, and drainage coefficient are relatively important. Overall standard deviation S_0, concrete strength and stiffness, modulus of subgrade reaction k, and terminal serviceability do not have a major effect on design thickness.

AASHTO 1998 Supplement

In 1998, AASHTO published a Supplement to the AASHTO Guide for Design of Pavement Structures, entitled *Part II – Rigid Pavement Design & Rigid Pavement Joint Design*. Use of this procedure is illustrated in Figures 8.2 and 8.3. This Supplement provided an alternate to the 1993 design procedure, and unlike the previous versions was not based on the AASHTO Road Test results. Instead, the 1998 Supplement procedure was based on the LTPP data base and NCHRP Project 1–30. Input values remain the same as for the 1993 procedure, with the addition of the joint spacing. The design is carried out using free spreadsheet software that may be downloaded from http://www.fhwa.dot.gov/pavement/ltpp/rigid.cfm.

Design inputs

The Supplement procedure provides a thickness design for JPCP, JRCP, or CRCP pavement, and also provides a joint faulting check for JPCP or JRCP. This reference is the source for the k-values provided in Table 4.1. Considerable information is provided on determining the k-value, particularly for moisture sensitive soils (AASHTO 1998: 2–16).

One additional key input required is the location of the project, and the associated mean annual wind speed, mean annual air temperature, and mean annual precipitation. These are provided for two to twelve cities per US state, in Table 15. For example, Cleveland, Ohio has a mean annual wind speed of 17.2 kph (10.7 mph), mean annual air temperature of 9.9 °C (49.6 °F), and mean annual precipitation of 900 mm (35.4 in).

Some further inputs are required:

- Joint spacing in feet. For JPCP, the actual joint spacing is used. For JRCP, the joint spacing is used up to a maximum of 9.1 m (30 ft). For CRCP, an arbitrary joint spacing of 4.6 m (15 ft) is used for slab thickness calculation only.
- Whether the traffic lane is conventional width (3.66 m or 12 ft), conventional width plus tied to a concrete lane, or widened to (4.27 m or 14 ft).
- Modulus of elasticity, thickness, and friction factor for the base.

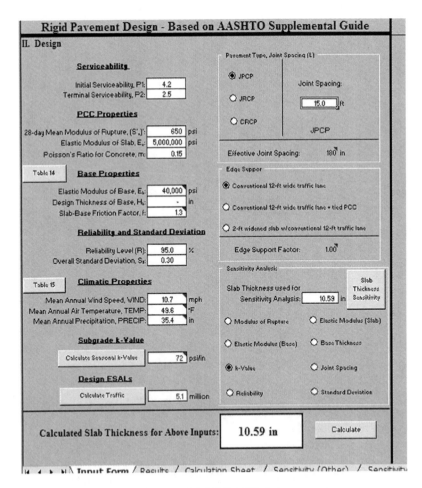

Figure 8.2 Spreadsheet solution to AASHTO 1998 design example.

Design examples

With the design example discussed above and a Cleveland location, the required pavement thickness for a 4.6 m (15 ft) joint spacing is 267 mm (10.59 in). Therefore, the thickness produced by the AASHTO 1998 Supplement procedure is a little greater than that from the AASHTO 1993 Design Guide. Changing the design to CRCP does not produce a change in pavement thickness. Thickness design is shown in Figure 8.2.

The next step in the process is to determine the required dowel diameter. For slabs less than 7.6 m (25 ft) in length, the critical mean joint

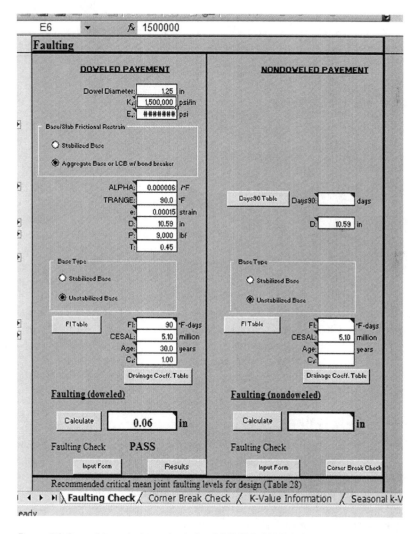

Figure 8.3 Spreadsheet faulting check for AASHTO 1998 design.

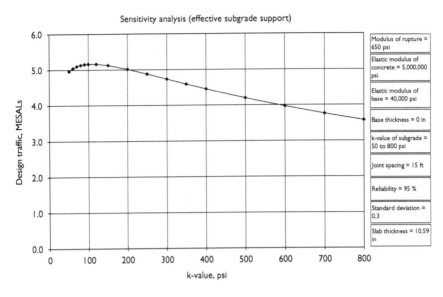

Sensitivity analysis (effective subgrade support)

Figure 8.4 Sensitivity analysis for AASHTO 1998 design.

faulting limit is 1.5 mm (0.06 in). This also requires the input of the annual temperature range in degrees, and mean annual freezing index in Fahrenheit degree-days. For Cleveland, these may be estimated as 90 degrees and 656 F degree-days. With a drainage coefficient of 1 and a pavement age of 25 years, 25 mm (1 in) dowels fail but 32 mm (1¼ in) dowels barely pass. The joint faulting check is shown in Figure 8.3.

Slab thickness sensitivity may also be performed automatically by the spreadsheet for modulus of rupture, elastic modulus of the slab, elastic modulus of the base, base thickness, *k*-value, joint spacing, reliability, and standard deviation. Sensitivity analysis for *k*-value is shown in Figure 8.4.

The supplement also provides a design example, with three different solutions. Solutions are provided for an undoweled JPCP with an aggregate base, an undoweled JPCP with ATB, and doweled JPCP with aggregate base.

PCA 1984 design

Unlike the AASHTO 1993 and 1998 procedures, the PCA 1984 procedure is mechanistically based. The charts and tables are based on finite element calculations of pavement stresses and deflections.

The original guide was set up for hand calculations. The hand calculations are described in detail by standard pavement textbooks such as Chapter 12 of Huang (2004) and Chapter 21 of Garber and Hoel (2002). The procedure is outlined below, but the tables and charts are not provided. It is simpler to use PCAPAV or StreetPave software, discussed below. Because the method is iterative, multiple design trials quickly become tedious with hand calculations. A hand solution is discussed below, but it is very similar to the StreetPave calculation presented in Chapter 9.

The PCA 1984 design procedure is also listed as an alternate method in the 1993 AASHTO Design Guide (AASHTO 1993: C-1). Although the PCA 1984 procedure may no longer be widely used for highway pavement design, it remains relevant for a number of reasons. The procedure provides insight into the role of various design variables – pavement thickness, dowels, and tied shoulders – into performance. It is particularly useful for lighter duty pavements, and has been used to develop design tables for parking lots (ACI Committee 330: 2001) and for streets and local roads (ACI Committee 325: 2002). An SI version of the Design Guide has been published by the Canadian Portland Cement Association (CPCA 1999).

Design criteria

Pavements are designed to prevent failure due to either fatigue or erosion. First, a trial pavement thickness is selected, along with joint and shoulder details. Then, the fatigue and erosion damage caused by the projected traffic is calculated, using equation 6.4. If both damage percentages are less than 100 percent, the design is satisfactory. It may, however, be overly conservative, so a thinner pavement may be tried. If either fatigue or erosion damage exceeds 100 percent, a thicker pavement is tried.

Design inputs

The PCA 1984 method design inputs are similar to those for other procedures. These include:

- The flexural strength of concrete, as measured through the third point MOR.
- Subgrade and subbase support (k-value). The PCA guide includes a relatively simple chart, similar to Table 4.1, for determining the k-value based on the ASTM soil classification, the AASHTO soil classification, resistance value, bearing value, or CBR. The k-value may then be adjusted upward for an untreated subbase or CTB. These adjustments are shown in Tables 4.2 and 4.3.
- Doweled or undoweled (aggregate interlock) joints are selected.
- Tied concrete shoulders or no shoulders/asphalt are selected.
- Traffic over the life of the pavement is divided into single and tandem, axles, and then divided into increments of 8.9 kN (2,000 lb) for singles and 17.8 kN (4,000 lb) for tandems.

- Load safety factor – reliability is not addressed specifically in the PCA 1984 procedure. Instead, the axle loads are multiplied by a load safety factor (LSF) of 1.2 for interstate and other multilane highways, or 1.1 for highways or arterial streets with moderate volumes of truck traffic.

Hand solution

First, a trial pavement thickness is assumed. Then fatigue and erosion analysis are carried out to find the percent damage.

Fatigue analysis

Based on whether or not the pavement has tied concrete shoulders, one of two charts is selected to determine the equivalent stress for single and tandem axles. Because tied concrete shoulders convert the loading from a near edge condition to a near interior condition, they reduce the stress considerably.

An example is provided in the 1984 PCA Guide. For example, with a 241-mm (9.5-in) doweled pavement and a subgrade modulus of 35 MPa/m (130 psi/in), the equivalent stresses for single and tandem axles are 1.42 MPa (206 psi) and 1.32 MPa (192 psi).

The equivalent stresses are divided by the concrete flexural strength of 4.48 MPa (650 psi) to get stress ratio factors of 0.317 and 0.295 for single and tandem axles, respectively. Next, the stress ratio is used for each single and tandem axle load to get an allowable number of repetitions for that axle.

In the example, the pavement is projected to carry 6,320 133 kN (30 kip) single axles. For a major highway, these are multiplied by an LSF of 1.2 to get 160 kN (36 kip). A nomograph is used to determine that the allowable number of load repetitions for this axle is 27,000. Therefore, these axles consume (6,320/27,000) or 23.3 percent of the pavement fatigue life.

This process is repeated for each single and tandem axle. At some point, the number of allowable repetitions becomes unlimited and there is no need to consider this axle or any lighter axles. For this pavement design, an unlimited number of 98 kN (22 kip) single axles or 214 kN (48 kip) tandem axles can be carried. Adding up the damage from all of the projected traffic shows 62.8 percent fatigue damage for this design, less than 100 percent, so the pavement is satisfactory for fatigue.

Erosion analysis

Erosion analysis is carried out in a similar manner. Four charts are provided for doweled or aggregate interlock joints, with and without concrete shoulders. In the example provided in the 1984 PCA Guide, the pavement has doweled joints, but no concrete shoulders. The erosion factors are 2.59 and 2.79 for single and tandem axles, respectively.

Using the nomograph, the allowable number of repetitions of a 160 kN (36 kip) single axle is 1,500,000 for erosion. Therefore, 6,320 repetitions of this axle weight consumes 0.4 percent of the pavement's erosion life. Adding up the damage from all of the projected traffic shows 38.9 percent erosion damage for this design, less than 100 percent, so the pavement is satisfactory for both fatigue and erosion.

PCAPAV and StreetPave software

If the original assumed design thickness is not satisfactory, the process must be repeated, and hand calculations can become tedious. PCA developed a DOS program, PCAPAV, to perform the designs. PCAPAV follows the 1984 PCA procedure exactly. The program automatically iterates until a satisfactory solution is found. More recently, ACPA has developed the Windows StreetPave program. This program is discussed in more detail in Chapter 9.

Introduction to the mechanistic-empirical pavement design guide

Under NCHRP Project 1-37A, research was carried out to develop a new pavement design guide to supersede the 1993 guide. At present, the guide has been completed and a research version is available on the web at http://www.trb.org/mepdg/guide.htm.

The new design guide is based on mechanistic-empirical principles. The pavement is analyzed with numerical models, which were calibrated with pavement performance data from the LTPP program. In essence, the LTPP program is the "Road Test" on which the new guide has been based. The procedure is completely software based and cannot be carried out manually.

Some specific points about the guide were made in a memorandum distributed by the AASHTO Task Force overseeing its development (AASHTO 2004). Some key excerpts from this three page memorandum are quoted below.

- The M-E pavement design guide uses mechanistic-empirical numerical models to analyze data for traffic, climate, materials, and proposed structure and to estimate damage accumulation over service life. It is applicable to designs for new, reconstructed, and rehabilitated flexible, rigid, and semi-rigid pavements. Performance predictions are made in terms of distress and smoothness. The predicted distresses . . . for rigid pavement designs (are) faulting, cracking, and continuously reinforced concrete pavement (CRCP) punchouts.

Design performance values can be compared with threshold values, or comparisons of performance may be made for alternate designs with varying traffic, structure, and materials.

- It must be understood that because the software is a tool for pavement *analysis* it does not provide structural thickness as an output. Nor, in its present form, does the M-E pavement design guide lend itself directly to use as a tool for routine, day-to-day production work.

The design guide is available only in US customary units, and does not address JRCP. The second point implies that the guide will be used primarily to develop catalogs or tables of pavement designs for most applications. Therefore, the guide will probably be used directly only for important heavy duty highway pavements, or as a forensics tool to analyze pavement performance problems.

There are a number of key features:

- It is an evaluation process like the PCA 1984 procedure, although vastly more sophisticated. A pavement structure is proposed, and then the projected distress and performance over the pavement service life under the expected traffic and environmental conditions is modeled. If necessary, the proposed design is adjusted until a satisfactory solution has been achieved.
- Much more detailed input data is required than for earlier design procedures. Separate input modules are used for traffic and materials. The software has a built-in climate model based on the location of the project.

Huang (2004: 716–727) provides additional details about the proposed procedure in appendix F of his textbook, entitled "A Preview Of 2002 Pavement Design Guide":

- The basic concrete pavement analysis module is the finite element program ISLAB2000, a successor to ILLI-SLAB.
- Whenever possible, inputs for flexible and rigid pavements are the same to make it easier to compare alternative pavement types.
- A hierarchical approach for determining design inputs is used. Level 1 inputs are very detailed and project specific, and more accurate, whereas level 3 inputs will often be default values. Therefore, the designer will use different amounts of data gathering effort based on the importance of the project.
- The Guide eliminates the ESAL approach and calculates damage directly from axle spectra, in a similar manner to the PCA 1984 method and StreetPave.

- The Enhanced Integrated Climate Model, developed by the FHWA, is used to provide site specific environmental conditions.
- Pavement roughness in IRI is determined based on pavement distress over time – for example, cracking, spalling, joint faulting, and other factors.

The implementation of the M-EPDG by state agencies and the FHWA will be a massive effort. At the time of this writing, the FHWA is sponsoring workshops, generally one full day, for state highway agencies across the United States. The workshops are described at http://www.fhwa.dot.gov/pavement/dgit/index.cfm. Workshop topics include:

- climate considerations;
- traffic inputs;
- local calibration of performance models; and
- materials characterization.

Furthermore, because the models were developed based on national LTPP data, many state agencies will be conducting research to further refine the models based on their local conditions. As a result, full implementation of the M-EPDG will probably take many years, and earlier design methods (e.g. AASHTO 1993) are likely to remain in general use for some time to come.

Chapter 9

Light duty pavement design

Concrete pavements have been widely used for important, heavy duty highways and airports. However, they also have an important place in lighter duty applications, such as city streets, local roads, and parking lots. For these applications, the long-term durability and reduced maintenance costs of concrete pavements may be highly attractive, as well as reducing lighting requirements at night and reducing heat buildup in warm climates. Concrete pavements with textured, patterned, or colored concrete may also be integrated into the landscape and architecture of neighborhoods to improve aesthetics.

However, for lighter duty pavements the complex methods used for highways and airports would be overkill. Moreover, although the AASHTO 1993 design procedure could be used for these pavements, light duty pavements were not well represented in the AASHO Road Test data used to develop the method. Furthermore, Huang (2004: 582–583) notes that the use of the AASHTO 1993 method may produce dangerously unconservative results for light duty pavements.

In order to develop appropriate light duty pavement designs, ACI Committees 325 and 330 applied the PCA 1984 method to develop simple design tables. Also, the ACPA developed StreetPave design software to design these types of pavements. ACPA has also developed *Design of Concrete Pavements for City Streets, Information Series IS184P* (ACPA 1992b).

This chapter contains pavement design tables for parking lots, streets, and local roads, generated using StreetPave. These are similar to the ACI 325.12R-02 and ACI 330R-01 tables. Because light traffic pavements are reasonable applications for pervious concrete pavements, design tables for pervious concrete are also provided.

Concrete intersections are also discussed. These are often used to replace asphalt intersections that exhibit rutting and other deterioration. Rutting is more likely with stopped or slowly moving heavy vehicles. Concrete bus stops may be used for the same reason.

Project quality control and opening the pavement to traffic may be on the basis of compressive strength, even though the pavement is designed on the basis of flexural strength. Various correlations between flexural and compressive strength are proposed by ACI Committee 325 (2002), ACI Committee 330 (2001), and the ACPA (1997). For intersections, the pavement may generally be opened to traffic once the concrete has achieved a flexural strength of 3 MPa (450 psi) or a compressive strength of 17 MPa (2,500 psi) (ACPA 1997: 23).

ACI 325.12R and ACI 330 design guides

ACI Committee 325 published the *Guide for Design of Jointed Concrete Pavements for City Streets and Local Roads* in 2002, and ACI Committee 330 published the *Guide for Design and Construction of Concrete Parking Lots* in 2001. Both documents are written in dual US customary and SI units. Maximum average daily truck traffic (ADTT) in two directions is 700 for parking lots and 1,500 for streets and local roads. In addition, ACI Committee 330 has published the *Specification for Unreinforced Concrete Parking Lots, ACI 330.1-03* (ACI Committee 330 2003). It includes "requirements for materials, placing, texturing, curing, jointing, and opening to traffic" (ACI Committee 330 2003: 1).

Because of the lighter traffic loads, undoweled JPCP pavements are often used. Undoweled, unreinforced pavements also have the advantage of not being susceptible to steel corrosion from deicing salts and other materials, because there is no steel to corrode.

Reinforcement is generally only necessary for irregularly shaped panels with aspect ratios of 1.7:1 or greater. Although reinforcement has been used in the past with long joint spacing, it is generally better practice to reduce the joint spacing. ACI 330R-01 states "The use of distributed steel reinforcement will not add to the load-carrying capacity of the pavement and should not be used in anticipation of poor construction practices" (ACI Committee 330 2001: 8–9).

Typical design features

Minimum pavement thickness is generally 100 mm (4 in). Pavement thickness for city streets and local roads may be as much as 200 or 225 mm (8 or 9 in). Dowels are often used for pavements more than 200 mm (8 in) thick. Dowels 32 mm (1.25 in) in diameter may be used for pavements up to 250 mm (10 in) thick, as noted in Table 7.1.

City streets often have curbs and gutters tied to the pavement edge and or placed integrally with the pavements. This reduces edge-stresses, and makes it possible to build thinner pavements. Curbs and gutters can also be built first and used as side forms to construct city street or parking lot pavements.

ACI 325.12R suggests that "use of dowel bars or stabilized bases is typically not recommended for low-volume design applications. Design options such as unbound bases, thickened edges, widened outside lanes, or tied curbs and gutters can be very cost effective" (ACI Committee 325 2002: 7).

With the traffic weights, volumes, and speeds typically carried by low-volume pavements, pumping is generally not a problem if drainage is adequate. Positive surface drainage of 2 or 2.5 percent also helps prevent moisture problems (ACI Committee 325 2002: 7). In urban areas, drainage for city streets and parking lots is often provided by subsurface storm sewer systems.

Traffic characterization

As with all concrete pavements, the heaviest axle loads tend to control design and performance. Therefore, it is important to accurately estimate the number of heavy trucks to be carried by the pavement, particularly any in excess of legal load limits (ACI Committee 325 2002: 01). These may be particularly important near industrial facilities. ACI Committee 325 (2002: 11) and ACI Committee 330 (2001: 6) provide the following facility classifications (Table 9.1).

Design tables for these traffic levels are provided later in this chapter.

Table 9.1 Street and parking lot traffic classifications from ACI 325.12R-11 and ACI-330R-6 (ACI Committee 325 2002: 11, ACI Committee 330 2001: 6)

Street/parking lot classification	VPD or ADT, two-way[1]	Heavy commercial vehicles (two axle, six tire, and heavier)		Category for StreetPave, Tables 9.3–9.5
		Percent	Trucks per day	
Car parking only		0	0	Residential
Truck access lanes			1–10	Residential
Light residential	200	1–2	2–4	Residential
Residential	200–1,000	1–2	2–4	Residential
Shopping center entrance and service lanes, bus, truck parking			25–300	Collector
Collector	1,000–8,000	3–5	50–500	Collector
Bus, truck parking			100–700	Minor arterial
Minor arterial	4,000–15,000	10	300–600	Minor arterial
Major arterial	4,000–30,000	15–20	700–1,500	Major arterial
Business	11,000–17,000	4–7	400–700	Major arterial
Industrial	2,000–4,000	15–20	300–800	Major arterial
Heavy truck parking			700	Major arterial

[1] Vehicles per day or average daily traffic

Joint layout

For parking lots, city streets, and local roads, jointing becomes an important consideration. Unlike mainline highway pavements, which are straight for long distances, light traffic facilities have to accommodate intersections, driveways, drainage structures, and utilities. Most of the joints are contraction joints, but construction joints are also necessary, as well as isolation or expansion joints to protect adjacent infrastructure. As a result, laying out joint patterns for these pavements often requires considerable planning.

Appendix C of ACI 325.12R-02 (ACI Committee 325 2002: 30–32) and appendix C of ACI 330R-01 (ACI Committee 330 2001: 27–29) provide examples of joint layouts for various types of facilities. Joint layout is also discussed in *Design and Construction of Joints for Concrete Streets, Information Series IS061P* (ACPA 1992c). Care should be taken that roof drains from buildings do not drain directly onto parking lot joints, because the large volumes of water moving at high speed can erode the supporting layers away rapidly through the joint opening.

Typical joint spacing for pavements in this thickness range is 3–5 m (12–15 ft). The joint spacing limits of Table 7.3 should not be exceeded.

One advantage of concrete pavements for city street and parking lot applications is that pavement markings delineating lanes or parking spaces may be incorporated into the joint pattern, making these features easier for drivers to follow.

For low-speed facilities such as parking lots, joints may be created using grooving tools while the concrete is still plastic, in the same manner as pavements. This technique produces a hard rounded edge at the joint that is easy to maintain, and any bump produced at the joint will not be noticeable at low speeds. Joints may also be sawn, in the same manner as high speed highway pavements.

ACPA StreetPave design software

The StreetPave design program allows the engineer to design light duty concrete pavements using either SI or US customary units. The program is available from the ACPA (www.pavement.com) for US $100.

The calculation follows the PCA 1984 procedure with a few exceptions. Probably the most significant exception is the use of a variable fatigue curve, based on the desired reliability and the allowable percentage of cracked slabs. The fatigue relationship is shown in equation 6.5 and Figure 6.1.

Two other important changes are the addition of a comparable asphalt pavement design and the addition of a life cycle cost module. Some of the other features and modules are discussed below. Help modules describe the input parameters in detail.

Global settings

The program's global settings are intended to be changed rarely, if at all. One is the system of units to be used, SI or US customary. Another is the mean annual air temperature (MAAT), which is based on the project location and is used only for the asphalt pavement design, using the Asphalt Institute (AI) design method. The AI design method is discussed at length in various transportation and pavement design textbooks, such as Chapter 20 of Garber and Hoel (2002) and Chapter 11 of Huang (2004) and will not be discussed in detail here. An additional global input is the terminal serviceability index, from the AASHTO design method, which is solely used to estimate the number of ESALs for the AI design. This input defaults to 2.0 and has little effect on the pavement design.

One important global input is the percent of pavement slabs cracked at the end of the service life. This ranges from 5 percent for interstate highway type pavements, to 15 percent for state roads, arterials, collectors, and county roads, to 25 percent for residential streets. The default value is 15 percent.

The percent of allowable cracked slabs is separate from reliability. With 25 percent cracked slabs and 85 percent reliability, there is no more than a 15 percent chance that the project will have more than 25 percent cracked slabs at the end of its service life. Setting either a low percentage of cracked slabs or a high reliability will make the design more conservative and more costly. It will not adjust the asphalt design, so if high or low values of cracked slabs or reliability are selected, the concrete and asphalt designs will no longer be comparable.

Project-level inputs

The project-level inputs are specific to a given project. These include the project information – project name, project description, route, owner/agency, location, and design engineer. Two types of project may be selected – either design of a new pavement, or analysis of an existing pavement. For a new design, in addition to determining the required concrete thickness, the designer may also elect to determine an equivalent asphalt pavement thickness and compare life cycle costs.

Two additional key inputs are the pavement design life and the desired reliability. Design life ranges from 10 to 40 years, with a default value of 20 years. The reliability recommendations of StreetPave are taken directly from the AASHTO design procedure, as shown in Table 8.1.

Traffic inputs

Probably the single most important traffic input is the traffic category, which ranges from residential to collector to minor arterial to major arterial. At each category, the maximum value of the single and tandem axle weights is different. Table 9.2 shows the average daily traffic (ADT), average daily truck traffic (ADTT), and maximum single and tandem axle assumed in each category.

Figure 9.1 and Table 9.3 show the number of single and tandem axles of various weights (in kN) per 1,000 trucks. Designers may also define their own traffic spectra, which may include tridem (triple) axles.

Other traffic inputs are the total number of lanes (in both directions), directional distribution, design lane distribution, either ADTT or ADT and percent trucks, and truck traffic growth per year. Directional distribution is normally assumed to be 50 percent in each direction, unless there is some reason truck traffic will always be heavier in one direction than the other.

Design lane distribution is the percent of trucks that travel in the outer lane, and recommendations are provided in a help file and appear automatically as defaults when the number of lanes is selected. For a four lane highway, the default assumption is 90 percent of heavy trucks in the outer lane and 10 percent in the inner or passing lane.

ADTT or ADT and percent trucks and estimated growth are typically estimated by the designer. Default ranges are also provided for ADTT or ADT and percent trucks, based on the traffic category. For example, the default ADTT of 1,000 for a major arterial is shown in Figure 9.1. The truck traffic growth rate default value is 2 percent, with a typical range of 1–3 percent.

Table 9.2 StreetPave traffic categories from help file

Category	Traffic ADT	ADTT		Maximum axle loads – kN (kip)	
		Percent	Per day	Single axles	Tandem axles
Residential	200–800	1–3	Up to 25	98 (22)	160 (36)
Collector	700–5,000	5–18	40–1,000	116 (26)	196 (44)
Minor arterial	3,000–12,000 lane, 3,000–50,000 4+ lane	28–30	500–5,000+	133 (30)	231 (52)
Major arterial	3,000–20,000 lane, 3,000–150,000 4+ lane	28–30	1,500–8,000+	151 (34)	267 (60)

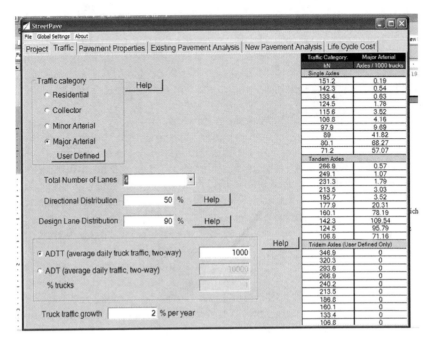

Figure 9.1 StreetPave traffic major arterial traffic category.

Pavement properties

The pavement properties must be input for the concrete pavement, and also for the asphalt pavement if a comparison between concrete and asphalt pavement designs is desired. A submodule may be used to compute the modulus of subgrade reaction. Other pavement property inputs include:

- Average 28-day concrete flexural strength (MOR), which may be estimated from the 28-day compressive strength.
- Concrete modulus of elasticity, which is computed automatically from the MOR.
- Whether load transfer dowels are used (yes/no flag).
- Whether edge-support is provided, either through a tied concrete shoulder, curb and gutter, or widened lane (yes/no flag).

The modulus of subgrade reaction (k-value) under the pavement may be input directly. Alternatively, the resilient modulus of the subgrade may be used, or it may be estimated from the CBR or R-value.

Up to three base/subbase layers may be placed between the slab and the subgrade. These layers may be econocrete/lean concrete base, CTB, hot-mix asphalt base, bituminous stabilized mixture, lime-stabilized

Table 9.3 Traffic categories – axles per 1000 trucks

Single axles

Axle load, kN	Axle load, kip	Residential	Collector	Minor arterial	Major arterial
151	34				0.19
142	32				0.54
133	30			0.45	0.63
125	28			0.85	1.78
116	26		0.07	1.78	3.52
107	24		1.6	5.21	4.16
98	22	0.96	2.6	7.85	9.69
89	20	4.23	6.63	16.33	41.82
80	18	15.81	16.61	25.15	68.27
71	16	38.02	23.88	31.82	57.07
62	14	56.11	47.76	47.73	
53	12	124	116.76	182.02	
44	10	204.96	142.7		
36	8	483.1	233.6		
27	6	732.28			
18	4	1693.31			

Tandem axles

Axle load, kN	Axle load, kip				
267	60				0.57
249	56				1.07
231	52			1.19	1.79
214	48			2.91	3.03
196	44		1.16	8.01	3.52
178	40		7.76	21.31	20.31
160	36	4.19	38.79	56.25	78.19
142	32	69.59	54.76	103.63	109.54
125	28	68.48	44.43	121.22	95.79
107	24	39.18	30.74	72.54	71.16
89	20	57.1	45	85.94	
71	16	75.02	59.25	99.34	
53	12	139.3	91.15		
36	8	85.59	47.01		
18	4	31.9			

subgrade/subbase, unbound compacted granular materials (sand/gravel, crushed stone), or fine graded or natural subgrade. The resilient modulus and thickness of each layer must be input – recommended resilient modulus ranges are provided. The *k*-value is adjusted for these layers

If an asphalt pavement design is also required, the necessary inputs include

- Resilient modulus of the subgrade, which is automatically calculated from the composite modulus of subgrade reaction submodule.
- The subgrade resilient modulus coefficient of variation, which ranges from 0.28 to 0.38 for projects with homogeneous soils and good quality control, to 0.59–0.70 for projects with little quality control of soils, such as municipal or industrial paving with little or no inspection. The default value is 0.38, taken from the AASHTO Road Test.

Existing and new pavement analysis

Once all of the inputs have been completed, either an existing pavement analysis or a new pavement analysis is performed. For an existing pavement analysis, the concrete thickness is input and either total erosion and fatigue damage are calculated, or the year in which the pavement will theoretically fail is calculated.

The new pavement analysis calculates the minimum required thickness of the concrete pavement, and recommends a thickness rounded up to the nearest 13 mm (1/2 in). Also, a maximum transverse joint spacing in meters (feet) is calculated. If dowel bars were selected, the required dowel diameter is also provided. If an asphalt pavement comparison is required, the thickness of the asphalt pavement is also calculated.

The new pavement analysis module also provides sensitivity analysis of the effects of k-value, reliability, concrete strength, percent slabs cracked, or design life on the required pavement thickness. A six-page design and analysis summary report is provided for each design.

Life cycle cost

The life cycle cost analysis module allows the comparison of the present worth of future maintenance costs as well as the initial construction cost between the concrete and asphalt designs. The comparison is only valid, of course, if both designs are equally conservative.

The life cycle cost module requires a number of specific inputs:

- Project information – project length, lane width, analysis period, and either interest and inflation rates or discount rate.
- Densities of the asphalt surface, asphalt base, and aggregate base.
- Initial material costs for concrete material and placement, aggregate base, and asphalt pavement layers.
- Concrete pavement maintenance and rehabilitation costs – annual maintenance, joint sealant, full depth repairs, partial depth repairs, and diamond grinding.

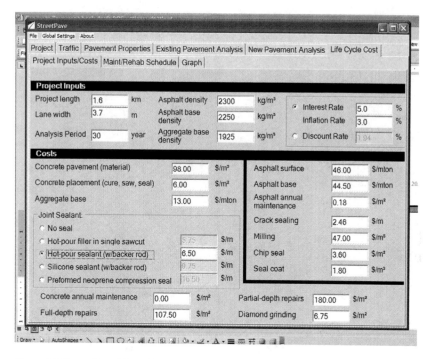

Figure 9.2 StreetPave life cycle cost module input screen.

- Asphalt pavement maintenance and rehabilitation costs – annual maintenance, crack sealing, milling, chip sealing, and seal coat.
- Maintenance and rehabilitation schedule – for each item of Maintenance, Preservation, and Rehab, a year and quantity are selected.

Project inputs/costs are shown in Figure 9.2. Once these have been specified, the schedule is developed. The program then produces graphs of life cycle costs.

Design examples

This design example produced using the StreetPave software generally follows the example from the PCA 1984 Design Guide. The design life is 20 years, and the reliability is 85 percent with 15 percent cracked slabs.

The traffic inputs are shown in Figure 9.3. The pavement is four lane, with 50 percent directional distribution and 90 percent of the trucks in the design lane. ADTT is 2,000. The maximum single axle load is 151.2 kN (34 kip), and the maximum tandem axle load is 266.9 kN (60 kip). These are higher than the PCA 1984 loads without the LSF, but lower than with the 1.2 LSF, so they are roughly comparable.

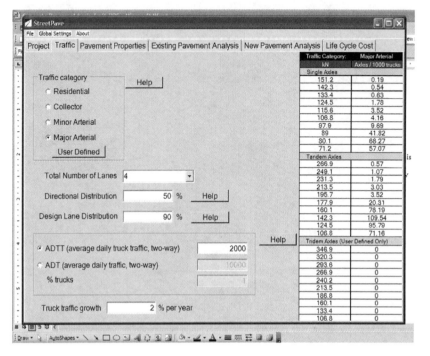

Figure 9.3 Design example traffic inputs.

Pavement properties are shown in Figure 9.4. To match the 1984 PCA design example, the subgrade and subbase *k*-value is set to 35.1 MPa/m (130 psi/in). Concrete flexural strength is 4.5 MPa (650 psi) and the concrete modulus of elasticity is 30.4 GPa (4.4 million psi). Dowels are used, but not tied shoulders. For this example, a comparable asphalt pavement design is not developed.

The pavement design is shown in Figure 9.5. The design concrete thickness is 223.8 mm (8.8 in), which rounds up to 229 mm (9 in). This agrees closely with the 141 mm (9.5 in) thickness of the 1984 PCA example. The recommended maximum transverse joint spacing is 4.57 m (15 ft) and 32 mm (1¼ in) diameter dowels should be used.

Figure 9.6 shows the fatigue/erosion table. The pavement has 98.9 percent fatigue life consumed, but only 11.76 percent of the erosion life consumed. Dowels could actually be omitted for this pavement, and the erosion life consumption would become 99.81 percent. However, given the magnitude and number of the traffic loads, it is better to use dowels.

Most of the fatigue consumption is due to the heaviest single axles – the 1,517 151.2 kN (34 kip) and the 4,310 142.3 kN (32 kip) together consume

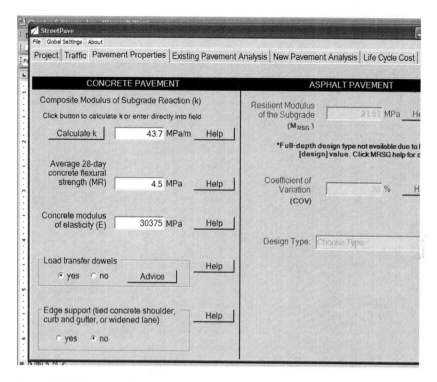

Figure 9.4 Design example pavement inputs.

over 78 percent of the fatigue, with lighter single axles and tandem axles contributing little. For doweled pavements, then, the highest single axle loads are the most significant. There are unlimited allowable repetitions for single axles 89 kN (18 kip) and lighter and tandem axles 195.7 kN (44 kip) and lighter. In contrast, virtually all of the axles do a small amount of erosion damage, with more erosion due to the tandem axles.

The design results also provide a discussion of rounding considerations. These are shown in Figure 9.7. Rounding the design thickness up increases the projected life to 33 years, or increases the 20-year reliability to 88.9 percent. Rounding the thickness down reduces the projected life to 8 years, or reduces the 20-year reliability to 75.4 percent.

StreetPave can calculate sensitivity analysis for *k*-value, reliability, concrete strength, percent slabs cracked, or design life. Figure 9.8 shows the sensitivity analysis for *k*-value. A decrease in *k*-value below approximately 27 MPa/m (100 psi/in) would require a thicker pavement, and an increase to 47.5 MPa/m (175 psi/in) or higher would allow a thickness reduction.

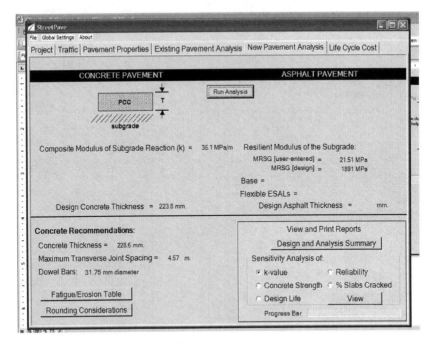

Figure 9.5 Concrete pavement design.

For a city street, the curb and gutter would provide edge-support. Repeating this design example with edge-support reduces the pavement thickness from 229 mm (9 in) to 203 mm (8 in). Joint spacing and dowel bar diameter are the same. If a 7.3 m (24 ft) four-lane pavement is widened to 7.9 m (26 ft) to widen the truck lane by 0.6 m (2 ft) and provide edge-support, the widening and reduction of pavement thickness cancel out so that no additional concrete is required.

Light traffic pavement design tables

The following design tables were prepared using StreetPave software, following the format of ACI 330R-01 and ACI 325.12R-02. These are based on 85 percent reliability and 15 percent cracked slabs allowed. These tables apply to parking lots, city streets, and local roads. These tables should produce design thicknesses close to those found in ACI 330R-01 and ACI 325.12R-02.

Table 9.4 provides design thicknesses for pavements with aggregate interlock joints and no edge-support, given concrete strengths of 3.45–4.5 MPa

Traffic Category:	Major Arterial		Fatigue Analysis			Erosion Analysis		
Axle Load, kips	Axles per 1000 Trucks	Expected Repetitions	Stress Ratio	Allowable Repetitions	Fatigue Consumed	Power	Allowable Repetitions	Erosion Consumed
Single Axles								
151.2	0.19	1517	0.647	3753	40.4	39.695	1271063	0.12
142.3	0.54	4310	0.611	11410	37.77	35.16	1826769	0.24
133.4	0.63	5028	0.575	43979	11.43	30.899	2715220	0.19
124.5	1.78	14207	0.539	230978	6.15	26.914	4212679	0.34
115.6	3.52	28096	0.503	1829339	1.54	23.204	6919975	0.41
106.8	4.16	33204	0.467	24461940	0.14	19.806	12234190	0.27
97.9	9.69	77343	0.43	726127261	0.01	16.642	24618442	0.31
89	41.82	333794	0.393	unlimited	0	13.754	61717535	0.54
80.1	68.27	544910	0.356	unlimited	0	11.141	262692430	0.21
71.2	57.07	455515	0.319	unlimited	0	8.803	unlimited	0
Tandem Axles								
266.9	0.57	4550	0.528	404290	1.13	47.033	774268	0.59
249.1	1.07	8540	0.495	2993295	0.29	40.969	1157982	0.74
231.3	1.79	14287	0.462	36379192	0.04	35.323	1801382	0.79
213.5	3.03	24185	0.428	865838808	0	30.095	2948617	0.82
195.7	3.52	28096	0.395	unlimited	0	25.286	5171864	0.54
177.9	20.31	162108	0.361	unlimited	0	20.896	10031761	1.62
160.1	78.19	624088	0.327	unlimited	0	16.923	22913349	2.72
142.3	109.54	874314	0.293	unlimited	0	13.369	72334824	1.21
124.5	95.79	764566	0.258	unlimited	0	10.234	669213478	0.11
106.8	71.16	567977	0.223	unlimited	0	7.531	unlimited	0
Tridem Axles								
346.9	0	0	0.34	unlimited	0	54.83	498524	0
320.3	0	0	0.315	unlimited	0	46.744	788181	0
293.6	0	0	0.291	unlimited	0	39.276	1311650	0
266.9	0	0	0.266	unlimited	0	32.457	2331506	0
240.2	0	0	0.241	unlimited	0	26.288	4549033	0
213.5	0	0	0.215	unlimited	0	20.769	10257526	0
186.8	0	0	0.19	unlimited	0	15.898	30126311	0
160.1	0	0	0.164	unlimited	0	11.679	176931115	0
133.4	0	0	0.138	unlimited	0	8.108	unlimited	0
106.8	0	0	0.112	unlimited	0	5.197	unlimited	0
				Total Fatigue Used:	98.9		Total Erosion Used:	11.76

Figure 9.6 Design example fatigue and erosion table.

Rounding Considerations

Specified Design Life	20	Recommended Thickness	228.6
Specified Reliability	85	Rounded-Down Thickness	215.9
Design Thickness	223.8		

Theoretical Life of Recommended Concrete Design

33 years @ 85 % reliability

Reliability of Recommended Concrete Design

88.9 % reliability for 20 -year design

Theoretical Life of Rounded-Down Concrete Design

8 years @ 85 % reliability

Reliability of Rounded-Down Concrete Design

75.4 % reliability for 20 -year design

Figure 9.7 Rounding considerations.

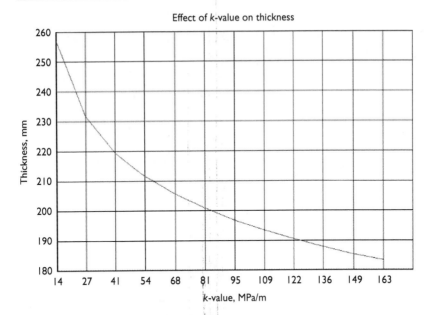

Figure 9.8 Sensitivity of slab thickness to k-value.

(500–650 psi) and k-values of 13.5–135 MPa/m (50–500 psi/in). Local roads often lack edge-support.

Table 9.5 is similar to Table 9.4, but for pavements with edge-support provided. This is generally the case for parking lots, city streets with concrete curb and gutter, and pavements with wider edges. The edge-stresses are reduced, so design thicknesses are decreased.

The use of dowels for all pavements in these tables was investigated. In only two cases, shown in Table 9.5 for low subgrade support, high concrete strength, and heavy traffic, was it possible to decrease the thickness. In all other cases, the required pavement thickness did not change.

For low speed, relatively low truck volume pavements of this type, joint faulting and pumping do not control the design. Therefore, strictly speaking dowels are not necessary. They are often included for pavements more than 203 mm (8 in) thick to reduce faulting potential. In harsh climates where corrosive deicing materials are used, it makes sense to eliminate the dowels and to instead attempt to reduce faulting by using a stabilized base.

For the design example discussed earlier in this chapter, the ADTT of 2,000 is higher than the traffic provided in the design tables, and the k-value of 35.1 MPa/m (130 psi/in) is also not in the tables. However, a comparison may be made with an ADTT of 1,500 and a k of 27 MPa/m (130 psi/in). For a concrete strength of 4.5 MPa (650 psi), Table 9.4 indicates a design thickness of 241 mm (9.5 in) for unsupported edges and Table 9.5 indicates

Table 9.4 Pavement thickness design tables, aggregate interlock joints, no edge-support

CBR = 2	13.5 MPa/m 50 psi/in	Concrete flexural strength							
		kPa	psi	kPa	psi	kPa	psi	kPa	psi
		4,500	650	4,150	600	3,800	550	3,450	500
Traffic	ADTT	Required pavement thickness							
		mm	in	mm	in	mm	in	mm	in
Residential	1	152	6	152	6	165	6.5	178	7
	10	165	6.5	178	7	191	7.5	191	7.5
Collector	25	191	7.5	191	7.5	203	8	216	8.5
	300	203	8	216	8.5	229	9	241	9.5
Minor	100	216	8.5	229	9	241	9.5	254	10
Arterial	300	229	9	241	9.5	254	10	267	10.5
	700	241	9.5	254	10	267	10.5	280	11
Major	700	254	10	267	10.5	280	11	292	11.5
Arterial	1500	267	10.5	280	11	292	11.5	305	12

CBR = 3	27 MPa/m 100 psi/in	Concrete flexural strength							
		kPa	psi	kPa	psi	kPa	psi	kPa	psi
		4,500	650	4,150	600	3,800	550	3,450	500
Traffic	ADTT	Required pavement thickness							
		mm	in	mm	in	mm	in	mm	in
Residential	1	140	5.5	140	5.5	152	6	165	6.5
	10	152	6	165	6.5	165	6.5	178	7
Collector	25	165	6.5	178	7	191	7.5	203	8
	300	191	7.5	203	8	203	8	216	8.5
Minor	100	203	8	216	8.5	216	8.5	229	9
Arterial	300	216	8.5	216	8.5	229	9	241	9.5
	700	216	8.5	229	9	241	9.5	254	10
Major	700	229	9	241	9.5	254	10	267	10.5
Arterial	1500	241	9.5	254	10	267	10.5	280	11

CBR = 10	54 MPa/m 200 psi/in	Concrete flexural strength							
		kPa	psi	kPa	psi	kPa	psi	kPa	psi
		4,500	650	4,150	600	3,800	550	3,450	500

Table 9.4 (Continued)

Traffic	ADTT	Required pavement thickness							
		mm	in	mm	in	mm	in	mm	in
Residential	1	127	5	127	5	140	5.5	152	6
	10	140	5.5	152	6	152	6	165	6.5
Collector	25	152	6	165	6.5	178	7	178	7
	300	178	7	178	7	191	7.5	203	8
Minor	100	191	7.5	191	7.5	203	8	216	8.5
Arterial	300	191	7.5	203	8	216	8.5	229	9
	700	203	8	203	8	216	8.5	229	9
Major	700	216	8.5	216	8.5	229	9	241	9.5
Arterial	1500	216	8.5	229	9	241	9.5	254	10

CBR =26	81 MPa/m 300 psi/in	Concrete flexural strength							
		kPa	psi	kPa	psi	kPa	psi	kPa	psi
		4,500	650	4,150	600	3,800	550	3,450	500

Traffic	ADTT	Required pavement thickness							
		mm	in	mm	in	mm	in	mm	in
Residential	1	114	4.5	127	5	127	5	140	5.5
	10	140	5.5	140	5.5	152	6	152	6
Collector	25	152	6	152	6	165	6.5	178	7
	300	165	6.5	178	7	178	7	191	7.5
Minor	100	178	7	178	7	191	7.5	203	8
Arterial	300	178	7	191	7.5	203	8	216	8.5
	700	191	7.5	203	8	203	8	216	8.5
Major	700	203	8	216	8.5	216	8.5	229	9
Arterial	1500	203	8	216	8.5	229	9	241	9.5

CBR =38	108 MPa/m 400 psi/in	Concrete flexural strength							
		kPa	psi	kPa	psi	kPa	psi	kPa	psi
		4,500	650	4,150	600	3,800	550	3,450	500

Traffic	ADTT	Required pavement thickness							
		mm	in	mm	in	mm	in	mm	in
Residential	1	114	4.5	114	4.5	127	5	127	5
	10	127	5	140	5.5	140	5.5	152	6
Collector	25	140	5.5	152	6	152	6	165	6.5
	300	165	6.5	165	6.5	178	7	191	7.5

		mm	in	mm	in	mm	in	mm	in
Minor	100	165	6.5	178	7	191	7.5	191	7.5
Arterial	300	178	7	191	7.5	191	7.5	203	8
	700	178	7	191	7.5	203	8	216	8.5
Major	700	191	7.5	203	8	216	8.5	229	9
Arterial	1500	203	8	203	8	216	8.5	229	9

CBR = 50	135 MPa/m 500 psi/in	Concrete flexural strength							
		kPa	psi	kPa	psi	kPa	psi	kPa	psi
		4,500	650	4,150	600	3,800	550	3,450	500

Traffic	ADTT	Required pavement thickness							
		mm	in	mm	in	mm	in	mm	in
Residential	1	114	4.5	114	4.5	127	5	127	5
	10	127	5	127	5	140	5.5	140	5.5
Collector	25	140	5.5	140	5.5	152	6	165	6.5
	300	152	6	165	6.5	165	6.5	178	7
Minor	100	165	6.5	178	7	178	7	191	7.5
Arterial	300	165	6.5	178	7	191	7.5	203	8
	700	178	7	191	7.5	191	7.5	203	8
Major	700	191	7.5	191	7.5	203	8	216	8.5
Arterial	1500	191	7.5	203	8	216	8.5	229	9

a design thickness of 203 mm (8 in). These are close to the values found earlier.

Pervious concrete pavement

The ACI A330R-01 and ACI 325.12R-02 design tables as well as Tables 9.4 and 9.5 are limited to concrete with a flexural strength of at least 3.4 MPa (500 psi). Pervious concrete often has a lower flexural strength than this. Therefore, the following design tables were developed to address the design of pervious concrete pavements. Flexural strengths of 2.1–3.1 MPa (300–450 psi) may be used (Table 9.6).

As with the earlier tables, a truck traffic growth rate of 2 percent per year is assumed. The concrete modulus of elasticity E is assumed to be 6,750 times the flexural strength or modulus of rupture. The pervious concrete pavement has no dowels, and is assumed not to have edge-support. If edge-support were provided, the pavement thickness could be reduced.

At this time, the fatigue relationship for pervious concrete is not known. As a result, these tables have been developed using 90 percent reliability rather than 85 percent, to add conservatism to the design until better information on the fatigue performance of pervious concrete becomes available.

Table 9.5 Pavement thickness design tables, aggregate interlock joints, supported edges

CBR =2	13.5 MPa/m 50 psi/in	Concrete flexural strength							
		kPa	psi	kPa	psi	kPa	psi	kPa	psi
		4,500	650	4,150	600	3,800	550	3,450	500
Traffic	ADTT	Required pavement thickness							
		mm	in	mm	in	mm	in	mm	in
Residential	1	127	5	140	5.5	140	5.5	152	6
	10	140	5.5	152	6	165	6.5	165	6.5
Collector	25	165	6.5	165	6.5	178	7	191	7.5
	300	178	7	191	7.5	191	7.5	203	8
Minor	100	191	7.5	203	8	203	8	216	8.5
Arterial	300	203	8	203	8	216	8.5	229	9
	700	203	8	216	8.5	229	9	241	9.5
Major	700	216	8.5	229	9	241	9.5	254	10
Arterial	1500	241[1]	9.5[1]	241[1]	9.5[1]	241	9.5	267	10.5

CBR =3	27 MPa/m 100 psi/in	Concrete flexural strength							
		kPa	psi	kPa	psi	kPa	psi	kPa	psi
		4,500	650	4,150	600	3,800	550	3,450	500
Traffic	ADTT	Required pavement thickness							
		mm	in	mm	in	mm	in	mm	in
Residential	1	114	4.5	127	5	127	5	140	5.5
	10	127	5	140	5.5	140	5.5	152	6
Collector	25	152	6	152	6	165	6.5	165	6.5
	300	165	6.5	165	6.5	178	7	191	7.5
Minor	100	178	7	178	7	191	7.5	203	8
Arterial	300	178	7	191	7.5	203	8	216	8.5
	700	191	7.5	191	7.5	203	8	216	8.5
Major	700	203	8	203	8	216	8.5	229	9
Arterial	1500	203	8	216	8.5	229	9	241	9.5

CBR =10	54 MPa/m 200 psi/in	Concrete flexural strength							
		kPa	psi	kPa	psi	kPa	psi	kPa	psi
		4,500	650	4,150	600	3,800	550	3,450	500
Traffic	ADTT	Required pavement thickness							
		mm	in	mm	in	mm	in	mm	in

Traffic	ADTT	mm	in	mm	in	mm	in	mm	in
Residential	1	102	4	114	4.5	114	4.5	127	5
	10	114	4.5	127	5	127	5	140	5.5
Collector	25	140	5.5	140	5.5	152	6	152	6
	300	152	6	152	6	165	6.5	178	7
Minor	100	152	6	165	6.5	178	7	191	7.5
Arterial	300	165	6.5	178	7	178	7	191	7.5
	700	178	7	178	7	191	7.5	203	8
Major	700	178	7	191	7.5	203	8	216	8.5
Arterial	1500	191	7.5	191	7.5	203	8	216	8.5

CBR = 26	81 MPa/m 300 psi/in	Concrete flexural strength							
		kPa	psi	kPa	psi	kPa	psi	kPa	psi
		4,500	650	4,150	600	3,800	550	3,450	500

Traffic	ADTT	Required pavement thickness							
		mm	in	mm	in	mm	in	mm	in
Residential	1	102	4	102	4	114	4.5	114	4.5
	10	114	4.5	114	4.5	127	5	127	5
Collector	25	127	5	127	5	140	5.5	152	6
	300	140	5.5	152	6	152	6	165	6.5
Minor	100	152	6	152	6	165	6.5	178	7
Arterial	300	152	6	165	6.5	178	7	178	7
	700	165	6.5	165	6.5	178	7	191	7.5
Major	700	178	7	178	7	191	7.5	203	8
Arterial	1500	178	7	191	7.5	191	7.5	203	8

CBR = 38	108 MPa/m 400 psi/in	Concrete flexural strength							
		kPa	psi	kPa	psi	kPa	psi	kPa	psi
		4,500	650	4,150	600	3,800	550	3,450	500

Traffic	ADTT	Required pavement thickness							
		mm	in	mm	in	mm	in	mm	in
Residential	1	102	4	102	4	102	4	114	4.5
	10	114	4.5	114	4.5	114	4.5	127	5
Collector	25	127	5	127	5	140	5.5	140	5.5
	300	140	5.5	140	5.5	152	6	152	6
Minor	100	140	5.5	152	6	165	6.5	165	6.5
Arterial	300	152	6	165	6.5	165	6.5	178	7
	700	152	6	165	6.5	178	7	178	7
Major	700	178	7	178	7	191	7.5	191	7.5
Arterial	1500	178	7	178	7	191	7.5	203	8

Table 9.5 (Continued)

CBR =50	135 MPa/m 500 psi/in	Concrete flexural strength							
		kPa	psi	kPa	psi	kPa	psi	kPa	psi
		4,500	650	4,150	600	3,800	550	3,450	500
Traffic	ADTT	Required pavement thickness							
		mm	in	mm	in	mm	in	mm	in
Residential	1	102	4	102	4	102	4	114	4.5
	10	102	4	114	4.5	114	4.5	127	5
Collector	25	114	4.5	127	5	127	5	140	5.5
	300	127	5	140	5.5	140	5.5	152	6
Minor	100	140	5.5	152	6	152	6	165	6.5
Arterial	300	152	6	152	6	165	6.5	178	7
	700	152	6	165	6.5	165	6.5	178	7
Major	700	165	6.5	165	6.5	178	7	191	7.5
Arterial	1500	178	7	178	7	191	7.5	191	7.5

1 Reduce pavement thickness 12.5 mm (1/2 in) if doweled

Table 9.6 Pavement thickness design tables for pervious concrete

CBR =2	13.5 MPa/m 50 psi/in	Concrete flexural strength							
		kPa	psi	kPa	psi	kPa	psi	kPa	psi
		3,100	450	2,750	400	2,400	350	2,100	300
Traffic	ADTT	Required pavement thickness							
		mm	in	mm	in	mm	in	mm	in
Residential	1	191	7.5	203	8	216	8.5	241	9.5
	10	216	8.5	229	9	254	10	276	10.5
Collector	25	241	9.5	254	10	280	11	305	12
	300	267	10.5	280	11	305	12	330	13
Minor	100	280	11	305	12	330	13	356	14
Arterial	300	292	11.5	318	12.5	343	13.5	381	15
	700	305	12	330	13	356	14	381	15
Major	700	330	13	356	14	381	15	419	16.5
Arterial	1500	330	13	356	14	394	15.5	432	17
CBR =2.5	20.3 MPa/m 75 psi/in	Concrete flexural strength							
		kPa	psi	kPa	psi	kPa	psi	kPa	psi
		3,100	450	2,750	400	2,400	350	2,100	300

Traffic	ADTT	Required pavement thickness							
		mm	in	mm	in	mm	in	mm	in
Residential	1	178	7	191	7.5	203	8	229	9
	10	203	8	216	8.5	229	9	254	10
Collector	25	229	9	241	9.5	254	10	280	11
	300	254	10	267	10.5	292	11.5	318	12.5
Minor	100	267	10.5	280	11	305	12	330	13
Arterial	300	280	11	292	11.5	318	12.5	356	14
	700	292	11.5	305	12	330	13	356	14
Major	700	305	12	330	13	356	14	381	15
Arterial	1500	318	12.5	330	13	368	14.5	394	15.5
CBR =3	27 MPa/m 100 psi/in	Concrete flexural strength							
		kPa	psi	kPa	psi	kPa	psi	kPa	psi
		3,100	450	2,750	400	2,400	350	2,100	300

Traffic	ADTT	Required pavement thickness							
		mm	in	mm	in	mm	in	mm	in
Residential	1	178	7	191	7.5	203	8	216	8.5
	10	191	7.5	216	8.5	229	9	241	9.5
Collector	25	216	8.5	229	9	254	10	280	11
	300	241	9.5	254	10	280	11	305	12
Minor	100	254	10	280	11	292	11.5	318	12.5
Arterial	300	267	10.5	280	11	305	12	343	13.5
	700	280	11	292	11.5	318	12.5	343	13.5
Major	700	292	11.5	318	12.5	343	13.5	368	14.5
Arterial	1500	305	12	318	12.5	343	13.5	381	15
CBR =6	40.5 MPa/m 150 psi/in	Concrete flexural strength							
		kPa	psi	kPa	psi	kPa	psi	kPa	psi
		3,100	450	2,750	400	2,400	350	2,100	300

Traffic	ADTT	Required pavement thickness							
		mm	in	mm	in	mm	in	mm	in
Residential	1	165	6.5	178	7	191	7.5	203	8
	10	191	7.5	203	8	216	8.5	229	9
Collector	25	203	8	216	8.5	241	9.5	254	10
	300	229	9	241	9.5	267	10.5	292	11.5

Table 9.6 (Continued)

		mm	in	mm	in	mm	in	mm	in
Minor	100	241	9.5	254	10	280	11	305	12
Arterial	300	254	10	267	10.5	292	11.5	318	12.5
	700	267	10.5	280	11	305	12	330	13
Major	700	280	11	292	11.5	318	12.5	356	14
Arterial	1500	280	11	305	12	330	13	356	14

CBR = 10	54 MPa/m 200 psi/in	Concrete flexural strength							
		kPa	psi	kPa	psi	kPa	psi	kPa	psi
		3,100	450	2,750	400	2,400	350	2,100	300

Traffic	ADTT	Required pavement thickness							
		mm	in	mm	in	mm	in	mm	in
Residential	1	152	6	165	6.5	178	7	203	8
	10	178	7	191	7.5	203	8	229	9
Collector	25	203	8	216	8.5	229	9	254	10
	300	216	8.5	241	9.5	254	10	280	11
Minor	100	229	9	254	10	267	10.5	292	11.5
Arterial	300	241	9.5	267	10.5	280	11	305	12
	700	254	10	267	10.5	292	11.5	318	12.5
Major	700	267	10.5	292	11.5	305	12	343	13.5
Arterial	1500	280	11	292	11.5	318	12.5	343	13.5

Additional conservatism is introduced into the design because the thickness is rounded up to the nearest even 12.5 mm (1/2 in) increment for ease of construction.

One efficient design concept is to combine pervious and conventional concrete pavement on the same site. Consider, for example, a four-lane city street with curb parking. The two center lanes, which carry most of the traffic, are made of conventional concrete and crowned. The curb lanes are made of pervious concrete, and are able to handle both the water that runs off of the conventional concrete and the rain that falls onto them. They handle less traffic than the two inner through lanes. This combines the proven durability of conventional concrete streets with the drainage capacity of pervious concrete.

Concrete intersections

Along asphalt city streets, intersections often exhibit considerable rutting and shoving due to the starting, stopping, and turning motions of heavy vehicles.

> At busy intersections, the added load and stress from heavy vehicles often cause asphalt pavements to deteriorate prematurely. Asphalt surfaces tend to rut and shove under the strain of busses and trucks stopping and turning. These deformed surfaces become a safety concern for drivers and a costly maintenance problem for the roadway agency.
>
> (ACPA 1997: 1)

Asphalt is a viscoelastic material – therefore, rutting progresses more rapidly under stopped or slowly moving heavy vehicle loads than under faster moving loads.

Bus stops often exhibit similar distress. Also, because intersections receive traffic from four directions, they carry up to twice the traffic as the city streets that they connect. Replacing deteriorated asphalt intersections with concrete can considerably reduce maintenance costs. A detailed 28-page guide to concrete intersections has been published by the ACPA as TB019P, entitled *Concrete Intersections: A Guide for Design and Construction* (ACPA 1997).

In addition to paving the physical area of the intersection, it is generally necessary to include the entire functional area where vehicles slow, stop, and accelerate. This may include 30–60 m (100–200 ft) on either side of the intersection, or it may be offset in the direction of approaching traffic to provide 60–120 m (200–400 ft) on the approach, but only 15 m (50 ft) past the intersection. With moderate traffic volumes and a low percentage of heavy vehicles, 15–30 m (50–100 ft) on either side of the intersection may be sufficient (ACPA 1997: 2).

Intersection pavement design thickness ranges from 100–125 mm (4–5 in) for light residential up to 150–225 mm (6–9 in) for minor arterial and 175–275 mm (7–11 in) for major arterial streets. Traffic from all directions coming into the intersection must be considered for the physical area, which may require increasing the pavement thickness 13–25 mm (1/2–1 in) over that in the functional area of the intersection (ACPA 1997: 3–4). Pavement thickness may be determined using the design tables above, or by using StreetPave.

Jointing considerations are important, and are similar to those for city street pavements, with a few additional considerations. Isolation joints are generally necessary for skew intersections and for the perpendicular street at T-intersections. They are also needed at manholes, drainage fixtures, pavements, aprons, and structures. It is important to develop a good jointing plan in advance, and make field adjustments as necessary. It is not uncommon for the actual physical location of some features, such as manholes, to vary somewhat from what is shown on the plans (ACPA 1997: 4–9). An isolation joint is shown in Figure 2.4.

Sample joint plans for typical intersection configurations are shown on page 10 of TB019P, *Concrete Intersections: A Guide for Design and Construction* (ACPA 1997). Additional details and configurations are provided in *IS006.01P, Intersection Joint Layout*, which provides a ten-step example (ACPA 1996b).

It is often necessary to maintain traffic through intersections during construction, so project phasing may be important. If possible, it is preferable to detour traffic away from the intersection to simplify and speed construction. This depends on having available alternate routes to handle the traffic. To maintain traffic through the intersection, construction may be divided into four phases by lane, or into four phases by quadrant (ACPA 1997: 9–11). Figure 9.9 shows quadrant construction at an urban intersection in downtown Cleveland, Ohio.

Intersections must often accommodate vehicle detector loops. Inductive loop detectors are used to magnetically detect the presence of vehicles in order to start or extend traffic signal green time. They require wire loops embedded in the pavement. Often, 6-mm (1/4-in) wide saw cuts 50 mm (2 in) deep are made in the pavement for wires, and then sealed. Loops may also be cast into the pavement, although in this case it is important that

Figure 9.9 Quadrant construction at an urban intersection in Cleveland, Ohio (photo by the author).

they be placed a minimum of 50 mm (2 in) above any reinforcing steel and offset as much as possible (ACPA 1997: 25).

RCC composite pavements

In some areas, including Columbus, Ohio, and Quebec, Canada, composite pavements consisting of an RCC structural base with a thin asphalt overlay as a riding surface have been used. An RCC composite pavement base under construction is shown in Figure 2.10. In the Columbus area, this type of pavement has proven to be more economical than either conventional concrete or asphalt pavements.

A relatively simple design method has been developed for this type of pavement (Delatte 2004). However, since the asphalt contributes little to the structural capacity of the composite pavement, the designs provided in this chapter in Table 9.4 are conservative and may be safely used.

Short, thin slab pavements

Covarrubias (2007) has recently developed a patented short, thin slab pavement system capable of carrying heavy truck loads. In this system, small square slabs are used, no more than half a lane in width. Slab dimensions are similar to those used for thin whitetopping, as discussed in Chapter 18.

Key features of the Covarrubias system include:

- Shorter slab dimensions, approximately 1.8 m (6 ft) square.
- Due to the small slab size, curling and warping stresses are very low. The slabs therefore have full bearing on the base and are not at a high risk of corner breaks.
- Due to the small slab size, only one wheel load is on a slab at a time, so there are no negative bending stresses due to multiple wheel loads. Slabs must be shorter than the distance between truck front and rear axles.
- Joints cut using a thin 2 mm (0.08 in) saw blade, with an early entry saw.
- Due to the small slab size and tight joints, no dowels or tie bars are used.
- Pins are used along the pavement edges to hold the transverse joints tightly together.
- Drainable base is used to prevent distress from water penetrating through the narrow joints. CTBs, ATBs or granular materials with less than 6 percent fines may be used.

With this system, a short slab pavement only 150-mm (6-in) thick has the same stress, and thus the same fatigue life, as a conventional 230-mm

(9-in) thick pavement. This technology has been demonstrated with trial sections in Puerto Montt and Temuco in Chile in January 2005 and 2006, respectively. At these sites, slabs as thin as 80 mm (3.1 in) over granular base have carried 70,000 standard axle loads to date. The concrete flexural strength was 4.8 MPa (700 psi).

Chapter 10

Airport pavement design

There are important differences between highway and airport pavements. Yoder and Witczak (1975: 559) note

> The number of repetitions of load is considerably less, but, on the other hand, gross load on the airfield pavement is greater than on the highway. Pumping can be a major problem on highways but is of lesser importance for rigid airfield pavements. In addition, a major portion of the load is applied just several feet from the edge of the rigid highway pavement, but loads are primarily in the center of airfield slabs.

The FAA notes that

> The design of airport pavements is a complex engineering problem that involves a large number of interacting variables... An airfield pavement and the aircraft that operate on it represent an interactive system that must be addressed in the pavement design process. Design considerations associated with both the aircraft and the pavement must be recognized in order to produce a satisfactory design. Producing a pavement that will achieve the intended design life will require careful construction control and some degree of maintenance. Pavements are designed to provide a finite life, and fatigue limits are anticipated. Poor construction and a lack of preventative maintenance will usually shorten the service life of even the best-designed pavement.
>
> (FAA 2004: 23)

In the past, some runway pavements were built with keel sections. Keel sections use thicker pavement near the runway centerline, and thinner pavement along the edges, because few aircrafts are expected at or near the edges. To prevent abrupt changes and stress concentrations, the pavement thickness is often tapered (Packard 1973: 20–21). Yoder and Witczak (1975: 592) suggested that the outer edges could be reduced by 30 percent.

Variable or keel cross-sections are now less common. The FAA notes

> Airport pavements are generally constructed in uniform, full width sections. Runways may be constructed with a transversely variable section, if practical. A variable section permits a reduction in the quantity of materials required for the upper paving layers of the runway. However, more complex construction operations are associated with variable sections and are usually more costly. The additional construction costs may negate any savings realized from reduced material quantities.
>
> (FAA 2004: 27)

Design procedures for airfield pavements have been developed in the United States by the FAA for civilian airfields and by the USACE for military airfields. The FAA advisory circulars that pertain to design and construction may be downloaded free from http://www.faa.gov/airports_airtraffic/airports/resources/advisory_circulars/. The PCA also developed procedures, which have later come under the ACPA purview. Packard (1973) authored *Design of Concrete Pavement*, which was supplemented by design charts for various aircraft types. The FAA allows the use of PCA procedures, which have now been updated by the ACPA with AIRPave software.

Military airfields are not covered in this book. In the United States, military airfields are designed and built using UFC 3-250-02 "Standard Practice Manual for Rigid Pavements," (UFC 2001a) and UFC 3-260-02 "Pavement Design for Airfields" (UFC 2001b). These documents are available on the web at http://65.204.17.188/report/doc_ufc.html.

It is important to understand that both civilian and military airport pavement design procedures assume an interior loading condition. In rare circumstances where aircraft wheels travel near an unsupported edge, these procedures are not valid. For example, a 30 m (100 ft) diameter circular power check pad was built at a military airfield. These pads are intended to anchor high performance jet aircrafts in order to test engines. As the aircraft exits the pad, the main gear traverses the outside slab edges, so the interior loading condition design assumption was violated. The edge-slabs failed rapidly with widespread cracking (Rollings 2005: 167–168).

Airport pavement bases and subbases

Base and subbase layers, including drainage layers, are widely used in civilian and military airport pavements. Hall et al. (2005: 1–2) note

> When specified, designed, and properly constructed, stabilized and permeable bases have a positive impact on pavement performance. However, to ensure success, the selection and specification of these layers

should be considered in the overall context of the rigid pavement design and construction process.

Stated in another way, mere inclusion of these bases in the typical section is not adequate to guarantee pavement performance under all situations. On the contrary, if careful attention is not paid to how these layers alter the early-age and long-term performance of pavements, they could even prove not as beneficial as anticipated. Design details of the PCC layers (thickness, joint spacing, etc.), PCC mixture properties, ambient conditions at the time of paving, and curing and jointing of the PCC layer interact with the base layer's as-constructed stiffness, thickness, and friction properties to create a unique set of circumstances for each project which need to be carefully accounted for to ensure a successful end product.

The Innovative Pavement Research Foundation report *Stabilized and Drainable Base for Rigid Pavement: A Design and Construction Guide, Report IPRF-01-G-002-021(G)* (Hall et al. 2005) discusses the higher risks of uncontrolled early-age cracking with certain base types and how to reduce those risks. Some of these factors should be addressed during design, and some during construction. Construction aspects are addressed in Chapter 13.

Hall et al. (2005: 13) suggest that stabilized base layers should be 152–203 mm (6–8 in) thick and permeable bases should be 102–152 mm (4–6 in) thick for airfield pavements. In this thickness range, the bases may be placed without segregation and permeable bases will have adequate permeability.

For airports, rigid pavement stabilized layers typically include Item P-304, CTB, Item P-306 Econocrete Subbase Course or LCB, and Item P-401 or P-403 ATB (FAA 2005). These materials are discussed in Chapter 4 of this book, and their construction is discussed in Chapter 13.

Design considerations

Standard FAA design procedures are discussed in Chapter 3 of FAA Advisory Circular AC 150/5320-6D (FAA 2004). These procedures are intended to address airports serving aircraft with gross weights of 13,000 kg (30,000 lb) or more. The same general procedures and considerations apply to both flexible and rigid pavements, although the specific design methods are different. For airports with only lighter aircraft, a separate and simpler design method has been developed, which is discussed in a later section.

The FAA standard design procedures are intended to provide pavements with structural lives of 20 years, although during that time there may be a need for some rehabilitation and renewal of skid resistance (FAA 2004: 23–24).

Determining equivalent annual departures of design aircraft

Standard gear configurations are assumed for single gear, dual gear, dual tandem gear, and wide body aircraft. Triple dual tandem gear aircraft are addressed separately by the new Chapter 7 – layered elastic design procedures. Tire pressures vary between 515 and 1,380 kPa (75 and 200 psi) (FAA 2004: 24).

The design is based on a single design aircraft type.

> The forecast of annual departures by aircraft type will result in a list of several different aircraft. The required pavement thickness for each aircraft type in the forecast should be checked using the appropriate design curve and the forecast number of annual departures for that aircraft. The design aircraft is the aircraft type that produces the greatest pavement thickness. It will not necessarily be the heaviest aircraft in the forecast... Since the traffic forecast is a mixture of aircraft having different landing gear types and different weights, the effects of all traffic must be accounted for in terms of the design aircraft. First, all aircraft must be converted to the same landing gear type as the design aircraft. The FAA has established factors to accomplish this conversion. These factors are constant and apply to both flexible and rigid pavements. They represent an approximation of the relative fatigue effects of different gear types.
>
> (FAA 2004: 24–25)

The conversion factors used to convert from one landing gear type to another are shown in Table 10.1.

After converting the aircraft to the same landing gear configuration, convert them to equal annual departures of the design aircraft:

$$\text{Log } R_1 = \text{Log } R_2 \times \sqrt{\frac{W_2}{W_1}} \qquad (10.1)$$

where:

R_1 = equivalent annual departures by the design aircraft;
R_2 = annual departures of the aircraft in question;
W_1 = wheel load of design aircraft;
W_2 = wheel load of the aircraft in question.

Wide body aircraft have significantly different landing gear assembly spacings, so wide body aircraft are treated as 136,100 kg (300,000 lb) dual tandem aircraft for computing equivalent annual departures, even when they are the design aircraft (FAA 2004: 25).

As an example, Table 10.2 shows the annual departures for eight different types of aircraft. The B727-200 is determined to be the design aircraft because its 9,080 annual departures would require the greatest pavement thickness. Therefore, the aircraft are all converted to dual wheel type by using the appropriate factor from Table 10.1, and then the equivalent annual departures are determined using equation 10.1. The equivalent annual departures of a B727-200 are shown as 16,241 in Table 10.2.

For example, the 3,050 average annual departures of the B707-320B with dual tandem gear type is multiplied by 1.7 to get 5,185 equivalent dual gear departures. Next, apply equation 10.1 with $W_1 = 20,520$ kg (45,240 lb) and $W_2 = 17,610$ kg (38,830 lb):

$$\text{Log } R_1 = \text{Log } (5,185) \times \sqrt{\frac{17,610}{20,520}} = 3.44$$

Table 10.1 Landing gear type conversion factors (FAA 2004: 25)

To convert from	To	Multiply departures by
Single wheel	Dual wheel	0.8
Single wheel	Dual tandem	0.5
Dual wheel	Single wheel	1.3
Dual wheel	Dual tandem	0.6
Dual tandem	Single wheel	2.0
Dual tandem	Dual wheel	1.7
Double dual tandem	Dual tandem	1.0
Double dual tandem	Dual wheel	1.7

Table 10.2 Forecast traffic and equivalent annual departures (FAA 2004: 26)

(a) Forecast departures

Aircraft	Gear type	Average annual departures	Maximum takeoff weight	
			kg	lb
727-100	Dual	3,760	72,580	160,000
727-200	Dual	9,080	86,410	190,500
707-320B	Dual tandem	3,050	148,330	327,000
DC-9-30	Dual	5,800	49,000	108,000
CV-880	Dual tandem	400	83,690	184,500
737-200	Dual	2,650	52,390	115,500
L-1011-100	Dual tandem	1,710	204,120	450,000
747-100	Double dual tandem	85	317,520	700,000

Table 10.2 (Continued)

(b) Equivalent annual departures

Aircraft	Equiv. dual gear departs.	Wheel load		Wheel load of design aircraft		Equiv. annual departs. design aircraft
		kg	lb	kg	lb	
727-100	3,760	17,240	38,000	20,520	45,240	1,891
727-200	9,080	20,520	45,240	20,520	45,240	9,080
707-320B	5,185	17,610	38,830	20,520	45,240	2,764
DC-9-30	5,800	11,630	25,650	20,520	45,240	682
CV-880	680	9,940	21,910	20,520	45,240	94
737-200	2,650	12,440	27,430	20,520	45,240	463
L-1011-100	2,907	16,160[1]	35,625[1]	20,520	45,240	1,184
747-100	145	16,160[1]	35,625[1]	20,520	45,240	83
					Total:	16,241

[1] Wheel loads for wide-body aircraft are taken as the wheel load for a 136,100 kg (300,000 lb) dual tandem aircraft for equivalent annual departure calculations

Finally, $10^{3.44} = 2,764$ annual departures of the design aircraft (B727-100). As noted previously, the wide bodied aircraft in the table (L-1011-100 and B747-100) are treated as 136,100 kg (300,000 lb) dual tandem aircraft.

This is a very straightforward calculation to set up with a spreadsheet. The weights of many aircraft are listed in FAA *Advisory Circular (AC) 150-5300-13 Airport Design* (FAA 1989) or may be found in the Aircraft Characteristics Data Base online http://www.faa.gov/airports_airtraffic/airports/construction/aircraft_char_database/.

Design for frost and permafrost

Airfield pavement designs must consider climatic conditions, particularly where adverse effects of seasonal frost or permafrost apply. Two approaches are available for designing for seasonal frost. The first approach is to ensure that the combined thickness of pavement and non-frost-susceptible material is sufficient to eliminate or limit frost penetration into the subgrade. The second approach is to provide a pavement strong enough to carry load during the critical frost melting period. Three design procedures are available:

1   Complete frost protection – the depth of frost penetration is developed using procedures in FAA AC 150/5320-6D Chapter 2, and the design thickness of pavement is compared to this depth. If frost will penetrate

deeper than the pavement thickness, subgrade material must be removed and non-frost-susceptible material provided. This is the most positive and costly method of frost protection. In any case, frost protection is normally limited to a maximum of 1.8 m (6 ft).

2   Limited subgrade frost protection – this method is similar to that above but frost protection is only provided to 65 percent of the depth of frost penetration. This method will not eliminate frost heave but will generally keep it to a tolerable level.

3   Reduced subgrade strength – based on the soil frost groups defined in FAA AC 150/5320-6D Chapter 2, the subgrade modulus is reduced to 6.8–13.5 MPa/m (25–50 psi/in). This reduced value is then used to determine the pavement thickness. This approach ignores the effects of frost heave (FAA 2004: 27–29).

> The design of pavements in permafrost regions must consider not only the effects of seasonal thawing and refreezing, but also the effects of construction on the existing thermal equilibrium. Changes in the sub-surface thermal regime may cause degradation of the permafrost table, resulting in severe differential settlements [and] drastic reduction of pavement load carrying capacity.
>
> (FAA 2004: 29)

Design approaches for permafrost include the complete protection method, the reduced subgrade strength method, and the less common insulating panel method. The first two methods are similar to those for ordinary frost protection discussed in the previous paragraph (FAA 2004: 29–30). There is no difference in the FAA design procedures for frost and permafrost between rigid and flexible pavements.

FAA rigid pavement design

The FAA pavement design procedure may be carried out using design curves. The curves

> are based on the Westergaard analysis of edge loaded slabs. The edge loading analysis has been modified to simulate a jointed edge condition. Pavement stresses are higher at the jointed edge than at the slab interior. Experience shows practically all load-induced cracks develop at jointed edges and migrate toward the slab interior. Design curves are furnished for areas where traffic will travel primarily parallel or perpendicular to joints and where traffic is likely to cross joints at an acute angle. The thickness of pavement determined from the curves is for slab thickness only. Subbase thicknesses are determined separately.
>
> (FAA 2004: 23)

Determining foundation modulus (k-value)

The k-value is established for the subgrade and then corrected for the subbase to a higher value. A plate bearing test is preferred to establish the subgrade k-value, but correlations such as those shown in Chapter 4 may be used instead. The FAA notes that "Fortunately, rigid pavement is not overly sensitive to k value, and an error in estimating k will not have a large impact on rigid pavement thickness" (FAA 2004: 56).

For both granular and stabilized subbases, charts and tables from various sources may be used to determine the composite k based on the subgrade k and the thickness and type of subbase. Tables 4.2 through 4.5 from Chapter 4 provide reasonable values.

Determining concrete slab thickness

Once the composite k under the concrete slab has been determined, the concrete flexural strength is used to determine the concrete slab thickness for a given aircraft type, gross weight of design aircraft, and number of annual departures. The concrete flexural strength is based on the strength that the concrete is required to have when it is opened to traffic, tested by ASTM C78 (ASTM C78 2002). In other words, if the pavement will be 3 months old when opened to traffic, the acceptance of the concrete is on the basis of 28-day flexural strength but the design should be on the basis of 90-day strength.

Separate design curves or nomographs are provided for each aircraft type on pages 59 through 83 of FAA AC 150/5320-6D. To use each chart, the concrete flexural strength is the first input and a horizontal line is carried across to the line representing the appropriate modulus of subgrade reaction k. From that intersection point, a line is drawn vertically downward to the line representing the gross aircraft weight. Finally, from that intersection point a line is drawn horizontally to the right to intersect with the number of annual departures and determine a design slab thickness. A dashed line is provided on each chart to indicate the procedure (FAA 2004: 58).

With eight different aircraft in the forecast traffic, this procedure must be carried out eight times – once for each aircraft, with its appropriate design chart. This will produce eight different required slab thickness values. The thickest value determines the design aircraft. Once the design aircraft has been determined, the procedure outlined above is used to determine the numbers of equivalent annual departures of the other seven aircraft types.

Since this manual procedure is obviously quite tedious, a spreadsheet computer program R805faa.xls has been developed and may be downloaded free from the FAA website, http://www.faa.gov/airports_airtraffic/airports/construction/design_software/. The use of the spreadsheet is illustrated in a design example below.

Critical and non-critical areas

FAA defines critical and non-critical areas for runway pavement. Critical areas include taxiways and the center sections of runways, as shown in Figure 3.1 of FAA AC 150/5320-6D. Non-critical areas include outer edges of runways. In non-critical areas, the pavement thickness may be reduced 10 percent (FAA 2004: 58).

High traffic volumes

The design charts go up to 25,000 annual departures. Some airports have more air traffic. With high traffic volumes, a stabilized subbase should be used, and pavement thickness may need to be slightly increased. Joint spacing may also be decreased to minimize joint movements and improve load transfer, and high quality joint sealants should be specified (FAA 2004: 84–85).

Jointing

Joint spacing and configuration is an important element of the FAA design procedure. Joints are classified as expansion, contraction, and construction joints, as discussed in Chapter 2. If an unstabilized subbase is used under the pavement, the maximum joint spacing listed in Table 10.3 may be used. Pavements with extreme temperature differentials during placement or in service should use shorter spacings (FAA 2004: 85–87).

If a stabilized subbase is used, the higher subbase k-values increase curling and warping stresses. Therefore, the joint spacing should be limited to five times the radius of relative stiffness of the slab (FAA 2004: 87). The radius of relative stiffness is calculated using equation 7.1, and the maximum slab length may be determined using Figure 7.3.

Joints are generally doweled, although the FAA AC 150/5320-6D allows keyed joints for slabs 230 mm (9 in) thick or more. Keyed joints are not

Table 10.3 Maximum joint spacing for unstabilized base (FAA 2004: 87)

Slab thickness		Transverse and longitudinal joint spacing	
mm	in	m	ft
150	6	3.8	12.5
175–230	7–9	4.6	15
230–305	9–12	6.1	20
>305	>12	7.6	25

Table 10.4 FAA dowel bar recommendations (FAA 2004: 88)

Slab thickness		Dowel diameter		Dowel length		Dowel spacing	
mm	in	mm	in	mm	in	mm	in
150–180	6–7	20	3/4	460	18	305	12
210–305	8–12	25	I	480	19	305	12
330–405	13–16	30	1 1/4	510	20	380	15
430–510	17–20	40	1 1/2	510	20	460	18
535–610	21–24	50	2	610	24	460	18

recommended when wide-body aircraft use the pavement and the foundation modulus is 54 MPa/m (200 psi/in) or less (FAA 2004: 87).

FAA recommendations for dowel bar diameter, length, and spacing are
based on the slab thickness. These recommendations are shown in Table 10.4.
These dowel diameters are smaller than those given in Table 7.1, so
the more conservative recommendations of Table 7.1 may be used instead.

Dowels are generally solid bars, but may be high-strength pipe plugged on
each end with a tight fitting cap or with a bituminous or mortar mixture. Tie
bars should be deformed bars 16 mm (5/8 in) in diameter spaced 760 mm
(30 in) on center (FAA 2004: 88).

Reinforced and CRC pavements

The FAA AC 150/5320-6D allows JRC and CRC airfield pavements in
addition to JPC. Reinforcement allows longer joint spacing, up to 23 m
(75 ft). Naturally joint openings will be larger and will require more care
to seal and maintain (FAA 2004: 89–93). CRC steel design for airport
pavements is addressed in Chapter 12.

Layered elastic pavement design

FAA airport pavement design procedures were substantially updated in
2004 to provide for new, larger aircraft, particularly the Boeing 777
and the Airbus A-380. Substantial changes were made to Section 3, rigid
pavement design, of Chapter 3, pavement design, and Chapter 7, layered elastic pavement design. Previously, design relied on curves developed
using the Westergaard stress equations, modified to model a jointed edge
condition.

The revised FAA AC 150/5320-6D states on p. 141 that

the design procedure presented in this chapter provides a method of
design based on layered elastic analysis developed to calculate design
thicknesses for airfield pavements. Layered elastic design theory was

adopted to address the impact of new gear and wheel arrangements such as the triple dual tandem (TDT) main gear.

The Boeing 777 and the Airbus A-380 are examples of aircraft that utilize this gear geometry. The TDT gear produces an airport pavement loading configuration that appears to exceed the capability of the previous methods of design, which incorporate some empiricism and have limited capacity for accommodating new gear and wheel arrangements. This design method is computationally intense and is thus in the form of a computer program called LEDFAA.

(FAA 2004: 141)

LEDFAA may be downloaded free from the FAA website, http://www. faa.gov/airports_airtraffic/airports/construction/design_software/. Some of the features of LEDFAA include:

- Designs may be carried out using either SI or US customary units.
- An extensive aircraft library is included and designers may adjust aircraft parameters.
- The basis of the design is layered elastic theory rather than the Westergaard solutions.
- A pavement design life other than 20 years may be considered, although this is a deviation from FAA standards.
- The aircraft traffic mix is not converted to equivalent departures of a single design aircraft. Instead, for rigid pavements, the fatigue damage caused by each aircraft is calculated. This is similar to the PCA/ACPA 1984 and StreetPave design method for highway pavements.
- Pavement materials are characterized by thickness, modulus of elasticity, and Poisson's ratio. Materials may be identified using their FAA specification designations, such as P-209 for crushed stone base course.
- The program does not automatically check that layer requirements are met – for example, the requirement for stabilized subbase layers under rigid pavements with wide body aircraft in the traffic mix. This is left to the designer (FAA 2004: 141–142).

Because of differences in philosophy between layered elastic design and the previously used FAA design procedure, there are also limitations placed on the use of LEDFAA by the FAA:

- If the aircraft traffic mix does not include aircraft with the TDT gear configuration, the previously used procedure should be used (i.e. Chapter 3 of FAA 2004) and the minimum pavement thickness determined using that procedure applies.

- The procedure should not be used to determine thickness requirements for an individual aircraft – the entire traffic mixture must be used.
- The program may not be used to evaluate existing pavements that were designed using the previous FAA design procedure (FAA 2004: 141–142).

The concrete pavement must be at least 152 mm (6 in) thick. Stabilized subbases are required under rigid pavements serving aircraft with TDT gear, with a minimum layer thickness of 102 mm (4 in). The subgrade is assumed to be elastic and infinite in thickness and characterized by an elastic modulus or k-value. The program converts an input k-value to an elastic modulus. In US customary units, the subgrade $E = 26\ k^{1.284}$ (FAA 2004: 143–144). Since the layered elastic design requires software, the procedure is discussed in the following section.

Airfield pavement design software

The FAA provides software for the standard and layered elastic design procedures. The ACPA has also developed a computer program, ACPA AirPave, which is based on the PCA design procedure (PCA 1966).

AirPave can only determine stresses and allowable repetitions for a single aircraft gear configuration at a time. However, the number of allowable repetitions for each aircraft in the traffic stream may be determined using AirPave, and then equation 6.4 may be used to compute the cumulative damage for the pavement. If necessary, the design can then be adjusted. Therefore, AirPave is convenient when designing an airfield pavement for a single aircraft type, but the FAA procedures are more useful for mixed traffic. The use of AirPave is very similar to RCCPave, which is discussed in detail in Chapter 11.

Standard FAA design procedure (Chapter 3)

As noted above, the FAA has developed spreadsheet design software, the program R805faa.xls, which may be downloaded free from the FAA website, http://www.faa.gov/airports_airtraffic/airports/construction/design_software/. The instructions file for the spreadsheet may be downloaded from the same website.

The basic program screen is shown in Figure 10.1. The eight design steps are carried by activating the various input screens in order. Step 1 is the basic airport and design information and does not affect the calculations.

Step 2 requires the input of the subgrade k-value, the subbase type and thickness, and the necessary information for subgrade frost design if it is required. Any of the frost design procedures discussed above may be used.

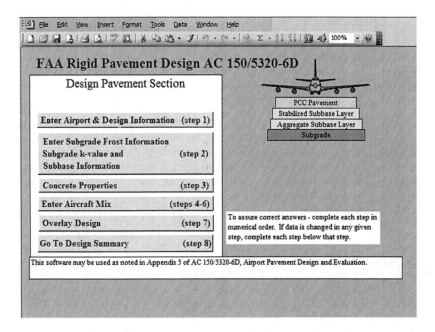

Figure 10.1 FAA R805FAA.xls rigid airport pavement design spreadsheet.

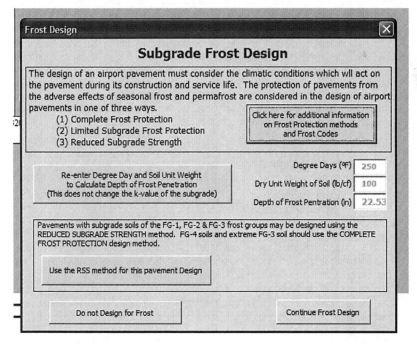

Figure 10.2 FAA R805FAA.xls rigid airport pavement design spreadsheet frost design.

The frost design input screen is shown as Figure 10.2. Frost design may be omitted.

If the pavement is to be designed for frost, a number of additional inputs are required, including the air freezing index in degree–days, the dry unit weight of the soil, and the subgrade soil frost code (non-frost or F-1 through F-4). These are used to calculate the depth of frost penetration. Alternatively, the reduced subgrade strength may be used.

Continuing with step 2, the subgrade/stabilized subbase/aggregate subbase values are input. These are shown in Figure 10.3. The input subgrade k-value of 27 MPa/m (100 psi/in) is automatically adjusted by the program for the 150 mm (6 in) of stabilized base to 57 MPa/m (215 psi/in).

The concrete properties are input in step 3. The main input is the concrete flexural strength. The Poisson's ratio is assumed to be 0.15 and the concrete modulus of elasticity is assumed to be 27.6 GPa (4,000,000 psi), although these default values may be changed.

The projected aircraft forecast is entered in steps 4–6. Up to 21 different aircraft types may be included. Aircraft types are entered one at a time, as shown in Figure 10.4, which repeats the calculations shown in Table 10.2.

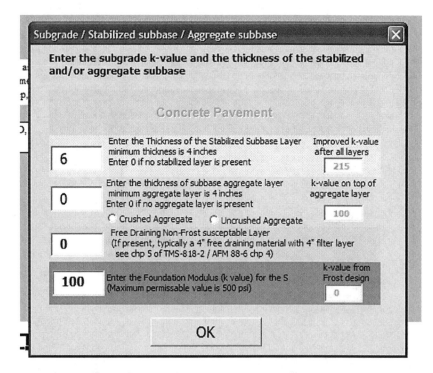

Figure 10.3 FAA R805FAA.xls rigid airport pavement design spreadsheet computation of composite k-value.

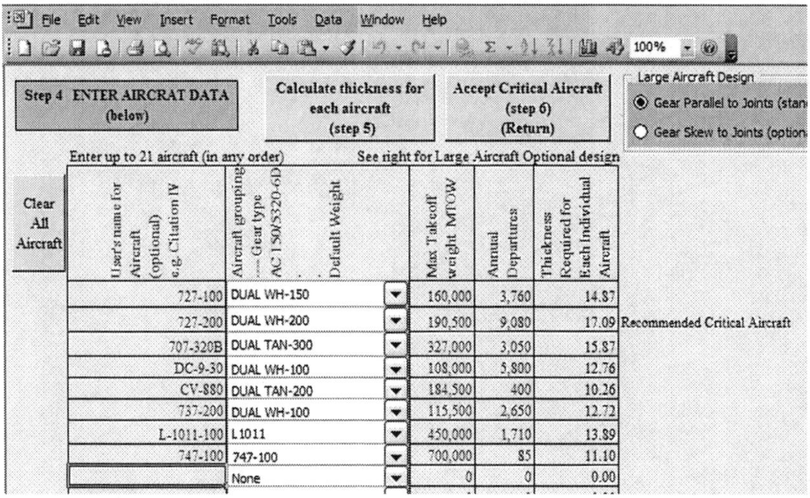

Clear All Aircraft	User's name for Aircraft (optional) e.g. Citation IV	Aircraft grouping — Gear type AC 150/5320-6D	Default Weight		Max Takeoff weight MTOW	Annual Departures	Thickness Required for Each Individual Aircraft	
	727-100	DUAL WH-150		▼	160,000	3,760	14.87	
	727-200	DUAL WH-200		▼	190,500	9,080	17.09	Recommended Critical Aircraft
	707-320B	DUAL TAN-300		▼	327,000	3,050	15.87	
	DC-9-30	DUAL WH-100		▼	108,000	5,800	12.76	
	CV-880	DUAL TAN-200		▼	184,500	400	10.26	
	737-200	DUAL WH-100		▼	115,500	2,650	12.72	
	L-1011-100	L1011		▼	450,000	1,710	13.89	
	747-100	747-100		▼	700,000	85	11.10	
		None		▼	0	0	0.00	

Figure 10.4 Determining critical aircraft.

For each aircraft type, the required inputs are the gear configuration, maximum takeoff weight, and annual departures (columns 2 through 4). A descriptive user name for each aircraft type (column 1) is optional. Some aircraft are already provided in the library, such as the L1011 and 747-100, and some are listed as gear configurations and an approximate gross aircraft weight in thousands of pounds, such as DUAL TAN-300 which is a dual tandem of approximately 136,000 kg (300,000 lb).

Once all of the aircraft have been entered, the pavement thickness for each aircraft is calculated by clicking the step 5 button. This provides the recommended critical aircraft, in this case the B727-200 as shown in both Table 10.2 and Figure 10.4. Next, the step 6 button is clicked to accept the critical aircraft and complete the design.

The program also offers an option to compute an overlay thickness, which is discussed in Chapter 18. Step 8 provides a printed design summary. As part of the summary, charts illustrate how required pavement thickness varies with the number of annual departures or with the pavement flexural strength. Unfortunately for international users, the spreadsheet is only available in US customary units, although the FAA design charts are presented in both SI and US customary units.

Layered elastic design (Chapter 7)

LEDFAA can be downloaded from the FAA website, http://www.faa.gov/airports_airtraffic/airports/construction/design_software/. The basic LED-FAA input screen is shown in Figure 10.5.

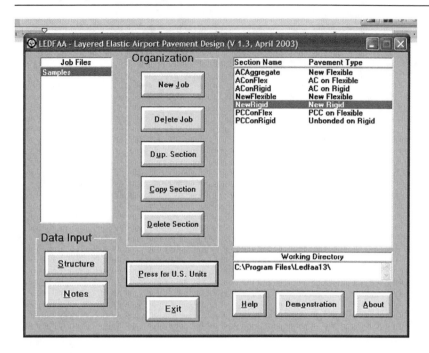

Figure 10.5 LEDFAA basic input screen.

The program may be used to design new flexible or rigid pavements and four types of overlays. Design examples are provided for each pavement type. The basic input screen is used to select the pavement type and units to be used (SI or US customary). An extensive help file is provided, as well as an embedded software demonstration.

The pavement structure is input first. A new rigid pavement design is copied into the design file from a file of samples stored in the program. A sample rigid pavement section is shown in Figure 10.6.

Layer thicknesses are in mm if the program is used with SI units. The concrete MOR is given in MPa, and the modulus of elasticity of each underlying layer is also given in MPa. One unusual aspect is that the modulus of subgrade reaction k is given in kg/cm^3 rather than MPa/m, but since the values are almost identical the traditional k in MPa/m should be input.

The basic rigid pavement structure stored within LEDFAA consists of 355.6 mm (14 in) of P-501 concrete pavement, 152.4 mm (6 in) each of P-306 Econocrete and P-209 crushed aggregate, with elastic moduli of 4.8 GPa (700,000 psi) and 517 MPa (75,000 psi), respectively. The modulus of subgrade reaction is 38.4 kg/cm^3 or MPa/m (141 psi/in), with an elastic modulus of 103 MPa (15,000 psi).

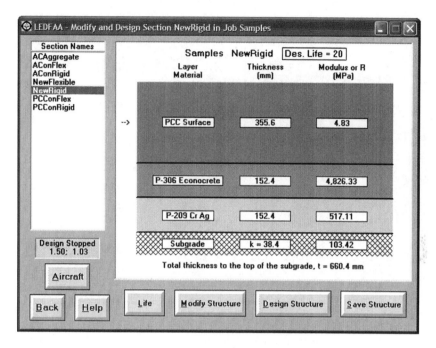

Figure 10.6 Sample rigid pavement section from the LEDFAA samples file.

Once the structure has been selected, the layers may be modified by changing the thickness or modulus, or by selecting a different type of material, such as replacing a P-306 econocrete subbase layer with P-304 CTB.

Once the structure has been modified, the designer selects the aircraft button from the structure screen to input the aircraft traffic mix. An example of an aircraft traffic mix is shown in Figure 10.7.

There are six built-in aircraft groups to select from – generic, Airbus, Boeing, McDonnell-Douglas, other commercial, and military. A library of aircraft is provided within each group. Once an aircraft type is selected from the library and added, the aircraft type, gross taxi weight (in metric tons or pounds), and a default number of 1,200 annual departures is added to the list. If only a single aircraft type is selected, the program notifies the user that this is a non-standard traffic mix.

The aircraft gross taxi weight may be modified, within limits. The number of annual departures may be set at any number up to 100,000. For some aircraft types, such as the DC-10-30 and MD-11, the wing and belly gear are treated separately and a line for each is included in the table. Finally, the aircraft traffic mix is saved, and the designer used the back button to return to the Structure screen.

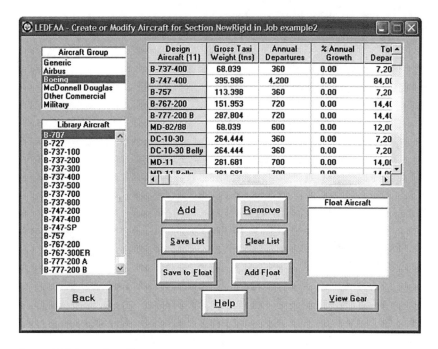

Figure 10.7 Aircraft traffic mix.

The design is completed by pressing the "Design Structure" button on the Structure screen. The thickness of the top layer is changed until the CDF for tensile stress at the bottom of the concrete is 1, based on the fatigue damage from each aircraft type as calculated using equation 6.4.

Design examples

FAA AC 150/5320-6D provides a rigid pavement design example using design curves. The design aircraft has a dual tandem gear configuration and a gross weight of 160,000 kg (350,000 lb). There are 6,000 annual equivalent departures of the design aircraft, including 1,200 annual departures of a B747 wide-body aircraft with a gross weight of 350,000 kg (780,000 lb).

The concrete flexural strength is 4.5 MPa (650 psi), and the subgrade is a low plasticity clay with a modulus of subgrade reaction of 27 MPa/m (100 psi/in). There is a 150-mm (6-in) thick stabilized subbase layer, which increases the *k*-value to 57 MPa/m (210 psi/in). Use of the design chart (Figure 3.19) provides a required thickness of 422 mm (16.6 in), which is rounded up to 460 mm (17 in) (FAA 2004: 84).

The design may also be carried out using the spreadsheet program R805faa.xls. The design aircraft may be input as either a DUAL TAN-300 or DUAL TAN-400 with the appropriate gross aircraft weight. The required pavement thickness is slightly less, 404 mm (15.9 in), which due to rounding will provide a savings of 25 mm (1 in) in the final pavement thickness. The required dowel diameter, length, and spacing, from Table 10.4, are 30 mm (1¼ in), 510 mm (20 in), and 380 mm (15 in), respectively. Because the subbase is stabilized, the joint spacing is limited to five times the pavement radius of relative stiffness.

If a pavement with the same flexural strength, support, and stabilized subbase is designed for the forecast traffic and equivalent annual departures shown in Table 10.2 and Figure 10.4, the required pavement thickness would be 450 mm (17.7 in). The required dowel diameter, length, and spacing, from Table 10.4, are increased to 40 mm (1½ in), 510 mm (20 in), and 460 mm (18 in), respectively.

LEDFAA provides a rigid pavement design example in its embedded help file. The structure shown in Figure 10.6 is modified so that the concrete flexural strength is 5.0 MPa (725 psi), the P-306 econocrete is replaced by P-304 CTB with an elastic modulus of 3.45 GPa (500,000 psi), and the modulus of subgrade reaction is changed to 32.6 MPa/m (120 psi/in). The thickness of the subbase layers is not changed.

The aircraft traffic mix is shown in Table 10.5 and Figure 10.7. With the TDT gear B-777-200 B in the traffic mix, the pavement cannot be designed using R805faa.xls.

To complete the design, the program iterates to find a new concrete pavement layer thickness of 406.7 mm (16 in). The two subbase layers are not adjusted. Results of the design are shown in Figure 10.8.

For comparison, the second traditional/R805faa.xls example shown above may be checked using LEDFAA. The sample rigid pavement structure

Table 10.5 Traffic mix for LEDFAA design example

No.	Name	Gross weight		Annual departures
		kg	lb	
1	B-737-400	68,000	150,000	360
2	B-747-400	396,000	873,000	4,200
3	B-757	113,000	250,000	360
4	B-767-200	152,000	335,000	720
5	B-777-200 B	288,000	634,500	720
6	MD-82/88	68,000	150,000	600
7	DC-10-30	264,000	583,000	360
8	DC-10-30 Belly	264,000	583,000	360
9	MD-11	282,000	621,000	700
10	MD-11 Belly	282,000	621,000	700
11	L-1011	225,000	496,000	320

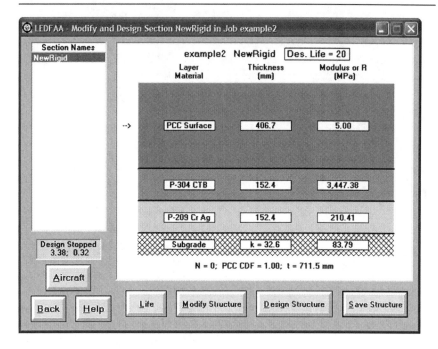

Figure 10.8 Example problem rigid pavement section as designed.

from Figure 10.6 is modified to change the concrete flexural strength to 4.5 MPa (650 psi), delete the P-209 crushed aggregate layer, and modify the modulus of subgrade reaction to 27 MPa/m (100 psi/in). The traffic mix is entered from the top half of Table 10.2.

The required pavement thickness calculated using LEDFAA is 461 mm (18 in), which is thicker than the design required by the traditional method. This aircraft traffic mix does not contain any TDT aircraft, so the result from R805faa.xls should be used rather than that from LEDFAA.

Airport pavements for light aircraft

FAA defines pavements for light aircraft as "those intended to serve aircraft weights of less than 30,000 pounds (13,000 kg). Aircraft of this size are usually engaged in nonscheduled activities, such as agricultural, instructional, or recreational flying" (FAA 2004: 125).

Whether subbases are required depends on the gross aircraft weight.

Rigid pavements designed to serve aircraft weighing between 12,500 pounds (5,700 kg) and 30,000 pounds (13,000 kg) will require a minimum subbase thickness of 4 inches (100 mm) except as shown in Table 3–4 of Chapter 3. No subbase is required for designs intended to serve aircraft weighing 12,500 pounds (5,700 kg) or less, except when soil types OL, MH, CH, or OH are encountered. When the above soil types are present, a minimum 4-inch (100-mm) subbase should be provided.

(FAA 2004: 129)

Thickness design is straightforward. "Rigid pavements designed to serve aircraft weighing 12,500 pounds (5,700 kg) or less should be 5 inches (127 mm) thick. Those designed to serve aircraft weighing between 12,501 pounds (5,700 kg) and 30,000 pounds (13,000 kg) should be 6 inches (150 mm) thick" (FAA 2004: 130). Although at first glance it seems unusual not to take the modulus of subgrade reaction into account, the subgrade and subbase construction requirements will ensure satisfactory support for the pavement.

Jointing of light load rigid pavements is also relatively simple.

The maximum spacing of joints for light-load rigid pavements should be 12.5 feet (3.8 m) for longitudinal joints and 15 feet (4.6 m) for transverse joints... Note that several differences exist between light-load and heavy-load rigid pavement joints. For instance, butt-type construction and expansion joints are permitted when an asphalt or cement stabilized subbase is provided. Also, half-round keyed joints are permitted even though the slab thicknesses are less than 9 inches (230 mm). Odd-shaped slabs should be reinforced with 0.05 percent steel in both directions. Odd-shaped slabs are defined as slabs that are not rectangular in shape or rectangular slabs with length-to-width ratios that exceed 1.25.

(FAA 2004: 130)

FAA also allows a special tension ring design for light duty pavements no more than 18.3 m (60 ft) wide. Interior joints may use keyways for load transfer.

The concept behind the jointing patterns shown is the creation of a "tension ring" around the perimeter of the pavement to hold joints within the interior of the paved area tightly closed. A tightly closed joint will function better than an open joint. The last three contraction joints and longitudinal joints nearest the free edge of the pavement are tied with [13 mm] #4 deformed bars, 20 inches (510 mm) long, spaced at 36 inches (1 m) center to center. At the ends of the pavement and in locations where aircraft or vehicular traffic would move onto or off the pavement, a thickened edge should be constructed. The thickened edge should be 1.25 times the thickness of the slab and should taper to the slab thickness over a distance of 3 feet (1 m).

(FAA 2004: 130)

Chapter 11

Industrial pavement design

Industrial pavement facilities include rail and truck terminals, industrial storage facilities, intermodal facilities, material handling yards, and ports. Design of industrial pavements is similar to that of other types of pavements, but there are important differences.

> For example, the wheel loads on container and trailer-handling vehicles may greatly exceed those of highway trucks and even of the heaviest modern aircraft; the required service life of the pavement may be shorter than that required for highway and airport pavements; and the pavement serviceability requirements may not have to be as strict as for highway and airport pavements carrying high-speed traffic.
>
> (PCA 1987: 1)

Industrial pavements may be made of conventional concrete pavement, although the heavy duty load-carrying capacity and economy of RCC pavements makes them ideal for this application. Vehicle speeds are generally low, so pavement smoothness is less important than for highways or airport runways. Owners are generally interested in keeping construction and maintenance costs low, so economical and durable pavements are favored. This accounts for the increasing popularity of RCC pavements for these applications.

The PCA has published IS234.01P, *Design of Heavy Industrial Concrete Pavements* (PCA 1988). This addresses some of the specialized aspects of industrial pavement design, and was adapted from design procedures for concrete airport pavements. The PCA has also published *IS233.01, Structural Design of Roller-Compacted Concrete for Industrial Pavements* (PCA 1987).

Like airport pavements, industrial pavements are designed to carry a relatively predictable number of heavy vehicle loads. The design vehicle may be heavy trucks parking at a facility, or heavy duty materials handling equipment moving freight containers or other large items. Loads may include large forklift trucks, straddle carriers, and log stackers, with wheel loads up to 578 kN (130 kip). Although the heavy point loads represent an unusual design situation, other aspects of conventional concrete paving still apply, such as joint design, subgrade and subbase preparation, and concrete quality (PCA 1988: 1).

The estimation of expected traffic is important. Required information includes the axle and wheel load magnitudes, wheel configurations in terms of the number of wheels and spacing between wheels, and the number of repetitions (PCA 1988: 5). It is also important to determine whether the vehicle traffic is heavily channelized, or relatively evenly distributed across the pavement.

> Usually the vehicle having the heaviest wheel load will control the design, but the design should also be checked for adequacy if other vehicle wheel loads are almost as heavy and travel the pavement frequently. The maximum wheel load is half of the heaviest axle load for the vehicle at its maximum loading. This information is usually available from the vehicle manufacturer.
>
> (PCA 1987: 3)

With rounding up of the required pavement thickness for the heaviest loads, the pavement will generally have sufficient fatigue capacity to handle any lighter loads. Total fatigue consumption due to the heavier and lighter loads can be checked using the pavement evaluation module for each vehicle type, and adding up the fatigue damage percentage due to each vehicle using equation 6.4 to determine the CDF. This is outlined in the RCCPave help file, and described in the PCA guide (PCA 1988: 10–11).

In some cases, it may be necessary to consider the weight of the loaded containers themselves. Individual containers may weigh more than 289 kN (65,000 lb) fully loaded, and may be stacked up to four high. In theory, this would require a pavement thickness of 457–610 mm (18–24 in). However, pavements in Halifax, Winnipeg, and Wando Terminal – Charleston have been able to handle containers four high with 275–400 mm (11–16 in) thick pavements. It is unlikely that all four containers would contain the maximum weight. It is also possible that under sustained container loads, some stress relief occurs due to creep (PCA 1988: 11–12). Therefore, the thickness design based on the container handlers is generally sufficient.

Point loads may also be important – if trailers are connected to dollies while parked, the dollies will not distribute loads as well as rear dual tires. This is only likely to control the design of lightly loaded industrial pavements. In some cases, it is desirable to provide thicker pads, with or without reinforcement, for smaller areas of the pavement subject to point loads.

Conventional industrial pavements

Because of the similarities to airport pavement design, the ACPA AirPave computer program may be used to design industrial pavements. Unfortunately, this program does not have a library of industrial vehicles, so a user-defined vehicle must be created. It is necessary to provide the gear/wheel configuration, the contact area, the tire pressure, the number of wheels, and the gross vehicle weight. Vehicle characteristics may be obtained from the RCCPave computer program, discussed in the next section.

If a single design vehicle is used, this will be a relatively straightforward procedure. Lighter vehicles can be checked to ensure that they will do little or no fatigue damage. Using the Packard and Tayabji (1985) fatigue relationships provided in equation 6.2, vehicles that impose a stress ratio (SR) <0.45 will do no fatigue damage to the pavement.

RCC industrial pavements

The popularity of RCC pavements for industrial applications has been increasing, because these pavements have low initial construction costs and low maintenance costs (ACI Committee 325 1995). In the southeastern US, automobile plant applications have included Saturn in Tennessee and Honda and Mercedes in Alabama (Delatte et al. 2003). A 400 mm (16 in) two-lift RCC pavement was constructed for the port of Norfolk, Virginia in 2006.

> The use of RCC for pavements at industrial facilities such as port and intermodal container terminals is particularly appropriate because of its ability to construct low-cost concrete pavements over large areas, allowing flexibility in terminal operations over time. Two basic pavement designs are used which incorporate RCC: 1) unsurfaced, where high-strength concrete is used as the surface layer, and 2) asphalt surfaced, where lower-strength concrete is used as a pavement base and an asphalt layer is used as the wearing surface ... RCC has been successfully used at the Port of Boston, Port of Los Angeles, and for container facilities at rail-truck intermodal yards in Denver, Colorado, and Calgary, Alberta.
>
> (PCA 2006)

Single lift RCC pavements are limited in thickness to about 200–250 mm (8–10 in). Thicker RCC pavements require two lift construction, preferably with the second lift placed within 1 hour of the first lift. This is termed a fresh joint or wet-on-wet construction. "Testing of core samples obtained from RCC paving projects indicates that properly constructed multiple-lift RCC pavements develop sufficient bond at the interface to be considered monolithic, except along edges that were unsupported during compaction" (ACI Committee 325 1995: 13–14).

The original design procedures for RCC pavements closely followed procedures for conventional concrete pavements. Both the PCA and the USACE developed RCC pavement design charts, which have been reproduced in ACI 325.10R (ACI Committee 325 1995: 12–15).

No dowels are used for RCC joints, which rely on aggregate interlock. When used, sawn transverse joints have been spaced 9.1–21.3 m (30–70 ft) apart. Often joints are not sawn and the pavement is allowed to crack naturally. Joints and cracks tend to remain tight. Maintenance costs for RCC pavements are generally low.

The ACPA has developed a computer program, RCCPave, to design RCC industrial pavements. It is computationally similar to AirPave, but uses a different fatigue relationship specifically developed for RCC. It also has a built-in library of industrial vehicles, including 80 kN (18 kip) single axles with single and dual tires, 178 kN (40 kip) tandem axles with dual wheels, and a variety of material handlers, as shown in Figure 11.1. User-defined vehicles may also be created. This library may also be used to determine vehicle characteristics for AIRPave to design a conventional concrete pavement.

The RCCPave program may be used to design the thickness of a new pavement or to evaluate an existing pavement. One important input is

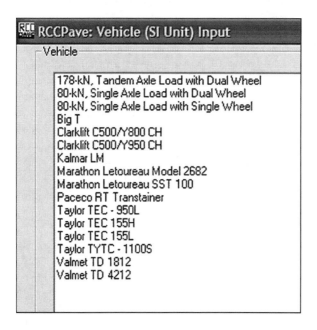

Figure 11.1 RCCPave vehicle library.

whether an interior or edge-loading condition should be used – in most cases, most vehicle traffic will be away from the pavement edge and it is reasonable to use the less conservative interior loading condition. This same assumption is typically used for airport pavement design. The PCA recommends increasing pavement thickness by 20 percent if vehicles will travel near or at the pavement edge (PCA 1987: 2). Rather than increasing pavement thickness, it makes more sense to extend the pavement width about 0.6 m (2 ft) outside of the traveled way.

The designer must specify the desired pavement life and number of load applications, along with the MOR and modulus of elasticity of the RCC and the modulus of subgrade reaction. If the concrete MOR and/or modulus of elasticity are not known, they may be estimated from compressive or splitting tensile strength. Modulus of subgrade reaction may be input directly or estimated based on a fine or coarse-grained soil with or without a subbase. The PCA recommends using the 90-day pavement flexural strength for design because the number of load repetitions in the first 3 months of pavement life is likely to be low.

Industrial pavement design examples (RCCPave and AIRPave)

These design examples have been modified from the PCA publication on design of heavy industrial concrete pavements (PCA 1988: 8–9), and redesigned using RCCPave software.

Design example 1 – container handler

The container handler has four wheels with a maximum single-wheel load of 125 kN (28,000 lb). There are 40 repetitions per day of the container handler, for a total of 292,000 repetitions over the 20-year life of the pavement. The pavement flexural strength is 4.6 MPa (660 psi), and the modulus of subgrade reaction is 27 MPa/m (100 psi/in). The design also requires a modulus of elasticity, which has a default value of 27.6 GPa (4 million psi). The modulus of elasticity has very little effect on the design, so it is generally reasonable to use the default value.

This requires a user-defined input vehicle, with tire pressure of 758 kPa (110 psi) and spacing between the tires of 300 and 1200 mm (12 and 48 in). Based on interior loading condition, the required pavement thickness is 200 mm (8 in) versus 250 mm (9.9 in) found using the PCA 1988 procedure. The design results are shown in Figure 11.2.

If edge-loading is selected instead, the required thickness becomes 300 mm (12 in). Therefore, the PCA 1988 result is halfway between the two conditions (PCA 1988: 8–9). If a 200 mm (8 in) RCC pavement is used, it may be built in a single lift.

RCCPave

(Click the "x" at the upper right corner to close the report and return to main window)

Thickness Design Report

GENERAL DESIGN INFORMATION

Project ID: Project 1   Operator:

Run Date: 8/17/2006

GENERAL DESIGN INPUT

Design Period: 20   Years

Unit: US Units   Total Design Traffic: 292000   Load Repetitions

USER DEFINED INPUT

Design Vehicle/Aircraft: Test Vehicle

Wheel/Axle Configuration: Single-Single   Pavement Type: RCC

Modulus of Elasticity (E):   4   million psi

Modulus of Rupture (MR):   660   psi

Modulus Subgrade Reaction (k):   100   pci

Computation Method: With Axle Rotation   Loading Condition: Interior Loading

Number of Wheels: 4   Contact Area: 64   sq. Inches

Contact Pressure: 110   psi

Total Load: 28160   lbf

COMPUTATION RESULT

X_Max: -2   inches

Y_Max: 0.5   inches

Maximum Angle: 27.2   degrees

Maximum Stress: 265   psi

Stress Ratio: 0.402

OUTPUT

Allowable Total Repetitions: 577270   Allowable Daily Repetitions: 79

Final Slab Thickness: 8.0   inches

WHEEL COORDINATES   inches

x: 0   12   0   12

y: 0   0   48   48

NOTE

A stress ratio (stress divided by design strength) greater than 0.75 may be too high to satisfy routine pavement design requirements (the thickness is inadequate), but may be used to evaluate the effect of unexpected heavy loads on an existing pavement. If the computed thickness for RCC pavement is less than 4" or 200 mm (20 cm), 4" or 200 mm should be used in design.

Figure 11.2 RCCPave design results for example 1.

AirPave produces a less conservative design for this example. The program does not calculate a pavement thickness – instead, the user checks different thickness values. For each thickness value, a number of repetitions is produced, which is compared to the desired number of repetitions. AirPave requires a pavement thickness of 178 mm (7 in) for this design example using interior loading.

Design example 2 – mobile gantry crane

Design example 2 uses a mobile gantry crane with a lift capacity of 40 tons. The vehicle has eight wheels (four duals) and a maximum dual-wheel load of 400 kN (90,000 lb). This closely matches one of the vehicles from the RCCPave library, so a Clarklift C500/Y800 CH is selected.

There are 50 repetitions per day of the container handler, for a total of 365,000 repetitions over the 20-year life of the pavement. The pavement flexural strength is 4.4 MPa (640 psi), and the modulus of subgrade reaction is 54 MPa/m (200 psi/in). The design modulus of elasticity is the default value of 27.6 GPa (4 million psi). The design inputs are shown in Figure 11.3.

The required pavement thickness based on interior loading is 457 mm (18 in), which is very close to the 431 mm (17 in) required thickness

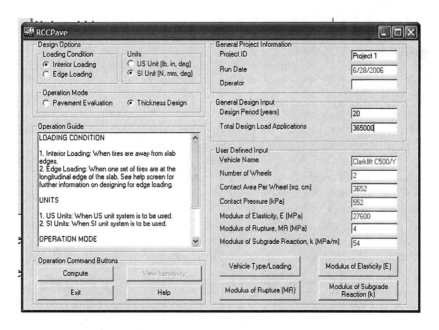

Figure 11.3 RCCPave design inputs for example 2.

calculated in the PCA 1988 Design Guide (PCA 1988: 9). For either thickness, the pavement must be built in two lifts.

If this pavement is designed using AirPave, it is necessary to use a user-defined vehicle to model the Clarklift C500/Y800 CH since only aircraft are provided in the AirPave library. Once again, the AirPave results are less conservative than the RCCPave results, with a 406 mm (16 in) pavement required for interior loading. The input screen for AirPave is similar to Figure 11.3, but only the pavement evaluation option is available, not thickness design.

Industrial pavement case studies

Two case studies are presented below illustrating conventional and RCC industrial pavement design and construction.

Conventional industrial pavement with shrinkage compensating cement

Keith et al. (2006) reported a case study of an industrial distribution center pavement made with shrinkage compensating concrete, long joint spacing, and diamond-shaped plate dowels. Shrinkage compensating concrete expands slightly during the first 1–7 days of curing, in order to offset subsequent later shrinkage. It may be made using either special type K cement, or with conventional type I, II, or V cement with an expansive admixture that has the same effect.

A first phase of construction with shrinkage compensating concrete at the Atlanta Bonded Warehouse Corporation (ABW) distribution center in Kennesaw, Georgia, began in 1993. Twelve years after construction, the pavement was still performing well. Despite joint spacing up to 32 m (105 ft), there were few random cracks and no observed edge-curl. Although the pavement cost more to build than conventional concrete pavements, long-term maintenance costs were projected to be much lower.

Therefore, a second phase was constructed 7 years later with approximately $27,000$ m^2 $(300,000$ ft$^2)$ of new pavement. One placement set a world record for more than $3,900$ m^2 $(42,000$ ft$^2)$ without interior joints. Joint spacing for the new construction varied from 26 to more than 104 m (85 to more than 340 ft).

There were some other unusual design features:

- Diamond square plate dowels, 115 by 115 by 6 mm (4½ by 4½ by 1/4 in), were used for load transfer. The dowels are encased in plastic sleeves that may be attached to the concrete form work to ensure accurate placement. With this spacing, all joints are construction joints and were built as expansion joints with polyethylene plank as filler. Joints were sealed with silicone.

- Joints were located to prevent reentrant slab corners. To minimize restraint against movement, catch basins and manholes were located near the center of slabs when possible, or isolated with compressible foam filler.
- Because shrinkage compensating concrete is very sensitive to curing, an innovative plastic backed wet curing sheet was used.
- The pavement was designed to carry tractor-trailer loads, and slab thickness was 150 mm (6 in).
- Two layers of plastic sheeting on top of a smooth base were used to reduce friction between the slab and base.
- A tight aggregate gradation was also used to minimize shrinkage.

Diamond plate dowels provide an advantage over conventional dowel bars in that they allow transverse movements parallel as well as perpendicular to joints. Therefore, they allow slabs to slip relative to each other, while providing load transfer. This is particularly important when new concrete is cast next to hardened concrete, because dowel bars provide restraint to the shrinkage of the new concrete and may initiate cracking. Although diamond plate dowels have been used for floor slabs and industrial pavements, they have not yet been adopted for highway or airfield pavements.

Norfolk international Terminal RCC port facility expansion

The Virginia Port Authority selected RCC for a large container storage and handling area at the Norfolk International Terminals in Norfolk, Virginia. The project is well documented in *Norfolk International Terminal Selects RCC for Port Facility Expansion* (PCA 2006) which provided the information for this case study. The port is the second largest in gross tonnage on the east coast of the United States and requires continuing expansion. The initial expansion included 10.5 hectares (26 acres) of pavement 420 mm (16.5 in) thick, for a total of 43,800 cubic meters (57,300 cubic yards) of concrete. RCC was estimated to be significantly cheaper than asphalt, conventional concrete, or concrete paver blocks.

The heavy container handling equipment has wheel loads of 133–266 kN (30,000–60,000 lb) per tire. The heavy wheel loads traveling continually along the same path had heavily rutted the asphalt pavement. To carry the traffic loads, the RCC was designed for a flexural strength of 3.1 MPa (450 psi) and a compressive strength of 17.2 MPa (2,500 psi) at 7 days.

Because of the pavement thickness it was necessary to place the RCC in two lifts. The second lift had to be placed within 1 hour of the bottom lift to ensure bond. Cores extracted after 7 days indicated that the interface was well bonded. The completed surface was water cured and then sealed with a bituminous tack coat before placing a 75 mm (3 in) asphalt overlay to allow adjustments for future differential settlement of the underlying soils.

The final project cost was $50 US per square meter ($42 per square yard) and the production rate was 5.4 days per hectare (2.2 days per acre). A large paver was able to place RCC 9.1 m (30 ft) wide in one pass. The Port was pleased with the work and planned a further 8.1 hectare (20 acre) expansion.

Chapter 12

Transitions, special details, and CRCP reinforcement

Transitions between concrete pavements and bridges and other structures and CRCP reinforcement require special designs and detailing. Details depend on whether movement is to be restrained or accommodated. If movement is to be restrained, enough reinforcement and anchorage must be provided to resist the forces that develop. If movement is to be allowed, the joints must be detailed and maintained so that they open and close freely.

Transitions

Transitions are necessary wherever concrete pavements abut bridges or asphalt pavements. Expansion and contraction of concrete pavements can impose substantial forces unless proper isolation is provided.

Jointed pavements (JPCP and JRCP) require relatively simple transition details. CRCP requires specialized details because it is necessary to anchor the pavement ends to allow the proper crack pattern to develop. Some special transition details have also been developed for overlays, and these are discussed in Chapter 18 along with the relevant overlay types.

Burke (2004a,b) provides a detailed discussion of the importance of providing expansion joints between jointed concrete pavements and bridges. A pavement growth/pressure phenomenon is defined – over time, incompressible materials infiltrate into joints as they expand and contract, and make it impossible for the joints to close completely. Therefore, the pavement attempts to increase in length. This can lead to damage to bridge abutments, or to blowups of the concrete pavement. This damage may occur 10–20 years after pavement construction.

Three bridge case studies are discussed in detail. In each case, the expanding pavement closed up expansion joints and intermediate movement joints within the bridge superstructure. "For a 24 ft × 9 in. (8.2 m × 230 mm) pavement, such pressures could result in a total longitudinal force of about 1300 ton (1200 tonne) or more than 25 times the forces usually assumed in the design of bridge abutments" (Burke 2004a: 57).

Unfortunately, often neither bridge engineers nor pavement engineers take appropriate note of this phenomenon. To prevent damage, Burke suggested that an addition be made to the AASHTO Bridge Specification: "Pavement Forces – Jointed rigid pavement approaches to bridges shall be provided with effective and long-lasting pressure relief joints. Otherwise, bridges to be constructed abutting such pavements without pressure relief joints shall be designed to withstand the potential longitudinal forces generated by the restrained growth of such pavements" (Burke 2004b: 89). However, if the bridge is already in place or if the bridge engineer has not designed a proper expansion joint, the pavement designer should address the problem and provide the joint.

Transitions between concrete pavements and bridges

Expanding concrete pavements can impose significant compressive forces on bridges unless expansion joints are provided. These forces are not accounted for in bridge design and have the potential to cause substantial damage.

Approach slabs may be used for pavements adjacent to bridges. Yoder and Witczak (1975: 618–619) suggested a 6.1-m (20-ft) reinforced approach slab adjacent to the bridge. One end of the approach slab rests on the bridge abutment and the other end rests on the subgrade. Typical reinforcement is 16 mm (#5 or 0.625 in) bars, 150 mm (6 in) on center in the longitudinal direction and 610 mm (2 ft) on center in the transverse direction. One or more doweled expansion joints should be provided where the approach slab connects with the concrete pavement.

Doweled expansion joints are similar to doweled contraction joints, but wider. The joint should be at least 19 mm (3/4 in) wide, with a non-extruding filler. Figure 2.9 shows an expansion joint detail. An expansion cap must be installed on one end of each dowel bar to allow for inward movement into the slab. In the past, expansion joints were used to prevent concrete pavement blowups, but they have proven to be difficult to maintain. Furthermore, because blowups are related to specific sources and types of coarse aggregates, it has been found to be easier to prevent blowups by avoiding those materials (Huang 2004: 179).

Transitions between concrete and asphalt pavements

There are two problems that may occur at intersections between concrete and asphalt pavements. One problem is that expanding concrete pavement can shove the asphalt, causing a hump in the pavement. A second problem is that asphalt adjacent to the concrete may rut, causing an impact load as vehicles strike the lip of the concrete pavement.

Although it would seem that these problems could be addressed by expansion joints, such joints would probably be more difficult to maintain than the problems they are intended to address. Therefore, for highways, joints between asphalt and concrete pavements are typically simple butt joints. For thin concrete pavements, a thickened slab abutting the asphalt as shown in Figure 18.9 is recommended to handle vehicle impact loads.

Yoder and Witczak (1975: 592–592) note that some airfields use concrete pavement for runway ends and asphalt pavement for the remainder of the runway. They proposed a detail using a concrete "sleeper slab" and a taper in the asphalt binder course. The top of the sleeper slab should be finished rough to promote bond with the asphalt and prevent it from sliding. This detail could also be modified for use in highway pavements.

Carefully constructed butt joints have also performed well. For butt joints, it is essential to make sure that the subgrade, subbase, and base under the asphalt pavement are stable and well compacted (Yoder and Witczak 1975: 592).

Kohn and Tayabji (2003: 85–86) discuss construction of butt joints:

- The asphalt pavement should be saw cut vertically to full depth, to minimize disturbance to the base layer.
- The base adjacent to the cut asphalt should be well compacted to prevent future faulting, using pole tampers and plate compactors.
- To minimize hand work at the interface, it is best to start paving from the butt joint and move away from it.
- If paving parallel to the cut asphalt pavement, do not allow the paver to track directly on the unsupported asphalt edge, and match elevations between the old and new pavement.

Thickness design for CRCP

For both highways and airports, the thickness design for CRCP is identical to that for JPCP. This is because the reinforcement does not carry load but merely holds cracks together. Therefore, the steel reinforcement does not reduce the flexural stress in the concrete, which is the basis for thickness design.

During the 1970's and early 1980's, CRCP design thickness was approximately 80 percent of the thickness of conventional jointed concrete pavement. A substantial number of the thinner pavements developed distress sooner than anticipated.

Attention to design and construction quality control of CRCP is critical. A lack of attention to design and construction details has caused premature failures in some CRCPs. The causes of early distress have usually been traced to: (1) construction practices which resulted in pavements which did not meet design requirements; (2) designs which resulted in excessive deflections under heavy loads; (3) bases of inferior quality, or; (4) combinations of these or other undesirable factors.

(FHWA 1990b)

In the AASHTO design procedure, the thickness of CRCP is likely to be less than that of JPCP because of the lower load transfer coefficient J, as shown in Table 8.3. However, this will not be a 20 percent reduction, and therefore is not likely to adversely affect performance.

Base or subbase design and friction are also important.

> The base design should provide a stable foundation, which is critical for CRCP construction operations and should not trap free moisture beneath the pavement. Positive drainage is recommended. Free moisture in a base or subgrade can lead to slab edge-pumping, which has been identified as one of the major contributors to causing or accelerating pavement distress. Bases that will resist erosion from high water pressures induced from pavement deflections under traffic loads, or that are free draining to prevent free moisture beneath the pavement will act to prevent pumping. Stabilized permeable bases should be considered for heavily traveled routes. Pavements constructed over stabilized or crushed stone bases have generally resulted in better performing pavements than those constructed on unstabilized gravel.
>
> The friction between the pavement and base plays a role in the development of crack spacing in CRCP. Most design methods for CRCP assume a moderate level of pavement/base friction. Polyethylene sheeting should not be used as a bond breaker unless the low pavement/base friction is considered in design. Also, States have reported rideability and construction problems when PCC was constructed on polyethylene sheeting.
>
> Continuously Reinforced Concrete Pavement is not recommended in areas where subgrade distortion is expected because of known expansive soils, frost heave, or settlement areas. Emphasis should be placed on obtaining uniform and adequately compacted subgrades. Subgrade treatment may be warranted for poor soil conditions.
>
> (FHWA 1990b)

CRCP steel design for highways – AASHTO

Reinforcing steel for CRCP is based on four different criteria:

1 Limiting crack spacing to between 1.1 and 2.4 m (3.5 and 8 ft) to reduce the risk of punchouts.
2 Limiting the width of cracks to 1 mm (0.04 in) to reduce the risk of water penetration or spalling.

3   Limiting the steel stress to no more than 75 percent of yield stress
    to guard against steel fracture. However, experience has shown that
    CRCP has performed satisfactorily in the field even when this limiting
    steel stress has been exceeded.

The design procedure was developed by McCullough and Elkins (1979)
and McCullough and Cawley (1981). Design equations and nomographs
are provided in Section 3.4.2 of the AASHTO Design Guide. All three
criteria should be checked and used to select the required percentage of steel
(AASHTO 1993: II-51–II-62). Although nomographs are available, these
equations may easily be programmed into a spreadsheet or some other type
of solver software. The pavement thickness is determined first, and then the
reinforcement is designed.

Steel percentage based on crack spacing

Huang (2004: 587) provides a version of the AASHTO equation (AASHTO
1993: II-57) to solve for percent steel P, in US customary units, based on
crack spacing:

$$P = \frac{1.062 \left(1 + \frac{f_t}{1000}\right)^{1.457} \left(1 + \frac{\alpha_s}{2\alpha_c}\right)^{0.25} (1 + \varphi)^{0.476}}{\left(\overline{X}\right)\left(1 + \frac{\sigma_w}{1000}\right)^{1.13} (1 + 1000Z)^{0.389}} - 1 \qquad (12.1)$$

where f_t = concrete indirect tensile strength, psi, α_s/α_c = ratio between the
thermal coefficients of the steel and concrete, respectively, ϕ = reinforcing
bar diameter, in, X = crack spacing, ft, σ_w = tensile stress due to wheel
load, psi, and Z = 28-day concrete shrinkage (in/in).
   These parameters are relatively straightforward to determine, except for
the tensile stress due to wheel load. The AASHTO Design Guide (AASHTO
1993: II-55) provides a nomograph, but no equation. The example given
in the design guide shows a wheel stress of 230 psi with a concrete splitting tensile strength of 550 psi, for a stress ratio of 0.42. Since pavement
thickness design generally keeps the stress ratio in this approximate range,
it may be satisfactory to estimate the wheel stress as 0.42 times the 28-day
splitting tensile strength of the concrete.
   The equation may be expressed in SI units:

$$P = \frac{1.062 \left(1 + \frac{f_t}{6894}\right)^{1.457} \left(1 + \frac{\alpha_s}{2\alpha_c}\right)^{0.25} (1 + 0.04\varphi)^{0.476}}{\left(0.328\overline{X}\right)\left(1 + \frac{\sigma_w}{6894}\right)^{1.13} (1 + 1000Z)^{0.389}} - 1 \qquad (12.2)$$

where f_t = concrete indirect tensile strength, kPa, α_s/α_c = ratio between the thermal coefficients of the steel and concrete, respectively, ϕ = reinforcing bar diameter, mm, X = crack spacing, m, σ_w = tensile stress due to wheel load, kPa, and Z = 28-day concrete shrinkage (mm/mm).

This equation should be solved twice, once for the lower crack spacing of 1.1 m (3.5 ft) and once for the higher crack spacing of 2.4 m (8 ft) to provide upper and lower bounds for the required steel percentage. This is an estimated mean crack spacing, and actual spacing will vary. Since the mean crack spacing is in the denominator, crack spacing decreases as the steel percentage increases. Therefore, the maximum crack spacing value will give a minimum percentage of steel and the minimum crack spacing will give the maximum percentage of steel.

Steel percentage based on crack width

Huang (2004: 588) provides a version of the AASHTO equation (AASHTO 1993: II-58) to solve for percent steel P, in US customary units, based on crack width:

$$P = \frac{0.358\left(1+\dfrac{f_t}{1000}\right)^{1.435}(1+\varphi)^{0.484}}{(CW)^{0.220}\left(1+\dfrac{\sigma_w}{1000}\right)^{1.079}} - 1$$

$$= \frac{0.727\left(1+\dfrac{f_t}{1000}\right)^{1.435}(1+\varphi)^{0.484}}{\left(1+\dfrac{\sigma_w}{1000}\right)^{1.079}} - 1 \qquad (12.3)$$

where CW = crack width in inches, and all other variables are as previously defined. This equation may also be expressed in SI units, with CW in mm:

$$P = \frac{0.358\left(1+\dfrac{f_t}{6894}\right)^{1.435}(1+0.04\varphi)^{0.484}}{(0.04 \times CW)^{0.220}\left(1+\dfrac{\sigma_w}{6894}\right)^{1.079}} - 1$$

$$= \frac{0.727\left(1+\dfrac{f_t}{6894}\right)^{1.435}(1+0.04\varphi)^{0.484}}{\left(1+\dfrac{\sigma_w}{6894}\right)^{1.079}} - 1 \qquad (12.4)$$

As noted above, the crack width CW should be set to 1 mm (0.04 in), which provides the second form of each equation.

Steel percentage based on steel stress

Huang (2004: 588) provides a version of the AASHTO equation (AASHTO) 1993: II-59) to solve for percent steel P, in US customary units, based on steel stress:

$$P = \frac{50.834 \left(1 + \dfrac{DT_D}{100}\right)^{0.155} \left(1 + \dfrac{f_t}{1000}\right)^{1.493}}{\sigma_s^{0.365} \left(1 + \dfrac{\sigma_w}{1000}\right) (1 + 1000Z)^{0.180}} - 1 \tag{12.5}$$

where DT_D is the design temperature drop in degrees F, σ_s is the allowable steel stress in psi, and all other variables are as previously defined. This equation may also be expressed in SI units, with DT_D in degrees C and σ_s in MPa:

$$P = \frac{50.834 \left(1 + \dfrac{DT_D}{55.6}\right)^{0.155} \left(1 + \dfrac{f_t}{6894}\right)^{1.493}}{(145\sigma_s)^{0.365} \left(1 + \dfrac{\sigma_w}{6894}\right) (1 + 1000Z)^{0.180}} - 1 \tag{12.6}$$

CRCP transverse reinforcement

The equations above provide the longitudinal reinforcement for the pavement. Transverse reinforcement should also be provided to prevent splitting along the longitudinal bars. The AASHTO Design Guide does not address transverse reinforcement, but it seems reasonable that the traditional temperature steel equation would be adequate for design. Transverse reinforcement is particularly important with frost susceptible or expansive soils (McCullough and Cawley 1981: 247).

Huang (2004: 167) provides a formula for calculating temperature steel which is half the quantity of steel required for tie bars as calculated using equation 7.15:

$$A_s = \frac{\gamma_c DL'f_a}{2f_s} = 0.065 DL'\, mm^2/m = 235 \times 10^{-6} DL'\, in^2/ft \tag{12.7}$$

where γ_c = unit weight of concrete (23.6×10^{-6} N/mm^3, or 0.0868 lb/in^3), D = slab thickness (mm, in), L' = distance from the longitudinal joint to a free edge (m, ft), f_a = average coefficient of friction between the slab and subgrade, usually taken as 1.5, and f_s = the allowable stress in the steel (two-thirds of the yield stress, conservatively 276 MPa or 40,000 psi with modern steel). To calculate the bar spacing, divide the cross-sectional area of a single bar by the required area of steel per unit length of the slab.

Table 12.1 AASHTO CRCP percent steel design example (AASHTO 1993 II-56–II-62)

Design input	SI	US customary
Concrete tensile strength f_t	3.8 MPa	550 psi
Concrete shrinkage Z	0.0004	0.0004
Wheel load stress σ_w	1.6 MPa	230 psi
Concrete thermal coefficient α_c	$2.1 \times 10^{-6}/°C$	$3.8 \times 10^{-6}/°F$
Steel thermal coefficient α_s	$2.8 \times 10^{-6}/°C$	$5 \times 10^{-6}/°F$
Design temperature drop DT_D	30.6 °C	55 °F
Allowable crack width	1 mm	0.04 in
Allowable stress in steel σ_s	427 MPa	62 ksi
Steel bar number (diameter)	16M (16 mm)	#5 (5/8 in)
Results		
Percent based on 1.1 m (3.5 ft) crack spacing	0.51	
Percent based on 2.4 m (8 ft) crack spacing	<0.4	
Percent based on allowable crack width	<0.4	
Percent based on allowable steel stress	0.43	
Final design steel percentage	0.43–0.51	

Design example

The AASHTO Design Guide provides an example, using the input values and results shown in Table 12.1. By modern standards, the steel percentage is somewhat low.

Software

The Concrete Reinforcing Steel Institute (www.crsi.org) sells design aids for CRCP, including the computer programs PowerPave, CRC-HIGHWAY PAVE and CRC-AIRPAVE. CRC-HIGHWAY PAVE and CRC-AIRPAVE are DOS programs.

CRCP highway field performance

Some problems occurred with early CRCP designs.

> Unfortunately, the demand for economy and overconfidence in the maintenance-free, "jointless" pavement led to some underdesigned projects on some questionable subbases, constructed to less than satisfactory standards. Longitudinal cracking, spalling, and punchouts have developed in many CRC pavements... Other states, the Federal Highway Administration, and AASHTO... are now recommending the same thickness as conventional designs.
>
> (Ray 1981: 7)

Tayabji et al. (1995) conducted a national survey of CRCP field performance. Some of the key findings were:

- Steel reinforcement percentages of 0.6–0.7 gave the best crack spacing for conventional strength concrete, and a minimum content of 0.65 percent is strongly recommended. With less than 0.6 percent steel, long crack spacing was coupled with some closely spaced adjacent cracks and a higher risk of punchouts.
- Although tied concrete shoulders did not seem to help CRCP performance, widened lanes appeared promising and should be considered.
- Subbase type did not seem to affect performance substantially. One test section of CRCP on permeable CTB had developed adequate crack spacing.
- Use of epoxy-coated reinforcing did not lead to any cracking problems, suggesting that there was no need to adjust the percentage of steel when using epoxy-coated bars.
- Some sections had very different cracking patterns from others, with no obvious cause, other than possible temperature and curing differences.
- Load transfer at CRCP cracks remained high, generally greater than 90 percent, even after many years of service.
- Crack spacing in CRCP is a complex problem that is not completely addressed by selecting the proper percentage of steel. Type of aggregate, curing method, concrete shrinkage, depth of steel cover, and rate of early strength gain are all factors that may affect the crack pattern but are generally not considered in the design process.

Darter (1994) reviewed the performance of CRCP in Illinois. The Dan Ryan Expressway in Chicago had carried very heavy traffic, approximately 66 million ESALs or 8,000 multiple unit trucks per day, over 19 years before rehabilitation, with a 7 percent growth rate. Some of the key findings were:

- D-cracking was the most prevalent distress on Illinois CRCP, followed by punchouts, wide transverse cracks caused by longitudinal steel ruptures, deterioration of longitudinal and construction joints, and end treatment failures.
- Even under very heavy traffic, the pavements generally lasted 17–22 years before an overlay was necessary. Pavements without D-cracking lasted about 5 years longer, on average, than the pavements with D-cracking.
- Putting steel closer to the top of the slab reduces crack width and punchouts.
- Failures were reduced if steel transverse reinforcement was increased from 0.6 to 0.8 percent.

CRCP steel design for airports – FAA

For airfield pavements, the support requirements and thickness design are identical to those determined for plain concrete pavements as discussed in Chapter 10. The steel percentage is selected to provide optimum crack spacing and crack width.

Three requirements need to be met – steel to resist subgrade restraint, steel to resist temperature effects, and a sufficient concrete to steel strength ratio. The highest of the three requirements should be selected. In any case, the steel percentage must be at least 0.5 percent. The FAA provides design nomographs, but the equations are easily solved with a calculator or spreadsheet (FAA 2004: 94).

Steel to resist subgrade restraint

The required percentage of steel to resist subgrade restraint P is:

$$P = 100(1.3 - 0.2F)(f_t/f_s) \tag{12.8}$$

where F = a friction factor between 1 and 2, and f_t and f_s are the tensile strength of the concrete and the allowable stress in the steel, respectively, which must be expressed in the same units, either MPa or psi. The tensile strength of concrete f_t may be taken as 67 percent of the flexural strength, and the allowable stress in the steel f_s may be taken as 75 percent of the yield strength (FAA 2004: 94–95).

Steel to resist temperature effects

The required percentage of steel to resist temperature effects P is:

$$P = \frac{50f_t}{f_s - 2.42T_C} = \frac{50f_t}{f_s - 195T_F} \tag{12.9}$$

where T_C and T_F are the maximum seasonal temperature differential in degrees Celsius and Fahrenheit, respectively. For equation 12.9 both stresses must be in MPa or psi (FAA 2004: 94).

Concrete to steel ratio

The required percentage of steel based on the ratio between concrete and steel strength P is:

$$P = 100(f_t/f_y) \tag{12.10}$$

where f_t is as previously defined and f_y is the yield strength of the steel (FAA 2004: 96).

Transverse steel

CRCP also requires transverse steel. The required amount of transverse steel P_t is:

$$P_t = 2.27 \frac{W_s F}{2 f_s} \text{(SI)} = 100 \frac{W_s F}{2 f_s} \text{(US)} \tag{12.11}$$

where W_s = slab width in meters for the SI equation and feet for the US customary units equation. Generally, slab widths are limited to 23 m (75 ft) and warping joints are used to divide the pavement into strips. Warping joints are similar to doweled contraction joints, but no dowels are used and the transverse steel is carried continuously through the joint (FAA 2004: 96–97).

Design example

A CRC pavement has been designed to handle DC-1010 aircraft with gross weight of 182,000 kg (400,000 lb) and 3,000 annual departures. The concrete flexural strength of 4.2 MPa (600 psi) and the foundation modulus is 109 MPa/m (400 psi/in) with a cement stabilized subbase. Based on the design method discussed in Chapter 10, the required pavement thickness is 305 mm (12 in). The next step is to design the steel.

The maximum specified yield strength of the steel is 414 MPa (60,000 psi) and the seasonal temperature differential is 56 °C (100 °F). The subgrade friction factor is F. The paving lane width is 7.6 m (25 ft). The longitudinal steel percentage is determined based on equations 12.8, 12.9, and 12.10. Working stress in the steel is 310 MPa (45,000 psi) and the tensile strength of the concrete is 2.8 MPa (400 psi).

The required percentage of steel to resist subgrade restraint P is:

$$P = 100(1.3 - 0.2F)(f_t/f_s) = 100(1.3 - 0.2 \times 1.8)(2.8/310) = 0.85 \text{ percent}$$

The required percentage of steel to resist temperature effects P is:

$$P = \frac{50 f_t}{f_s - 2.42 T_C} = \frac{50 \times 2.8}{310 - 2.42 \times 56} = 0.80 \text{ percent}$$

The required percentage of steel based on the ratio between concrete and steel strength P is:

$$P = 100(f_t/f_y) = 100(2.8/414) = 0.68 \text{ percent}$$

The required percentage of steel based on subgrade restraints is the largest value, and more than 0.5 percent, so $P = 0.85$ percent. Next, transverse steel is determined using equation 12.11. The required amount of transverse steel P_t is:

$$P = 2.27 \times \frac{W_s F}{2f_s} = 2.27 \times \frac{7.6 \times 1.8}{2 \times 310} = 0.05 \text{ percent}$$

CRCP reinforcement and end anchors

CRCP uses longitudinal reinforcement and usually transverse reinforcement also. End treatment details are also necessary.

Reinforcement details

Longitudinal reinforcement percentages traditionally range between 0.5 and 0.7 percent, with higher percentages used in northern climates. Bar sizes are usually 16 or 19 mm (5/8 in #5 or 3/4 in #6). Spacing between bars should be at least 100 mm (4 in) for satisfactory consolidation of the concrete, and should not exceed 229 mm (9 in). Bars are usually placed at or near the midpoint of the slab, with a minimum of 64 mm (2½ in) of cover on top to protect against corrosion (CRSI 1983: 7). Top cover of 75 mm (3 in) is preferable, and some US state agencies have used two layers of longitudinal steel for pavements thicker than 279 mm (11 in) (FHWA 1990b). Recommended minimum and maximum bar sizes, based on bond considerations, are provided in Table 12.2.

It is important that lap splices between bars be long enough, otherwise wide cracks may develop. Lap splices should be at least 25 bar diameters or at least 406 mm (16 in), whichever is greater. Splices must also be staggered across the pavement to avoid localized stress/strain concentrations. Several splice patterns are available, including a chevron or skewed pattern, staggering laps, or staggering laps alternating every third bar (CRSI 1983: 7–8).

Table 12.2 Recommended longitudinal reinforcement sizes (FHWA 1990b)

% steel	Pavement thickness											
	mm	in	mm	in	mm	in	mm	in	mm	in	mm	in
	203	8	229	9	254	10	280	11	305	12	330	13
0.60	13, 16	4, 5	16, 19	5, 6	16, 19	5, 6	16, 19	5, 6	16, 19	5, 6	19	6
0.62	16, 19	5, 6	16, 19	5, 6	16, 19	5, 6	16, 19	5, 6	16, 19	5, 6	19	6
0.64	16, 19	5, 6	16, 19	5, 6	16, 22	5, 7	16, 22	5, 7	19, 22	6, 7	19, 22	6, 7
0.66	16, 19	5, 6	16, 22	5, 7	16, 22	5, 7	16, 22	5, 7	19, 22	6, 7	19, 22	6, 7
0.68	16, 19	5, 6	16, 22	5, 7	16, 22	5, 7	19, 22	6, 7	19, 22	6, 7	19, 22	6, 7

Note: bars are uncoated deformed bars given as minimum, maximum

If a staggered splice pattern is used, not more than one-third of the bars should terminate in the same transverse plane and the minimum distance between staggers should be [1.2 m] 4 feet. If a skewed splice pattern is used, the skew should be at least 30 degrees from perpendicular to the centerline. When using epoxy-coated steel, the lap should be increased a minimum of 15 percent to ensure sufficient bond strength.

(FHWA 1990b)

Transverse bars may be used to help support and form the pattern for longitudinal bars, to hold any unplanned longitudinal cracks together, and to serve as tie bars between lanes. Satisfactory transverse reinforcement installations have included 13 or 16 mm (1/2 in #4 or 5/8 in #5) bars spaced 1.2 m (48 in) on center. Tie bars are needed if transverse reinforcement is not carried across the longitudinal construction joints (CRSI 1983: 8–9). Transverse bar spacing should be no closer than 0.9 m (36 in) and no further than 1.5 m (60 in). A combination chair and transverse steel detail is shown in Figure 12.1. (FHWA 1990b).

Reinforcing bars may be placed in two ways. They may be placed before paving on chairs, with the longitudinal and transverse bars secured in place by ties. Presetting the bars in this manner makes it easy to check that they are properly positioned and secure before paving. Alternatively, mechanical methods such as tube feeders are used. The steel is preplaced and lapped

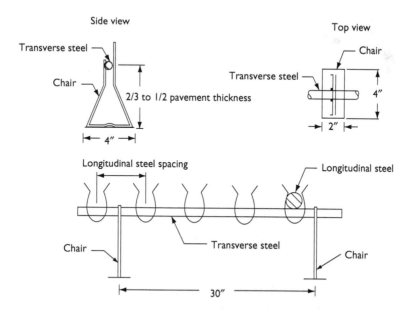

Figure 12.1 Combination chair and transverse steel detail (1 in = 25.4 mm) (FHWA 1990b).

on the subbase, and then raised on rollers and threaded on tubes within the spreader. Sometimes transverse reinforcement is omitted if mechanical methods are used (CRSI 1983: 9–12).

> Many CRCP performance problems have been traced to construction practices which resulted in a pavement that did not meet the previously described design recommendations. Because CRCP is less forgiving and more difficult to rehabilitate than jointed pavements, greater care during construction is extremely important. Both the contractor and the inspectors should be made aware of this need and the supervision of CRCP construction should be more stringent.
> Steel placement has a direct effect on the performance of CRCP. A number of States have found longitudinal steel placement deviations of ± [75 mm] 3 inches in the vertical plane when tube feeders were used to position the steel. The use of chairs is recommended to hold the steel in its proper location. The chairs should be spaced such that the steel will not permanently deflect or displace to a depth of more than 1/2 the slab thickness.
>
> (FHWA 1990b)

End anchors and terminal joints

CRCP requires special details at the beginning and end of the pavement, as well as at junctions with JPCP or bridges. The FAA notes that

> Since long slabs of CRCP are constructed with no transverse joints, provisions must be made to either restrain or accommodate end movements wherever the CRCP abuts other pavements or structures. Rather large end movements, up to 2 inches (50 mm), are experienced with CRCP due to thermal expansion and contraction. End movement is normally not a problem except where CRCP abuts another pavement or structure. Experience with highway CRCP shows that attempts to restrain end movement have not been too successful. More favorable results are achieved where end movement is accommodated rather than restrained. Joints designed to accommodate large movements are required where CRCP intersects other pavements or abuts another structures. Failure to do so may result in damage to the CRCP, pavement or other structure. Wide flange beam type joints or "finger" type expansion joints can accommodate the movements. The wide flange beam type joint is recommended due to its relatively lower costs.
>
> (FAA 2004: 96)

The most widely used system for restraining movement of CRCP is end anchors (CRSI 1983: 21). These are reinforced beams embedded into the subgrade, perpendicular to the pavement centerline. Figure 12.2 shows a typical anchor lug.

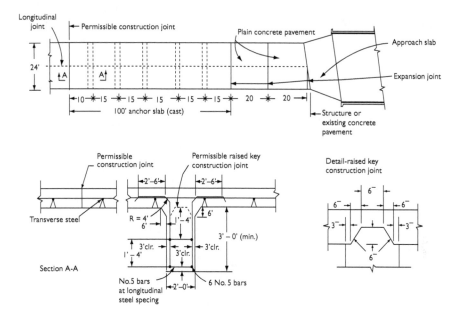

Figure 12.2 Lug anchor treatment (1 in = 25.4 mm, 1 ft = 305 mm) (FHWA 1990b).

The purpose of the end anchors is twofold. One is to help form the proper crack pattern in the CRCP. The other is to prevent the pavement ends from moving, because a CRCP pavement is essentially a single slab and the movement at the ends could be very substantial. Without restraint, end movements on the order of 50 mm (2 in) may be expected (McCullough and Cawley 1981: 248). McCullough and Cawley (1981: 248) note that end anchors can only reduce the movement by about 50 percent, so expansion joints are still necessary.

> The lug anchor terminal treatment generally consists of three to five heavily reinforced rectangularly shaped transverse concrete lugs placed in the subgrade to a depth below frost penetration prior to the placement of the pavement. They are tied to the pavement with reinforcing steel. Since lug anchors restrict approximately 50 percent of the end movement of the pavement an expansion joint is usually needed at a bridge approach. A slight undulation of the pavement surface is sometimes induced by the torsional forces at the lug. Since this treatment relies on the passive resistance of the soil, it is not effective where cohesionless soils are encountered.
>
> (FHWA 1990b)

An alternative approach to addressing CRCP end movement is to allow the movement to occur with special expansion terminal joints. Experience has shown that these generally cost less and provide a better ride than end anchors (CRSI 1983: 21).

The most commonly used terminal treatments are the wide-flange (WF) steel beam which accommodates movement, and the lug anchor which restricts movement... The WF beam joint consists of a WF beam partially set into a reinforced concrete sleeper slab approximately [3.05 m] 10 feet long and [254] 10 inches thick. The top flange of the beam is flush with the pavement surface. Expansion material, sized to accommodate end movements, is placed on one side of the beam along with a bond-breaker between the pavement and the sleeper slab. In highly corrosive areas the beam should be treated with a corrosion inhibitor. Several States have reported premature failures of WF beams where the top flange separated from the beam web. Stud connectors should be welded to the top flange... to prevent this type of failure.

(FHWA 1990b)

Table 12.3 and Figure 12.3 contain recommended design features, including the stud connectors. Any beam size heavier than that listed in Table 12.3 would be satisfactory, as long as web and flange width and thickness are equaled or exceeded.

McCullough and Cawley (1981: 248) note that

a wide flange beam joint is cast in a reinforced concrete sleeper slab that supports the ends of the abutting pavement. End movement is accommodated by a compressible material placed on the side of the web adjacent to the CRC pavement. This type of joint provides excellent load transfer, a smooth riding transition, and requires very little maintenance. The weight of the WF section should be selected to resist fatigue under heavy truck loadings.

Table 12.3 Recommended WF beam dimensions (FHWA 1990b)

CRCP thickness, mm (in)	Embedment in sleeper slab, mm (in)	US WF beam size	Flange		Web thickness, mm (in)
			Width, mm (in)	Thickness, mm (in)	
203 (8)	152 (6)	W 14 × 61	254 (10)	16 (5/8)	9.5 (3/8)
229 (9)	125 (5)				
254 (10)	152 (6)	W 15 × 58	216 (8.5)	16 (5/8)	11 (7/16)
280 (11)	125 (2)				

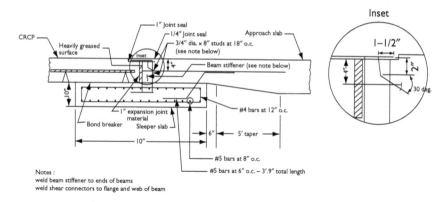

Figure 12.3 Recommended WF steel beam terminal joint design (1 in = 25.4 mm, 1 ft = 305 mm) (FHWA 1990b).

Transverse construction joints

A construction joint is formed by placing a slotted headerboard across the pavement to allow the longitudinal steel to pass through the joint. The longitudinal steel through the construction joint is increased a minimum of one-third by placing [0.9 m] 3-foot long shear bars of the same nominal size between every other pair of longitudinal bars. No longitudinal steel splice should fall within [0.9 m] 3 feet of the stopping side nor closer than [2.4 m] 8 feet from the starting side of a construction joint... If it becomes necessary to splice within the above limits, each splice should be reinforced with a [1.8 m] 6-foot bar of equal size. Extra care is needed to ensure both concrete quality and consolidation at these joints. If more than 5 days elapse between concrete pours, the adjacent pavement temperature should be stabilized by placing insulation material on it for a distance of [61 m] 200 feet from the free end at least 72 hours prior to placing new concrete. This procedure should reduce potentially high tensile stresses in the longitudinal steel. Special provisions for the protection of the headerboard and adjacent rebar during construction may be necessary.

(FHWA 1990b)

Seamless pavement – case study

One interesting alternative first tried in Australia is a seamless pavement, which combines CRCP and concrete bridges without expansion joints. The pilot project was the Westlink M7 in Sydney.

The seamless pavement is an enhancement that eliminates transverse joints and provides a reinforced concrete connection between the CRCP and the bridge deck. This results in improved ride quality for highway users and reduced maintenance costs. Additionally, it eliminates the need for pavement anchors behind each abutment, thereby reducing the pavement cost and minimizing construction activities in an area that is generally on the critical path. The seamless connection between CRCP and bridge deck must accommodate the stresses induced by shrinkage, creep, thermal strain, embankment settlement and traffic loads.

(Griffiths et al. 2005: 48)

The pilot project in Australia is 40 km (25 miles) long and 10.5 m (34 ft) wide, and crosses 68 bridges over the length of the project. There has been a long-term transition to multispan and integral abutment bridges to reduce the number of joints, which are difficult to maintain for bridges as well as pavements. With the seamless pavement concept, the CRCP and the bridges are connected with reinforced transition sections. This eliminates the bump that is often found at conventional pavement to bridge transitions. In the event of foundation settlement adjacent to the bridge abutments, the transition sections are designed to carry the traffic without any subgrade support. Complete description, details, and construction observations are provided by Griffiths et al. (2005: 48–65).

Chapter 13

Subgrade and subbase construction

The performance of the pavement system as a whole depends greatly on the construction of the supporting subgrade, subbase, and base (if any) layers. There is a saying in building construction that "there is nothing more expensive than a cheap foundation." This is also true of pavements. Quality construction of these layers ensures a well-drained, uniformly supported pavement, and makes it much easier to build a smooth pavement because the paver has a smooth working platform. In contrast, a weak subgrade generally requires time-consuming and costly full-depth reconstruction to correct.

Grade preparation

Construction begins with subgrade preparation.

Preparation of the subgrade includes:

1. Compacting soils at moisture contents and densities that will ensure uniform and stable pavement support.
2. Whenever possible, setting gradelines high enough and making side ditches deep enough to increase the distance between the water table and the pavement.
3. Crosshauling and mixing of soils to achieve uniform conditions in areas where there are abrupt horizontal changes in soil type.
4. Using selective grading in cut and fill areas to place the better soils nearer to the top of the final subgrade elevation.
5. Improving extremely poor soils by treatment with cement or lime, or importing better soils, whichever is more economical.

(PCA/ACPA 1991: 1)

Chapter 4 of the Kohn and Tayabji report (2003: 32–40) provides considerable information about grade preparation for airport pavements, but many of their recommendations are pertinent to all concrete pavements.

The Kohn and Tayabji report (2003: 32) notes that "important areas of subgrade preparation include:

1. Evaluation of subgrade suitability
2. Subgrade modification to improve stability
3. Evaluation of surface tolerances."

They further recommend consulting with a geotechnical engineer for large facilities such as airports if soils are variable, of low bearing strength (<96 kPa or 1 ton/sq ft), organic, swelling/expansive, or frost susceptible.

Surveying, excavation, and embankment construction procedures are similar for all pavement types and are well covered in typical transportation engineering textbooks, for example Garber and Hoel (2002) chapters 15 and 18. This encompasses cutting high points and filling in low points to achieve the desired final subgrade finished elevation.

Subgrade grading and compaction

Embankments must be placed in uniform layers and compacted to a high density. The strength of compacted soil depends on the dry density achieved during compaction. The optimum moisture content (OMC) and maximum dry density for compaction are determined using either the Standard Proctor test (ASTM D698 2000, AASHTO T99 2005) or the Modified Proctor test (ASTM D1557 2002, AASHTO 2005).

The Standard and Modified Proctor differ in the compaction energy applied to the soil sample. The Standard Proctor uses a 2.5 kg (5.5 lb) hammer with a 300 mm (12 in) free fall distance, while the Modified Proctor uses a 4.54 kg (10 lb) hammer with a 450 mm (18 in) free fall distance. Garber and Hoel (2002: 858–859) note that OMC and maximum dry density depend on the compacting effort. Their textbook example shows a Standard Proctor OMC of 16.5 percent and maximum dry density of 1,900 kg/m^3 (120 pcf) versus a Modified Proctor OMC of 15 percent and a maximum dry density of 2,180 kg/m^3 (136 pcf) for the same soil. Therefore, as the moisture content decreases, the compaction effort required to achieve a given density increases.

One typical US State Department of Transportation specification uses Standard Proctor for highway construction (ODOT 2005: 93). For airfield pavements, Modified Proctor is used for pavements that will handle aircraft heavier than 27,000 kg (60,000 lb), with Standard Proctor used for pavements carrying only lighter aircraft (FAA 2004).

Kohn and Tayabji (2003: 34) note typical compaction requirements for airport pavement construction with cohesive and cohesionless soils in fill and cut sections:

- For cohesive soils in fill sections, the entire fill should be compacted to 90 percent of maximum density.
- For cohesive soils in cut sections, the top 150 mm (6 in) should be compacted to 90 percent maximum density.
- For cohesionless soils in fill sections, the top 150 mm (6 in) should be compacted to 100 percent maximum density, and the remainder of the embankment to 95 percent maximum density.
- For cohesionless soils in cut sections, the top 150 mm (6 in) should be compacted to 100 percent maximum density and the next 450 mm (18 in) to 95 percent maximum density.
- The previous requirements are based on structural considerations. In order to provide a stable working platform for heavy construction equipment, it is considered good practice to compact all subgrade materials to 95 percent of Modified Proctor for airport pavements.

Kohn and Tayabji (2003: 34–35) also address the moisture requirements for obtaining a stable subgrade prior to compaction:

- The moisture contact is generally required by specifications to be within ±2 percent of the OMC.
- For expansive soils, the moisture should be 1–3 above OMC to reduce swell potential.
- For fine-grained non-swelling soils, the moisture content should be 1–2 percent dry of OMC.
- If cohesive soils are compacted wetter than OMC, they may become unstable under construction traffic even if target density is reached.
- Clayey sand soils are much more moisture sensitive than silty clay soils, and small changes in moisture content may result in compaction difficulty.

The selection of compaction and moisture control equipment is also important (Kohn and Tayabji 2003: 35):

- Cohesive soils are generally compacted with sheepsfoot rollers, with pads penetrating 70 percent of the lift thickness.
- Discing is necessary to control moisture in cohesive soils.
- Static steel drum rollers may be used after compaction to smooth the grade.

- Vibratory steel drum rollers are used for cohesionless soils. Use vibration with caution if the water table is close to the surface or if the subgrade is saturated.

Field compaction testing is generally performed with nuclear density gauges. Ideally, equipment and personnel are available throughout subgrade construction to provide quality control of density and allow for timely field adjustments to compaction procedures as necessary.

Proof rolling is useful for verifying grade preparation. A heavy pneumatic tired roller or fully loaded truck may be driven across the prepared surface, while observing whether any rutting or deformation occurs. For airport and other heavy duty pavements, the following guide is suggested for proof rolling:

- Grade is acceptable if rutting is less than 6 mm (1/4 in).
- The grade should be scarified and re-compacted if rutting is between 6 and 40 mm (1/4 and 1½ in).
- Removal and replacement is recommended if rutting exceeds 40 mm (1½ in).
- Deformation over 25 mm (1 in) which rebounds indicates a soft layer near the surface, and further investigation is necessary (Kohn and Tayabji 2003: 39).

If acceptance of the grade is part of the project specifications, surface deviations based on a 5 m (16 ft) straightedge are typically limited to 13 or 25 mm (1/2 or 1 in) and surface elevation should be within 15–30 mm (0.05–0.10 ft). Autograders or trimmers may be used to smooth the grade.

A traffic control plan should be implemented to keep heavy equipment off of the grade. Surface drainage is necessary to keep the subgrade from ponding and becoming saturated. If rain is expected, a rubber tired or steel drum roller should be used to seal the surface. If the subgrade becomes frozen, the surface should be scarified to a depth of 150 mm (6 in) and recompacted. Troubleshooting of subgrade problems is addressed by Kohn and Tayabji (2003: 39–40).

Controlling expansive soils

Some of the construction techniques and methods for controlling expansive soils are described in the PCA/ACPA Publication *Subgrades and Subbases for Concrete Pavements* and are outlined below (PCA/ACPA 1991: 4–6).

Soil swell can be controlled by surcharge loads, by placing the soils on the lower part of an embankment. Selective grading and soil mixing is also helpful. In deep cut sections, construction removes surcharge loads and allows soils to swell, so it is advisable to excavate deep cuts in advance of other grading work to allow the expansion to occur and stabilize.

Proper moisture and density controls during construction also play a considerable role in controlling soil volume changes. Highly expansive soils should be compacted at 1–3 percent above OMC as determined by Standard Proctor. Use of the OMC from the Modified Proctor will leave the soil too dry and more prone to future expansion. The soil should not be allowed to dry out excessively before placing the concrete pavement.

In areas with long dry periods, it may be necessary to cover highly expansive soils with a full width placement of non-expansive soil. This layer should be a soil that is not be susceptible to pumping.

If non-expansive soil is not readily available, it may be more economical to modify the existing soil with cement or lime. Laboratory tests should be used to determine the appropriate cement or lime content, generally in the range of 3 or 5 percent by mass.

Other methods that have been used to control expansion include ponding or preswelling the soil to allow the expansion to occur before it damages the pavement, and various membranes and moisture barriers. Chemical stabilization processes have also been used.

Controlling frost heave

The performance of older concrete pavements in frost-affected areas under today's increased traffic shows that extensive, costly controls are not needed to prevent frost damage. Surveys of those pavements indicate that control is needed only to reduce excessive heave and, more critically, prevent differential heave by achieving reasonably uniform subgrade conditions. As in the case of expansive soils, a large degree of frost-heave control is attained most economically by appropriate grading operations and by controlling subgrade compaction and moisture.
(PCA/ACPA 1991: 8–9)

If soils susceptible to frost heave are present and cannot be economically removed, drainage, moisture control, and compaction can mitigate the potentially harmful effects. Some of the construction techniques and methods for controlling frost heave are described in the PCA/ACPA publication *Subgrades and Subbases for Concrete Pavements* and are outlined below (PCA/ACPA 1991: 7–10).

One method is to raise the pavement grade high enough and cut side ditches deep enough to keep the water table well below the pavement. If possible, the grade should be kept 1.2–1.5 m (4–5 ft) above the ditch bottom. Alternatively, edge-drains may be used to lower groundwater tables.

When building up embankments, the more frost-susceptible soils should be placed near the bottom and less frost-susceptible soils near the top.

Mixing soils can help prevent differential frost heave. If localized pockets of silts are distributed among less susceptible soils, the pockets can be excavated and backfilled with better material.

Grading establishes reasonable uniformity, and proper compaction at the proper moisture content provides additional uniformity. Permeability is reduced by compacting the soil slightly wetter than Standard Proctor OMC. This is the same treatment as for expansive soils.

Subgrade stabilization

The need for subgrade stabilization, if any, depends in large part on the traffic loads to be carried by the pavement as well as on the soil type. Parking lots and streets and local roads are unlikely to require stabilization. On the other hand, airports and heavy industrial pavements may require stabilization of soils, particularly as in many areas of the world the large, flat areas available for these facilities consist of poor quality soils. Reasons for stabilizing subgrades include improving low strength soils, reducing swelling potential, and improving construction conditions (Kohn and Tayabji 2003: 36). The advantage of stabilizing the soil to provide a good working platform should be carefully considered.

The FAA design Advisory Circular states

> Subgrade stabilization should be considered if one or more of the following conditions exist: poor drainage, adverse surface drainage, frost, or need for a stable working platform. Subgrade stabilization can be accomplished through the addition of chemical agents or by mechanical methods . . . In some instances subgrades cannot be adequately stabilized through the use of chemical additives. The underlying soils may be so soft that stabilized materials cannot be mixed and compacted over the underlying soils without failing the soft soils. Extremely soft soils may require bridging in order to construct the pavement section. Bridging can be accomplished with the use of thick layers of shot rock or cobbles. Thick layers of lean, porous concrete have also been used to bridge extremely soft soils. Geotextiles should be considered as mechanical stabilization over soft, fine-grained soils. Geotextiles can facilitate site access over soft soils and aid in reducing subgrade soil disturbance due to construction traffic. The geotextile will also function as a separation material to limit long-term weakening of pavement aggregate associated with contamination of the aggregate with underlying fine-grained soils.
>
> (FAA 2004: 16)

The most common subgrade stabilization materials are lime and cement. Lime is used for cohesive or clayey subgrade soils with high moisture content. Soils with a plasticity index greater than 10 percent require 3–5 percent lime for stabilization. Lime may be placed dry or as a slurry. Cement is usually used to stabilize either coarse-grained soils or soils high in silt. The amount of cement to use is determined through a laboratory mix design process. Lime and cement stabilized soils are discussed in detail by Kohn and Tayabji (2003: 36–38).

Other subgrade considerations

Sometimes unsuitable soils may be encountered during excavation, and it is useful to have a contingency plan ready to deal with localized pockets as they are found. Possible alternatives include:

- Removing soft or disturbed material and replacing it from material from nearby areas, which generally only works for surface pockets.
- Remove soft soil and replace with crushed stone, using a geotextile if necessary to prevent contamination of the stone layer with subgrade material.
- Placing a geogrid and a 250 mm (10 in) lift of crushed stone over the soft area.

Environmental considerations are very important. Kohn and Tayabji (2003: 33) point out that "measures to control water pollution, soil erosion, and siltation that are shown on the plans or that are needed for the site should be provided. All pertinent local, State, and federal laws should be followed."

Subbase and base construction

Kohn and Tayabji (2003) discuss base and subbase construction in Chapter 5 of *Best Practices for Airport Portland Cement Concrete Pavement Construction (Rigid Airport Pavement)*. Although this guide focuses on airport pavements many of the recommendations also apply to highway and other concrete pavements.

The terminology can occasionally be confusing – the Kohn and Tayabji guide refers to a base layer over a subbase layer, while the FAA Pavement Design Guide (FAA 2005) refers to two layers of subbase, with the top layer meeting base course requirements. Highways generally have only a single layer of subbase under the concrete pavement, while airports have two layers between the subgrade and pavement, particularly if they handle heavy wide body aircraft.

Untreated subbases

"A wide variety of materials and gradations has been used successfully for untreated subbases by different agencies. These include crushed stone, bank run sand-gravels, sands, soil-stabilized gravels, and local materials such as crushed mine waste, sand-shell mixtures, and slag" (PCA/ACPA 1991: 12).

Important elements are:

- proper control of gradation – although many different gradations have proven satisfactory, it is important to keep the gradation relatively constant for a single project;
- limiting the amount of fines passing the number 200 sieve (0.075 mm);
- avoiding soft aggregates that will break down under the compacting rollers;
- compacting the subbase to a very high density, a minimum of 100 percent of Standard Proctor, to avoid future consolidation;
- Only a 75-mm (3-in) thick subbase is necessary for preventing pumping (PCA/ACPA 1991: 12–14).

Untreated subbase construction elements for airfields noted by Kohn and Tayabji (2003: 41–42) include:

- Start construction at the centerline or high point of the pavement in order to maintain drainage.
- Place using automated equipment or a stone box on a bulldozer.
- Control the moisture so that it is within 1 percent of OMC.
- Keep the minimum layer thickness to three or four times the maximum aggregate size.
- Repair any soft or unstable subgrade areas before placing the subbase.
- Maintain the finished subbase elevation within 12 mm (1/2 in) as measured using a 5 m (16 ft) straightedge. On larger projects, lasers or autotrimmers may be used for the final grade.
- Protect the finished subbase surface once it has been placed. Provide drainage so that water does not pond on the surface. If the weather is excessively dry, the subbase may need to be watered.
- Roll with vibratory drum rollers. Rubber tired rollers may be used to knead the surface if compaction is difficult.

Many of these elements, of course, apply to non-airfield pavements.

Mechanically stabilized base courses are similar to untreated subbases, but must meet higher quality requirements in terms of crushed aggregate content, deleterious material, and gradation. On the whole placement is similar to untreated subbase courses, with some additional considerations:

- Before placing the base course, the underlying subbase or subgrade should be checked and soft or yielding areas should be corrected.
- The base should not be placed if the underlying layer is wet, muddy, or frozen.

- Work should be suspended during freezing conditions.
- The base should be compacted with vibratory, rubber tired, or static wheel rollers.
- Grade tolerance is typically 10 mm (3/8 in) measured with a 5 m (16 ft) straightedge, which may require automated placing equipment (Kohn and Tayabji 2003: 42).

Cement treated subbases and bases/CTB

> Granular materials in AASHTO Soil Classification Groups A-1, A-2-4, A-2-5, and A-3 are used for cement-treated subbases. They contain not more than 35 percent passing the [0.075 mm] No. 200 sieve, have PI of 10 or less, and may be either pit-run or manufactured. Cement-treated subbases have been built with A-4 and A-5 soils in some nonfrost areas and are performing satisfactorily; generally, however, such soils are not recommended in frost areas or where large volumes of heavy truck traffic are expected. Use of A-6 and A-7 soils is not recommended. To permit accurate grading of the subbase, maximum size of material is usually limited to [25 mm] 1 in., and preferably to [19 mm] ³/₄ in.
>
> (PCA/ACPA 1991: 14–15)

A-1 soils are typically USCS/ASTM GW, GP, GM, SW, SP, or SM. A-2-5 and A-2-6 soils may be GM, GC, SM, or SC. A-3 corresponds to SP. A-4 and A-5 soils are typically ML, OL, MH, or OH. A-6 and A-7 soils correspond to CL, MS, CH, and OH (Garber and Hoel 2002: 847).

Important elements are:

- Dirty granular materials that do not meet subbase specifications can be used and may in fact require less cement than clean aggregates.
- Cement-treated subbases and bases are highly resistant to erosion, and ideally suited to preventing pumping.
- These layers may be constructed using road mix or central plant methods. For road mixing, the material may be processed in a blanket on the subgrade.
- The final step is to finish the cement-treated subbase to an accurate grade and crown it, and then to provide a light fog spray of water followed by bituminous curing material (PCA/ACPA 1991: 15–16).

Cement-treated subbase and base construction elements for airfields noted by Kohn and Tayabji (2003: 43–44) include:

- When using a central mixing plant, introduce water and cement into a pug mill mixer in the appropriate proportions, and make sure the cement does not ball in the mixer.

- Complete delivery and placement within 60 minutes of initial mixing.
- Complete final grading and compaction within 2 hours of mixing.
- Monitor density using a nuclear density gauge. The layer should be compacted to 97 or 98 percent of the maximum.
- Moisture content should be within 2 percent of optimum, and 2 percent over optimum in hot weather.
- Place the material with mechanical spreaders or bulldozers with automated blades.
- Minimize the number of longitudinal and transverse joints.
- Place a transverse header at the end of the day or if construction is interrupted for more than 60 minutes, using full depth saw cutting.
- Form longitudinal joints by saw cutting the free edge.
- Ensure that the temperature at the time of placement is at least 4 °C (40 °F) and suspend placement if temperatures are expected to fall below 2 °C (35 °F) within 4 hours of placement.
- Generally, limit the layer thickness to 200 mm (8 in) for compaction, although some compaction equipment can effectively compact 300 mm (12 in) layers.
- Compact using a combination of vibratory and rubber tired rollers.
- If multiple lifts are necessary, keep the lower lift moist until the next lift is placed.
- Maintain the finished subbase elevation within 10 mm (3/8 in) as measured using a 5 m (16 ft) straightedge. A trimmer should be used to maintain grade tolerance. CTB is a rigid material and may not be re-graded after compaction.

Many of these elements, of course, apply to non-airfield pavements. For larger projects, CTB should be produced in a continuous flow pugmill mixer central plant to ensure uniform quality. During haul, it may be necessary to use truck bed covers to protect the CTB from rain (Hall et al. 2005: 55–56).

Attention to potential adverse weather is important for CTB.

> The CTB material should not be mixed or placed while the air temperature is below 40 °F (4 °C) or when conditions indicate that the temperature may fall below 35 °F (2 °C) within 24 hours. The CTB should not be placed on frozen ground. Further, CTB should not be placed when rainfall is occurring. If an unexpected rain event occurs during placement, the layer should be quickly compacted and protected. CTB material that becomes wet by rain during transport or placement should be examined and rejected if excess water in the mixture changes its consistency and uniformity.
>
> (Hall et al. 2005: 57)

CTB is placed in 100–200 mm (4–8 in) lifts and compacted with rollers. If multiple lifts are necessary, the compacted lift should be kept moist until the next lift is placed over it. All lifts should be placed and compacted within a 12-hour period. CTB should be cured with a liquid membrane forming curing compound, which should be maintained and protected until the concrete pavement is placed. Before placing the concrete pavement, a single layer of choke stone should be applied as a bond breaker (Hall et al. 2005: 57–59).

Lean concrete subbases and bases/econocrete (LCB)

"Lean concrete subbase mixtures are made with a greater amount of cement and water than cement-treated subbases, but they contain less cement than conventional concrete. Having the same appearance and consistency of conventional concrete, lean concrete is consolidated by vibration" (PCA/ACPA 1991: 16). The mixture is often called Econocrete because it makes use of locally available low cost materials.

Important elements are:

- Aggregates that do not meet concrete specifications can be used and may in fact require less cement than clean aggregates.
- Recycled pavement material may be used as aggregate.
- Typical cement factors are 118–208 kg/m^3 (200–350 lb/cu yd), and typical slumps are 25–75 mm (1–3 in) (PCA/ACPA 1991: 16).

LCB production is similar to that of conventional concrete. It may be mixed at a stationary mixer or central batch plant, or it may be truck mixed. It is hauled in agitator trucks, truck mixers at agitating speed, or non-agitating trucks (Hall et al. 2005: 59).

Lean concrete subbase and base construction elements for airfields noted by Kohn and Tayabji (2003: 44–45) include:

- Place econocrete in the same way as conventional concrete.
- Keep seven-day strength below 5,200 kPa (750 psi) and 28-day strength below 8,300 kPa (1,200 psi) to prevent reflective cracking.
- Alternatively, cut joints in the Econocrete to match the future joints in the concrete pavement.
- Cure Econocrete with a double application of wax-based curing compound. This treatment cures the Econocrete and prevents bond and reflective cracking between the Econocrete and the concrete pavement.

Temperature and rainfall precautions are similar to those for CTB discussed above. It is important to avoid segregation of LCB. Either slipform or fixed form paving may be used to place LCB. No finishing is necessary, but the

LCB should be cured with a liquid membrane forming curing compound. If seven-day strength is less than 5.5 MPa (800 psi), a second layer of curing compound may be used for the bond breaker. If the strength is higher, choke stone should be used instead (Hall et al. 2005: 60–62).

Asphalt-treated subbases and bases

"The production, placement, control, and acceptance of ATB are similar to that of a high-quality asphalt paving layer" (Hall et al. 2005: 62). Kohn and Tayabji (2003: 45–46) discuss construction of asphalt treated bases:

- Layer thickness is typically limited to 100–125 mm (4–5 in) although some equipment can properly compact 150 mm (6 in) layers.
- Check compacted density using a nuclear density gauge.
- Determine layer thickness and air voids using cores.
- Stop placement if air temperature is below 4 °C (40 °F).
- Under summer conditions, the surface of the asphalt treated base can exceed 60 °C (140 °F) with potential to cause early-age cracking and high shrinkage in the concrete pavement. These layers should be white-washed with a lime-water solution to reduce the temperature before concrete placement.
- Do not trim the surface with a milling machine because the concrete will bond to that surface and have a higher risk of cracking.

Permeable bases

"The permeable bases are made of crushed aggregates with a reduced amount of fines. The materials fall into two categories: untreated and treated. Treated subbases, which provide a stable construction platform, are bound with either cement [118–177 kg/m^3] (200 to 300 lb. per cu yd) or asphalt (2 to 2.5 percent by weight)" (PCA/ACPA 1991: 18).

Important elements are:

- Permeable subbase material is usually placed and trimmed with a paver, a hopper converted auto-trimmer, or with a truck.
- Layers are lightly rolled, generally one to three passes of a 4 to 10 ton steel wheel roller in static mode.
- Cure cement-treated permeable bases with water misting or with polyethylene sheets for 3–5 days.

As noted in Chapter 4, drainage layers may be stabilized or unstabilized. Drainage layers stabilized with cement or asphalt provide better stability during construction. These layers are generally 100–150 mm (4–6 in) thick. Unstabilized drainage layers are not recommended for airport pavements serving wide body aircraft. Placing drainage layers deeper within the pavement structure reduces the stresses on the layer (Kohn and Tayabji 2003: 46).

CTPB is mixed in a central batch plant or stationary mixer, similar to conventional concrete. Because of the lack of fines, it is important to avoid segregation by minimizing handling and limiting drop heights. The material may be placed using a mechanical spreader or an asphalt paver with dual tamping bars (if the paver does not fracture the aggregate). To ensure compaction, layers should be 100–150 mm (4–6 in) thick. CTPB is usually compacted with static rollers, starting within 30 minutes of mixing the material and completing compaction within 60 minutes. Choke stone should be placed as a bond breaker and rolled with two passes of a vibratory roller. Test sections are required for both CTPB and ATPB layers to verify materials, methods, and properties. The material is placed and hauled like conventional asphalt, but a remixing system is recommended to avoid segregation. Pneumatic rollers should be used to compact the ATPB, and no curing or bond breaker is required. In summertime, the ATPB should be whitewashed before placing concrete pavement (Hall et al. 2005: 63–66). ATPB is mixed with batch or continuous mixing plants (Hall et al. 2005: 66–69).

Issues with stabilized bases

Stabilized bases provide stiff support under concrete pavements and reduce the flexural stresses due to loading. However, they increase stresses due to curling and warping unless the joint spacing is shortened. The window for joint sawing (discussed in Chapter 15) may also be shortened, increasing the risk of random cracking. Troubleshooting of base and subbase problems is addressed by Kohn and Tayabji (2003: 46–47).

Chapter 14

Paving

Quality of pavement construction is important. "A well-built but poorly-designed pavement is likely to outperform a poorly-built but well-designed pavement. Consequently, the impact of a capable contractor with skilled and experienced superintendents and laborers on achieving good pavement performance cannot be over-emphasized" (Rollings 2001: 292).

The two main types of concrete paving are slipform paving and fixed form paving. Both have advantages and disadvantages and each is better suited to some projects than to others. On most slipform paving projects, however, small amounts of fixed form paving or "handwork" are still necessary.

Agency specifications, such as the FAA P-501 which is part of *Standards for Specifying Construction of Airports, AC 150/5370-10B* (FAA 2005) and Ohio Department of Transportation Specifications (ODOT 2005), often address many aspects of paving. In some cases, the agency furnishes a "recipe" specification with mixture proportions for the concrete.

A detailed discussion of concrete paving operations, with photographs, is provided in the AASHTO/FHWA/Industry Joint Training Participant's Manual, *Construction of Portland Cement Concrete Pavements* (ACPA 1996a). A CD with PowerPoint presentations that accompanies this manual may be purchased separately from ACPA. Concrete plant operations are addressed in pages V-1–V-74. Slipform paving is covered on pages VI-1–VI-64, and fixed form paving on pages VI-65–VI-105.

Concrete production considerations

Concrete production and plant operations are beyond the scope of this book, but are nevertheless important for pavement quality. Perhaps the key consideration is that the material delivered be of consistent quality and also be delivered at a constant rate. This is particularly important for slipform paving, because continuous movement of the paver is critical for construction of a smooth pavement. Smoothness will suffer if the paver must be slowed or halted to wait for concrete because it is delivered too slowly, and concrete quality may suffer if delivery trucks are backed up at the project site because the material is delivered too quickly. It can be argued that the construction of a smooth, durable concrete pavement begins at the concrete plant.

One useful reference on concrete production is ACI Committee 304 *Guide for Measuring, Mixing, Transporting, and Placing Concrete, ACI 304R-00* (ACI Committee 304 2000), although this reference is not specific to paving concrete. Often ready-mixed concrete plants are inspected or certified to meet agency requirements, or to meet the National Ready Mixed Concrete Association (NRMCA) certification checklist (in the United States) (ACPA 1996a: V-22).

In many cases, truck or ready-mixed concrete is used for the project. Truck mixed concrete is discussed in detail in the ACPA course manual (ACPA 1996a: V-5–V-45). For large projects, a central mixed concrete plant may be set up. A central plant can produce up to 3,000 cubic meters (3,900 cubic yards) per ten-hour work day. Central plants are also called on-site mixers, stationary mixers, wet-mix, or batch plants. Central mixed plant operations are described in the ACPA course manual, including considerations for the plant location (ACPA 1996a: V-47–V-74).

Regardless of the equipment and production methods used, the ACPA training course manual stresses the importance of communication and preparation between the agency (specifier), paving contractor (purchaser), ready-mix plant (concrete producer), and testing personnel. Specifications must clearly state the quality of concrete required, and must be appropriate for the work. The contractor must arrange a realistic quantity and delivery rate for the concrete. The producer must assure that quality and production rate requirements can be met. These items should be arranged through a meeting of these parties before construction (ACPA 1996a: V-10–V-11).

For large projects that extend over several months, such as major highway and airport projects, it may be necessary to consider seasonal requirements and different mix designs of concrete. For hot weather concreting, a mixture with substitution of fly ash and/or GGBFS will help reduce heat of hydration. As construction extends into the fall, it may be necessary to reduce the content of these materials and increase cement in order to achieve set and strength gain without undue delay.

One important element is avoiding segregation during aggregate and concrete handling and batching operations. Segregation is defined as "the differential concentration of the components of mixed concrete, aggregate, or the like, resulting in nonuniform proportions in the mass" (ACI Committee 116 2000: 57). Segregation is avoided by properly managing aggregate stockpiles, and proper handling of materials throughout concrete batching, transportation, and placing (ACPA 1996a: V-24–V-25, V-33).

Another important consideration is the management of the water content of the concrete. As noted in Chapter 6, concrete mixture proportions are based on SSD conditions and must be adjusted for the actual moisture content of the aggregate. If the aggregate moisture content changes during the course of production, often due to weather conditions, the water added to the mixture must be adjusted to compensate or the slump of the concrete will be too high or too low. Moisture meters may be used to monitor the aggregate moisture content in bins. The moisture content of the fine aggregate is generally more critical than that of coarse aggregate (ACPA 1996a: V-15, V-35, V-38, V-68).

Slipform paving

Slipform construction means

> to consolidate, form into geometric shape, surface finish a concrete mass (vertical or horizontal) by a "slipping" or pulling the forms continuously through and surrounding the plastic concrete mass. In slipform paving of a roadway the forms for shaping the mass, the tools for consolidation, and the tools for surface smoothing are firmly mounted into a self-propelled machine.
>
> (ACPA 1996a: VI-6)

Slipform paving uses low slump concrete, allows high production paving, and can produce a very smooth pavement. Slipform paving is an extrusion process (ACPA 1996a: VI-6, VI-38). Slipform paving production on the order of 1.6 km (1 mile) per day can be achieved. Figure 14.1 shows

Figure 14.1 Slipform paving train (ACPA 1996a: VI-6).

Figure 14.2 Concrete pavement produced by slipforming (ACPA 1996a: VI-6).

a slipform paving train, and Figure 14.2 shows a slipformed pavement. Figures 1.2 and 1.3 also show slipform paving.

Kohn and Tayabji (2003: 68) note some common elements of slipform pavers:

- self-propelled by either two or four tracks;
- weight generally in the range of 3,000 kg/m (2,000 lb/ft) of paving lane width;
- typical paver width 8–11 m (25–37.5 ft), although some can pave as wide as 14–15 m (45–50 ft);
- an array of hydraulically controlled variable speed internal vibrators for consolidating the pavement;
- can carry a head of concrete in front of the leveling screed;
- use continuous augers or similar devices to evenly distribute concrete in front of the screed;
- finishing attachments.

A smooth slipform pavement begins with a good, well-compacted subgrade. The subgrade profile should be within tolerances. Even if subgrade stabilization is not necessary for the support of the constructed pavement, a stabilized subgrade will support the hauling units better and aid in the compaction of any base and subbase layers. Some subgrade trimming may be necessary to achieve the final tolerances. It is important that adequate site drainage be maintained throughout construction. Base and subbase materials must also be placed, compacted, and trimmed accurately. Treated base and subbase materials may be placed using pavers (ACPA 1996a: VI-8–VI-9, VI-17–VI-18).

The elevation of the final pavement surface is generally maintained through the use of stringlines. Sensors on the paver ride along the stringlines to set the final pavement elevation. Stringlines may be wire, cable, woven nylon, polyethylene rope, or another similar material. They are attached to rigid stakes and must be stretched tightly enough to eliminate appreciable sagging. Generally, the stakes should be at no more than 8 m (25 ft) intervals, and closer for horizontal and vertical curves. In very uniform conditions, it is permissible to increase the stake spacing to 16 m (50 ft). Either a single stringline on one side of the paver or dual stringlines on both sides may be used. It is important to accurately survey the stringline and check it just before paving to ensure that it has not been displaced. Changes in temperature and humidity may cause the stringline to sag (ACPA 1996a: VI-14–VI-16). Figure 14.3 shows the stringline and the paver sensing wand.

Recently, stringless pavers have been developed. These rely on global positioning systems to maintain the proper geometrics for the pavement.

The paver rides on tracks outside of the edge of the proposed finished pavement width. This path is generally called pad line, track line, or form line. The pad line is shown in Figure 14.4.

Figure 14.3 Stringline and paver sensing wand (ACPA 1996a: VI-15).

Figure 14.4 The pad line (ACPA 1996a: VI-20).

To support the paver, well-compacted base material should be extended 1 m (3.3 ft) past the outside edge of the pavement. It should be durable enough to provide for smooth passage of the entire paving train, including any texturing and curing equipment. If edge-drains are used, they should be installed after paving so that the paving train will not crush them. The pad line should be carefully maintained and kept free of debris (ACPA 1996a: VI-19).

It is important to carefully set up the paving equipment before paving and check the alignment of the components. A modern placer/spreader typically has an unloading belt, augers, a plow system, and/or a strike off. It is important to ensure that sensors operate properly and that there are no hydraulic oil leaks. Four track paving machines must be set up square to avoid skewed forward motion. The attitude or angle of attack of the paver will have an important effect on the smoothness of the pavement (ACPA 1996a: VI-21–VI-22).

Slipform pavers uses internal vibrators to consolidate the concrete, as shown in Figure 14.5. The vibrators consolidate the concrete mass to remove undesirable voids, and fluidize the mass to aid in forming the shape of the finished pavement.

Proper positioning and frequency of the vibrators is important. Typically, the vibrators should be set for 7,000–9,000 vibrations per minute. The vibrators should be spaced horizontally so that the zones of influence of the adjacent vibrators overlap by 2–3 in (50–75 mm). Too much vibration can

Figure 14.5 Vibrators (ACPA 1996a: VI-26).

segregate the concrete and drive out entrained air, causing visible vibrator trails behind the paver. Details of proper adjustment and maintenance of vibrators are provided in the ACPA course manual (ACPA 1996a: VI-22–VI-30).

For doweled JPCP pavements, dowel bars may be installed with basket assemblies or dowel bar inserters. Dowel baskets are heavy wire assemblies that are staked to the subgrade or base prior to paving. They must be rigidly staked so that they don't move during the paving. Figure 14.6 shows a dowel basket. The dowels must be lubricated with a thin film of grease to allow movement as the concrete slabs expand and contract (ACPA 1996a: VI-31).

When using dowel baskets, the baskets should be checked prior to placing the concrete to ensure that the dowels are properly aligned and that the dowel basket is securely anchored in the base. It is recommended that dowel baskets be secured to the base with steel stakes having a minimum diameter of [7.5 mm] 0.3 inch. These stakes should be embedded into the base a minimum depth of [100 mm] 4 inches for stabilized dense bases, [150 mm] 6 inches for treated permeable bases, and [250 mm] 10 inches for untreated permeable bases, aggregate bases, or natural subgrade. A minimum of 8 stakes per basket is recommended. All temporary spacer wires extending across the joint should be removed from the basket. Securing the steel stakes to the

Figure 14.6 Dowel basket (ACPA 1996a: VI-88).

top of the dowel basket, as opposed to the bottom, should stabilize the dowel basket once these spacer wires are removed.

(FHWA 1990a)

Dowels should be lightly coated with grease or other substance over their entire length to prevent bonding of the dowel to the concrete. This coating may be eliminated in the vicinity of the welded end if the dowel is to be coated prior to being welded to the basket. The traditional practice of coating only one-half of the dowel has frequently resulted in problems, primarily caused by insufficient greasing and/or dowel misalignment. The dowel must be free to slide in the concrete so that the two pavement slabs move independently, thus preventing excessive pavement stresses. Only a thin coating should be used, as a thick coating may result in large voids in the concrete around the dowels.

(FHWA 1990a)

Dowel bar inserters eliminate the need for dowel basket assemblies. The dowels are pressed into the fresh concrete just after placement, as shown in Figure 14.7. Regardless of whether baskets or inserters are used, it is important to accurately mark the joints for later sawing (ACPA 1996a: VI-32–V-33).

Figure 14.7 Dowel bar inserters (ACPA 1996a: VI-32).

The placement of dowels should be carefully verified soon after paving begins. If specified tolerances are not being achieved, then an evaluation of the dowel installation, concrete mix design, and placement techniques must be made. Appropriate corrections should be made to the paving process to ensure proper alignment of the load transfer devices.

(FHWA 1990a)

Tie bars are used across longitudinal joints. These may be placed on chairs ahead of the paver or inserted into the fresh concrete. Reinforcing mesh or bars for JRCP can be installed by using two concrete lifts, and placing the bars on the first lift. Alternatively, they may be pressed into the plastic concrete with a mesh depressor (ACPA 1996a: VI-34–VI-35). As noted in Chapter 2, lightly reinforced JRCP has fallen out of favor.

Steel for CRCP may be set on chairs or installed using tube feeders. Bar chairs assure proper location of the reinforcement and serve as horizontal reinforcement. Tube feeders are shown in Figure 14.8. It is important to ensure that the paver vibrators do not interfere with the steel (ACPA 1996a: VI-36–VI-37).

One essential element of building a smooth concrete pavement is maintaining constant movement of the paving equipment. In addition, uniform particle (aggregate) content, water content, and constant pressure and vibration

Figure 14.8 CRCP steel installation with tube feeders (ACPA 1996a: VI-37).

are important. In summary, key features for constructing a smooth slipform pavement are consistency of delivery, quantity, quality, and motion. Normal paving speeds are on the order of 1–2.5 m (3.3–8.2 ft) per minute. Tamper bars are used on some pavers to help consolidate the concrete (ACPA 1996a: V-21, VI-39–VI-41, VI-47).

Construction joints at the end of the work day require formed headers, or paving a short distance past the last contraction joint. Headers on CRCP require care to ensure continuity of the steel reinforcement (ACPA 1996a: VI-48–VI-49).

A checklist of critical slipform paving considerations is provided in the ACPA course manual (ACPA 1996a: VI-42–VI-43). Troubleshooting tips are provided in the same manual on pages VI-62 through VI-64.

Fixed form paving

Fixed form paving uses molds staked to the subgrade or base to hold the concrete in place. The top edge of the form establishes the grade and alignment, and serves as the tracks for the paving equipment. Fixed form paving is shown in Figure 14.9. Fixed form paving is used for streets, local roads, airports, and pavements with complicated geometry, short length, or variable width (ACPA 1996a: VI-68–VI-69).

Figure 14.9 shows dowel basket and tie bar assemblies, in addition to the forms and paving equipment.

Figure 14.9 Fixed form paving (ACPA 1996a: VI-68).

Advantages of fixed form paving include:

- ability to handle tight tolerances and side clearances;
- ease of changing the width of the pavement;
- blockouts are simple to handle;
- it is possible to pave intersections while keeping some quadrants open to traffic, and to handle other difficult traffic control situations;
- quality can be maintained when concrete delivery is variable;
- the pavement may be built with small, inexpensive equipment (ACPA 1996a: VI-71).

The main disadvantage of fixed form paving is that the production rate is typically much lower than that of slipform paving under ideal conditions. Of course, not all projects offer ideal conditions for slipform paving.

The form is the key to successful fixed form paving construction. A typical paving form is a complex steel section with the face the same height as the required pavement thickness. It has a wide, flat base to provide stability and a solid upper rail to support paving equipment. Holes in the form are provided for pins or stakes to firmly attach it to the subgrade or subbase. Forms are shown in Figure 14.10.

Figure 14.10 Steel forms for fixed form paving (ACPA 1996a: VI-72).

The ACPA course manual (ACPA 1996a: VI-73) provides typical specifications for paving forms:

- metal at least 5.6 mm thick (1/4 in) and 3 m (10 ft) long;
- form depth equal to pavement thickness;
- form depth may be increased up to 50 mm (2 in) by building up the form bottom;
- adequate base width to provide stability;
- flange braces extending outward on the base at least two-thirds of the depth of the form;
- tops of the form do not vary from true plane by more than 3 mm (1/8 in) in 3 m (10 ft) in vertical alignment;
- face of the forms do not vary from true plane by more than 6 mm (1/4 in) in 3 m (10 ft) in horizontal alignment;
- forms should lock together tightly and be flush when they are in position.

Forms must be cleaned and oiled before use, and bent, twisted, and broken forms should not be used. Each 3 m (10 ft) form should be secured with at least two iron pins. Minor adjustments may be made with wedges against the iron pins. Tight curves use flexible steel or wood forms, with similar requirements to the steel mainline forms (ACPA 1996a: VI-74).

Although the base is less critical for fixed form paving than for slipform paving, a good base will improve pavement performance. A stringline is typically used to set the forms. Once the forms are set, they must be checked for alignment and tolerance before paving, and joints between forms must be tight and smooth (ACPA 1996a: VI-75–VI-81). The stiffness and alignment of the forms will determine the smoothness of the pavement.

Special details for fixed form paving include tie bolt assemblies or tie bars, curves, integral curbs, and blockouts. Tie bars may be secured directly to the forms for longitudinal joints. For curves, straight metal forms may be used unless the radius of curvature is less than about 30 m (100 ft). Otherwise, special curved metal forms or wooden forms should be used. Integral curbs may be formed by hand or by using special devices called curb mules. Utilities and drainage structures typically require the construction of blockout forms before paving. When blockouts are used, it may necessary to provide special jointing to prevent random cracking (ACPA 1996a: VI-84–VI-86, VI-90–VI-91).

Typical blockouts are shown in Figure 14.11. If these provide internal restraint to the concrete as it shrinks, cracks will initiate from corners of the blockouts. Therefore, joints should be sawn or formed to control these cracks.

Uniformity and consistency of the concrete is the key to placing, consolidating, and finishing. A conventional form riding train includes a spreader

Figure 14.11 Blockouts for utility, drainage, and similar structures (ACPA 1996a: VI-91).

with embedded vibrators and a finishing machine to strike off and consolidate the pavement. Some small projects use only the finishing machine, while some larger projects use multiple spreaders or multiple finishing machines. For small, isolated areas, small machines or hand placing may be used, but care is required to place, spread, and vibrate the concrete uniformly. It is also important to ensure that curbs, if any, are finished to maintain water flow (ACPA 1996a: VI-93–VI-99).

Curing and sawing and sealing of fixed form concrete pavement is the same as that for slipform concrete pavement, except that the sides of the pavement must be sprayed with curing compound once the forms are removed. Forms may be removed as early as 6–8 hours after paving if enough care is used. After removal, the forms should be cleaned immediately and stored with care (ACPA 1996a: VI-100–VI-102).

According to the ACPA course manual (ACPA 1996a: VI-103–VI-104), the following critical factors for fixed-form paving affect ride quality:

- The grade must be uniform and compacted and moistened ahead of concrete placement.
- Forms must meet specifications and be well oiled and properly aligned.
- Blockouts and utility fixture adjustments must be carefully prepared.
- Dowel baskets and reinforcement (if any) must be placed properly.
- Concrete should be of consistent quality and delivered at a consistent rate.
- Extra care should be taken to achieve consolidation around dowel basket assemblies and blockouts.
- The pavement must be properly finished and cured.

Bridge deck pavers

Some airfield pavements are built with bridge deck pavers. Kohn and Tayabji (2003: 68–70) discuss some of their features:

- truss system with a suspended screw auger to spread concrete, an oscillating vibrator, and a roller which compacts and finishes the surface;
- may incorporate a texturing device following the roller assembly;
- rides on either forms or self-propelled wheels;
- has vibrators that move transversely in front of the screed and may also have fixed vibrators near form edges;
- usually weigh less than 1,500 kg/m (1,000 lb/ft) of paving width;
- production capacity much less than a slipform paver;
- most economical in paving widths of 12–15 m (40–50 ft).

Planning and troubleshooting

Careful preparations can go a long way toward ensuring trouble-free paving, which is the best way to produce a smooth pavement. Some important planning elements include verifying that:

- all equipment in the paving train is in operational condition;
- the grade is ready to start paving;
- any required approved test reports for project materials are available;
- backup testing equipment is available;
- all necessary concrete placement tools are on hand;
- communications between the batch plant and the construction site have been established;
- equipment to wet the grade, if needed, is available;
- the string line is straight and properly tensioned (and there is a plan to monitor the string line during construction);
- the header is in place for the day's work;
- the extreme weather management plan has been developed (and check the weather forecast);
- enough plastic covering is available to protect the pavement in case of sudden and unexpected rain (Kohn and Tayabji 2003: 48).

Equipment condition is important:

- Enough hauling trucks should be available, because they affect slipform production rates. If the paver has to be slowed or stopped while waiting on concrete, the pavement will be rough.
- Enough concrete saws need to be available, because joint sawing windows may be too short to obtain additional saws in case of breakdown. Saw blades must be appropriate for the aggregate used in the concrete.
- Slipform paver vibrators should have a frequency of 6,000–12,000 vibrations per minute and an amplitude of 0.6–1.3 mm (0.025–0.05 in) when not under load. They should be high enough not to interfere with dowel baskets or reinforcement.
- Curing application equipment should be checked to ensure that it will apply curing compounds uniformly at the proper rate (Kohn and Tayabji 2003: 52).

Accelerated concrete pavement construction

On some projects, it is necessary to accelerate pavement construction. Accelerated paving may involve changes in planning, design, and materials. ACI 325.11R *Accelerated Techniques for Concrete Paving* (ACI Committee 325 2001) provides extensive guidance. Traffic control issues are addressed in

Traffic Management – Handbook for Concrete Pavement Reconstruction and Rehabilitation, Engineering Bulletin EB213P (ACPA 2000c). For airfield pavements, a planning guide and detailed case studies in two volumes have been published by the IPRF (Peshkin et al. 2006a,b).

Mixture considerations for accelerated pavement construction are discussed in Chapter 6, and some specific mixtures are shown in Table 6.1. In many cases, conventional paving concrete will develop sufficient strength to carry traffic within a few days.

Maturity testing

When EOT or fast-setting concretes are used and must be opened to traffic at an early age, it is desirable to provide some sort of strength verification. The rate of strength gain of a given concrete mixture depends on the progress of the hydration reaction which is highly temperature dependent – concrete gains strength more rapidly in hot weather and more slowly in cold weather. When the concrete has a high cement content, an accelerator, and/or is covered by curing blankets, the strength gain rate increases further. Neville (1997: 305) states "as the strength of concrete depends on both age and temperature, we can say that strength is a function of $\sum$ (time interval × temperature), and this summation is called maturity." The temperature is measured from a datum of $-10\,°C$ or $14\,°F$, below which the concrete does not gain strength.

The maturity method provides a way to adjust a concrete strength prediction for the actual temperature history of the concrete, and requires the development of a strength–maturity relationship that is specific to the mixture. The procedure for developing the strength–maturity relationship is provided by ASTM C1074 (ASTM C1074 2004). A device is required to monitor and record the concrete temperature as a function of time. The procedure is outlined as follows:

- Prepare at least 15 specimens, with mixture proportions and constituents similar to the concrete that will be used in practice.
- Embed temperature sensors at the centers of at least two specimens and connect to maturity instruments for recording.
- Moist cure the specimens in water bath or moist room.
- Perform strength (compression) tests at ages of 1, 3, 7, 14, and 28 days. Test two specimens at each age and compute the average strength. If the range exceeds 10 percent of the average, test another specimen and average all three tests. If a low test result is due to an obviously defective specimen, discard the low test result.
- At each age, record the average maturity index for the instrumented specimens.

- On a graph paper, plot the average strength as a function of the average value of the maturity index. Draw a best-fit curve through the data, which provides the strength–maturity relationship.

The ASTM procedure is written in terms of compressive strength, but for pavements beam specimens may be more relevant. For the specific concrete mixture, the maturity index required to achieve the specified strength is determined. Various temperature measuring devices may be used in the field to record the temperature history of the concrete and determine when the required maturity index has been reached. It is useful to supplement maturity testing with an independent physical measurement to verify strength, such as ultrasonic pulse velocity.

Chapter 15

Finishing, texturing, curing, and joint sawing and sealing

The final operations in concrete pavement construction are finishing, texturing, curing, and (for jointed pavements) joint sawing and sealing. Proper planning and execution of these steps has an important impact on pavement performance.

Finishing and texturing

For slipform paving, the pavement is finished behind the paver with a burlap drag and tine texturing. A tube float, which is a self-propelled machine not attached to the paver, may be used to correct minor variations and seal small imperfections in the finished slab surface. A key consideration for finishing is water management. The concrete should not be finished until the bleed water has disappeared from the surface. Longitudinal or transverse floats may be used. A straight edge may be used to check the pavement for surface imperfections that may be corrected with hand operated floats. For the most part, no water should be added to aid in finishing beyond a mist from a hand pump sprayer (ACPA 1996a: VI-50–VI-52).

Edge-slump occurs when the concrete is not stiff enough to hold its edge after being extruded from the paver. While some limited edge-slump may be corrected by hand work, it is better to adjust the concrete mixture so that it does not slump.

The concrete surface is textured with tining machines to increase the friction of the surface. The tining may be longitudinal or transverse, with grooves 3–5 mm (1/8–1/5 in) deep, 3 mm (1/8 in) wide, and 12–20 mm (1/2–3/4 in) spacing. Tining provides macrotexture, while microtexture is achieved with burlap or Astroturf drags. These are illustrated in Figures 15.1 and 15.2, respectively (ACPA 1996a: VI-53–VI-54).

Curing

Typically, concrete pavements are cured by spraying white pigmented or clear curing compounds. When the surface has a tine texture, the curing

Figure 15.1 Transverse tining to provide macrotexture (ACPA 1996a: VI-53).

Figure 15.2 Burlap drag to provide microtexture (ACPA 1996a: VI-53).

compound should be applied in two coats, in the forward and backward directions. In case of rain, clear plastic sheets should be available to prevent damage to the new concrete pavement surface (ACPA 1996a: VI-55–VI-56).

The FHWA recently published the *Guide for Curing of Portland Cement Concrete Pavements, Volume I* (Poole 2005). This guide addresses the major events in curing pavements – curing immediately after placement, or initial curing, curing following final finishing, or final curing, and termination of curing and evaluation of effectiveness. Poole defines curing as "the process of deliberate control of moisture and temperature conditions within prescribed limits. The process allows concrete properties to develop and prevents damage as a result of drying and/or thermal gradients during the early history of the structure" (Poole 2005: 1).

Factors affecting curing requirements

Curing is particularly important for pavements because of their high surface to volume ratio. Poor curing can result in significant damage to concrete pavements. Damage occurs due to plastic shrinkage cracking, thermal stress, or drying shrinkage cracking. Poorly cured concrete pavements may also have poor abrasion resistance and may not resist deicing salts and other deterioration mechanisms as well.

> When PCC layers are placed over stabilized or permeable bases, the impact of shrinkage and curling and warping gradients due to improper curing will be exaggerated. Therefore, special care should be taken to ensure that adequate curing is provided to the PCC layers in the presence of certain stabilized and permeable bases (namely, CTB, LCB, and CTPB).
>
> (Hall et al. 2005: 27)

Concrete materials, mixture proportions, early-age properties, and probable climactic conditions during and in the few days following concrete placement are important for planning and preparation for curing the concrete. These topics are addressed in detail in Chapter 2 of Poole (2005: 3–16).

The type of cement used has significance, primarily from the standpoint of the strength gain of the cement. Curing is generally specified either for a particular period of time (number of days), or until a certain concrete strength has been achieved. Types I, II, and I/II cement have similar strength gain, with type V slower and type III faster. Fineness of cement affects bleeding and development of internal dessication with very low w/cm ratios. Very fine cements (Blaine values of more than 400 m^2/kg) may develop too little bleed water in dry climates or with w/cm less than 0.40. The fineness of pozzolans, particularly silica fume, may also contribute to this problem.

Pozzolans, particularly class F fly ash, delay setting time and retard strength gain. Thus, the optimum time for final curing is delayed and there is a greater opportunity for plastic shrinkage cracking to occur. Slow strength gain also prolongs the required curing time. Class C fly ash also delays setting, but does not retard strength gain as much as class F ash.

The amount of bleeding of concrete depends on the w/cm ratio. If the rate of evaporation from the concrete exceeds the bleeding rate, then plastic shrinkage cracking occurs. On the other hand, excessive bleeding leads to a low w/cm ratio at the surface of the concrete, a weak surface layer, and poor abrasion resistance. The rate of bleeding, BR, in kg of water/m^2/hr, may be calculated using an empirical equation (Poole 2005: 6–7)

$$BR = (0.51^*w/cm - 0.15)^*D \tag{15.1}$$

where w/cm = water/cementitious materials ratio and D = pavement thickness in mm (or in * 25.4). It is also possible to determine the bleeding for a specific concrete experimentally.

Poole (2005: 7) notes

> Paving concretes tend to have w/cm's between 0.38 and 0.48. For a 30-cm [12 in] thick pavement, bleeding would then range from about 0.13 to 0.28 kg/m^2/h [0.027 to 0.057 lb./ft.^{2}/hr]. These are lower average bleeding rates than found in more general-use concretes, which range from about 0.5 to 1.5 kg/m^2/h [0.1 to 0.3 lb./ft.^{2}/hr]. The result of this is that paving concretes are more susceptible to losing more-than-safe amounts of bleed water to evaporation. ACI 308 considers that drying conditions of less than 0.5 kg/m^2/h [0.1 lb./ft.^{2}/hr] represent a mild threat to most concrete. A safer upper limit for paving would be about 0.3 kg/m^2/h [0.061 lb./ft.^{2}/hr].

High cementitious materials content, particularly fine particles like type III cements or pozzolans, tends to reduce bleeding. These high contents also increase long-term drying shrinkage.

Time of initial setting is also important.

> In conventional concreting, final finishing is typically not executed until about the time of initial setting. In slip-form paving, final finishing is usually completed within a few minutes of placing the concrete, well before the time of initial setting and the end of the bleeding period. If bleeding rates are low relative to evaporation rates, then loss of surface sheen will appear rather soon after placing, suggesting that final curing should be initiated even though bleeding is continuing.
>
> (Poole 2005: 9–10)

Unfortunately, problems may occur with starting final curing before initial setting, as water continues to bleed to the surface. It may wash out fines and pond against sheet materials, or damage sprayed curing membranes.

Evaporation of bleedwater from the concrete surface depends on wind speed, concrete temperature, air temperature, and relative humidity. The evaporation rate has traditionally been determined using a nomograph published in ACI 308, *Guide to Curing Concrete, ACI 308R-01* and in ACI 306R, *Hot Weather Concreting, ACI 305R-99* (ACI Committee 308 2001: 3, ACI Committee 305 1999: 5). Poole (2005: 11) provides an equation that may be programmed into a calculator or spreadsheet, and yields the same result:

$$\text{ER} = 4.88 \left[0.1113 + 0.04224 \frac{\text{WS}}{0.447} \right] (0.0443)(e^{0.0302(\text{CT} \cdot 1.8) + 32})$$

$$- \left[\left(\frac{RH}{100} \right) (e^{0.0302(\text{AT} \cdot 1.8) + 32}) \right] \tag{15.2}$$

where

> ER is evaporation rate (kg/m^2/h), multiply by 0.2048 to get the rate in lb/ft^2/h,
> WS is the wind speed (m/s), multiply speed in mph by 0.447,
> CT is concrete temperature (°C), to convert temperatures in °F subtract 32 and then divide by 1.8,
> AT is air temperature (°C), and
> RH is relative humidity (percent).

Evaporation increases as wind speed and air or concrete temperature increase, and as relative humidity decreases. For concrete paving, only the concrete temperature may be easily and reliably controlled. It may be advisable, before paving, to estimate the evaporation rate that will occur under the anticipated environmental conditions.

The wind speed increases rapidly with the height above the surface. Therefore, for purposes of evaporation prediction, the wind speed should be measured 0.5 m (20 in) above the concrete surface. Measurements taken higher up will overestimate evaporation (ACI Committee 305 1999: 5).

> Standard guidance recommends that when evaporation exceeds bleeding, then something must be done to reduce evaporation rates. Standard remedies include use of fogging and wind breaks. Neither of these is particularly useful for large paving projects.
>
> Three practices are potentially useful in paving large areas. One is to shift paving operations to a time of day in which the drying conditions

are less severe. Night-time placement is often attractive because relative humidity is usually higher than during the day.

(Poole 2005: 14)

The other two practices are to reduce the concrete temperature at the time of placement, and to use evaporation reducers. Evaporation reducers can reduce the rate of evaporation by up to 65 percent (Poole 2005: 15).

Initial curing

Initial curing is the period between placing the concrete and application of final curing. Initial curing is addressed in detail in Chapter 3 of Poole (2005: 17–19). The major action items during initial curing are to verify environmental conditions and to make onsite adjustments. The proper time to apply final curing is at the time of initial setting of the concrete.

The onsite environmental conditions that affect the evaporation rate, as shown in equation 15.2, may be verified in the field using inexpensive instruments. Concrete temperature should be routinely measured in hot weather.

The two onsite adjustments that may be useful during paving are reducing the concrete temperature and applying evaporation reducers. Concrete temperature may be reduced by spraying aggregate stockpiles to cool them, cooling mixing water, or adding ice to mixing water. ACI 305R, *Hot Weather Concreting, ACI 305R-99* (ACI Committee 305 1999), provides equations for determining concrete temperature based on the temperature of the constituent materials, including ice, if used. In hot climates, the injection of liquid nitrogen directly into ready-mix trucks has proven very effective for reducing the concrete temperature.

> Evaporation reducers are a relatively new product developed to specifically address the condition of excessive evaporation rates. The approach is to apply evaporation reducers in sufficient quantity and frequency that the concrete does not ever lose critical amounts of water to evaporation. Application is made using the same or similar equipment to that used to apply curing compounds.
>
> Evaporation reducers are water emulsions of film-forming compounds. The film-forming compound is the active ingredient that slows down evaporation of water. There is also a benefit from the water fraction of the evaporation reducers in that it compensates to a small degree for losses of mixing water to evaporation.
>
> (Poole 2005: 19)

Depending on the environmental conditions, the evaporation reducer may need to be applied several times. The required frequency of application for a water reducer may be determined using the following equation (Poole 2005: 19):

$$F = \frac{AR}{ER(1 - 0.4) - BR} \tag{15.3}$$

where

F = frequency of application, h,
AR = application rate, kg/m^2,
ER = evaporation rate of bleed water, kg/m^2/h,
BR = bleed rate of concrete, kg/m^2/h.

This equation assumes that the evaporation reducer reduces evaporation by 40 percent, which is less than the value claimed by most manufacturers. The equation may be used in pounds per square foot without conversion.

A portion of the curing compound may be applied early to serve as an evaporation reducer. However, this may interfere with final curing.

Final curing

Final curing is the period between application of final curing and the end of deliberate curing. Final curing is addressed in detail in Chapter 4 of Poole (2005: 21–33).

Final curing methods are classified as curing compound methods, water added methods, and water retentive methods. For large paving projects, only curing compound methods are generally considered practical and economical. Water added methods (ponding water on concrete) and water retentive methods (waterproof sheets) may be considered for very small areas or for repairs.

> Curing compounds are normally the most economical method for curing large areas of paving because of the relatively low labor costs. Once the application is satisfactorily completed, little or no additional attention is required. The negative side to using curing compound methods is the relatively complicated selection and specification-compliance issues that are frequently encountered, and the skill required to apply the material correctly.
>
> (Poole 2005: 22)

The major issues associated with the use of curing compounds are (Poole 2005: 22):

• selection of the curing compound – generally based on water retention. Factors include pigments, drying time, type and amount of solids,

volatile organics (for environmental and safety considerations), compatibility with coatings, and viscosity. Many specifications address curing compounds;
- time of application;
- application rate;
- verification of application.

Timing of application is very important.

Curing compounds perform best if applied after time of initial setting. Typical guidance on paving is to apply curing compound when the surface sheen has disappeared. Taken literally this practice can lead to poor performance. Paving concretes tend to be made with a relatively low w/c, so under even relatively mild drying conditions, the surface sheen may disappear soon after placing, even though bleeding is continuing. Application of the curing compound then slows or stops the evaporation of bleed water, which then either accumulates under the membrane or dilutes the curing compound. In either case, the membrane is likely to be damaged and suffer reduced performance during final curing period. This damage is sometimes visible as cracks or tears in the membrane. In other cases it can only be seen with moderate magnification. If drying conditions are mild, (e.g. < 0.5 kg/m2/h [0.1 lb./ft.^{2}/hr]), this result may have no detrimental effect.

(Poole 2005: 25–26)

In paving, curing compounds are often applied directly following concrete placement, often by automated equipment following a slipform paver. The curing compound acts as an evaporation reducer.

The potential difficulty with this practice is that the curing compound will not then retain water during the final curing period to the level of performance expected from the job specification on the material for the reasons cited above. However, it may be reasonable practice to apply part of the curing compound early, for purposes of controlling evaporative losses during the initial curing period, then applying the remainder after time of initial setting to restore the integrity of the membrane from any damage suffered from the early application.

(Poole 2005: 26)

The application rate of the curing compound needs to be sufficient to form a continuous membrane to reduce moisture loss.

AASHTO guidance recommends a coverage rate of no more than 5 m^2/L [8.3 ft.^{2}/gal]. This is also common manufacturers' recommended guidance. As discussed above, many curing compounds cannot be applied at this rate in a single pass without serious running into low areas. Grooving patterns increase the effective surface area. This effect must be accounted for to maintain the target application rate.

(Poole 2005: 26)

Standard practice has often been to document the amount of curing compound used and the area of pavement covered, and calculate the application rate. Most curing compounds are pigmented to provide a visual indicator of the degree of coverage. For example, the inspector may look for areas where the gray of the concrete shows through the white curing compound, and direct an additional application. Portable reflectometers are more sensitive than the human eye.

Water added methods, primarily spraying, may be used for RCC. Pervious concrete pavement cannot be satisfactorily cured with either curing compounds or added water, and is cured for 7 days under plastic.

Temperature management

It is important to manage temperature changes in concrete pavements as well as moisture loss. As discussed in earlier chapters, concrete expands and contracts with temperature.

Concrete generates heat internally starting soon after placing due to the hydration of the cementitious materials. The most intense heating from this source occurs in the first 24 hours, reaching a peak approximately 6–8 hours after placing, depending on the chemistry of the cement. In thin pavements, this heat is usually dissipated to the environment about as fast as it develops and doesn't contribute significantly to the overall heating of the pavement. In thick highway pavements, some of the heat can accumulate.

Concrete can also warm if the air temperature is higher than the placement temperature and if there is significant solar radiation. Cool atmospheric conditions and evaporation of water from the surface of the concrete act against the warming.

The typical pattern in warm weather placements is for the concrete to warm up at least a little, but if the heat of hydration of the cement, peak air temperatures, and peak solar radiation occur at the same time, then temperatures as high at about 60 °C (140 °F) can be reached if measures are not taken to prevent this.

(Poole 2005: 32)

As the concrete subsequently contracts as it stiffens and becomes brittle, it is likely to crack if it is restrained. Proper jointing and detailing are intended to handle this cracking potential, but cracking may occur nevertheless if a substantial temperature drop occurs after paving. HIPER-PAV software may be used to assess this potential. It may be necessary to cover the concrete pavement with blankets to retain heat in cold weather.

Termination of curing and evaluation of effectiveness

When to terminate curing, and how to evaluate whether it has been effective, are addressed in detail in Chapter 5 of Poole (2005: 35–39). The major actions are to determine the proper length of curing, and to apply either prescriptive or performance measures to verify that curing is complete.

Because curing compounds are not removed from the pavement, curing is not really terminated. The requirements instead apply to water retentive and water added curing methods.

"The traditional prescriptive way of specifying length of curing is with fixed time periods. The requirement is usually accompanied by a minimum temperature during the specified time interval, typically 10 °C [50 °F]" (Poole 2005: 35). More than half of states require 3 days, some require 4, and about 25 percent require 7–14 days. ACI documents offer differing guidance but generally require 7 days.

Termination of curing may also be based on achieving a certain percentage of concrete strength, typically 70 percent. Particularly in cold weather, the strength should be verified using companion specimens cured next to the concrete pavement in the field, rather than specimens stored in a heated laboratory that will gain strength more quickly.

In order to account for the effects of time and temperature, the maturity method may be used (ASTM C1074 2004). This method is discussed in Chapter 14. "In actual field application, the maturity method normally takes temperature input from in-place thermocouples located at critical points in the pavement. Determining critical locations is an important part of the application. Pavement corners, sections of elevated pavement, and most recently placed pavements are particularly sensitive to low temperature events" (Poole 2005: 37).

Verification of the effectiveness of curing requires an actual measurement of the in-place strength of the concrete. Methods include the extraction of cores, the rebound hammer, ultrasonic pulse velocity, and abrasion resistance (Poole 2005: 37–39).

To supplement *ACI 308R-01, Guide to Curing Concrete* (ACI Committee 308 2001), ACI has also published a Specification, *ACI 308.1-98, Standard Specification for Curing Concrete* (ACI Committee 308 1998). This is a reference standard that may be cited in project specifications.

Considerations for thin overlays

Thin concrete overlays, either bonded, thin whitetopping, or UTW, are particularly sensitive to curing. Reasons include the high surface to volume ratio, the low w/cm ratio and high cement factor and heat generation of many overlay mixtures, and possibly heat retained by the base pavement layers. Improper curing of thin overlays can lead to high shrinkage and debonding. Potential adverse conditions include very hot or cold base pavement, low humidity, and high wind speeds. Typically, for thin overlays, curing compounds are applied at twice the manufacturer's recommended rate (ACI Committee 325: 2006: 10, 11).

Because of the difficulty of curing thin overlays, consideration should be given to internal curing using a partial replacement of a portion of the fine aggregate by saturated lightweight aggregate with absorption on the order of 15–20 percent. Background for internal curing and suggestions for mixture proportioning, as well as test results, are provided by Bentz et al. (2005), Lam (2005), and Mack (2006).

Joint sawing and sealing

A joint is a designed crack. Concrete pavements will crack due to shrinkage and temperature differences. The purpose of joint sawing and sealing is to ensure that cracks form at the proper locations and will be easy to maintain. Generally, transverse joints in JPCP or JRCP are initiated by sawing through the surface slab to create a plane of weakness.

Joints are often sealed to reduce the amount of water flowing through the pavement and to prevent incompressible material, such as small stones, from becoming trapped in the pavement joint. Incompressible materials can lead to chips or spalls when joints close as temperatures increase.

The initial saw cut is made with a thin bladed saw. This is often followed by a widening cut, which is not as deep as the initial cut, to provide a reservoir for sealant material. Special saw blades are available that make both cuts at the same time. Figure 15.3 shows the initial and widening cut, along with the crack that propagated through the slab from the initial cut (ACPA 1996a: VII-11). It also shows a tie bar at the transverse joint, which is not good practice because the bond between the tie bar and the concrete will be poor.

Two types of sawing may be used – wet sawing with diamond blades or dry sawing with abrasive or diamond blades. Wet sawing, with water to lubricate and cool the saw blade, is the most common method used. Water also helps control dust. Dry sawing is used with softer aggregates such as limestone. Small, medium, large, and high production saws are available, depending on the size of the project and the required production rate (ACPA 1996a: VII-12–VII-17).

Figure 15.3 Transverse joint initial and widening cut (ACPA 1996a: VII-11).

It is important to accurately locate joints before sawing so that the dowel bars will function properly. Control points should be established far enough away to not be disturbed by the paving operations.

There is a window of time during which a saw cut may be made. If the attempt is made too early, the pavement will not be able to support the sawing equipment and will be damaged or "raveled" by the saw. If the joint is not cut in time, a crack will form. The window will be earlier and shorter in hotter weather or if higher cement contents or accelerators are used with the concrete. The window is shorter and earlier with stiffer or open graded base materials, and extends longer with granular bases. With experience, it is possible to tell by scratching the concrete surface with a knife or a nail whether it is ready to saw yet. In addition to ambient temperature, the sawing window is affected by wind, humidity, cloud cover, and temperature of the base (ACPA 1996a: VII-21–VII-23).

Because of the difficulties in predicting joint sawing windows, the contractor must be ready to saw as soon as the concrete is ready, at any time of day or night. If sawing is too late, random cracking may occur. If spalling occurs during saw cutting, the concrete is not ready to saw yet. It is important to saw before the surface concrete temperature decreases significantly (Kohn and Tayabji 2003: 100–101).

In 1993 at an airbase in Texas, an apron was slipformed in warm, dry summer weather. It was wet cured with burlap for 1 day and then membrane cured. The contractor waited until the burlap was removed to begin sawing transverse contraction joints... over 25 percent of the slabs had cracked... Concrete generally has to be sawed the day it is placed. Delays until the next day or later almost invariably lead to cracks starting to develop on their own and leave the saw cuts untouched.

(Rollings 2005: 176)

The purpose of a joint sealant is to deter the entry of water and incompressible material into the joint and the pavement structure. It is recognized that it is not possible to construct and maintain a watertight joint. However, the sealant should be capable of minimizing the amount of water that enters the pavement structure, thus reducing moisture-related distresses such as pumping and faulting. Incompressibles should be kept out of the joint. These incompressibles prevent the joint from closing normally during slab expansion and lead to spalling and blowups.

(FHWA 1990a)

Once joints are cut and widened, a number of different joint sealing materials may be used. Various types of joint sealant materials are listed in Table 15.1.

Sealant behavior has a significant influence on joint performance. High-type sealant materials, such as silicone and preformed compression seals, are recommended for sealing all contraction, longitudinal, and construction joints. While these materials are more expensive, they provide a better seal and a longer service life. Careful attention should be given to the manufacturer's recommended installation procedures. Joint preparation and sealant installation are very important to the successful performance of the joint. It is therefore strongly recommended that particular attention be given to both the construction of the joint and installation of the sealant material.

(FHWA 1990a)

It is important to clean out the joint reservoir well. Joint sealants may be formed-in-place or preformed. Formed-in-place sealants include hot applied and cold applied materials, with a backer rod inserted into the joint before the material is poured. Backer rods are made of polyethylene foam and control the sealant shape while keeping it from flowing out of the bottom of the reservoir. Preformed sealants are pressed down into joints and do not require backer rods (ACPA 1996a: VII-31–VII-44).

Table 15.1 Joint sealing materials (ACPA 1996a: VII-32)

Type of material	Material types and properties	Common specifications
	Formed-in-place (hot applied)	
Polymeric asphalt based	Self-leveling	AASHTO M0173, ASTM D3405, SS-S-1401C, ASTM D1190
Polymeric sealant	Self-leveling	ASTM D3405
Low modulus	Self-leveling	
Elastomeric sealant	Self-leveling	SS-S-1614
Coal tar, PVC	Self-leveling	ASTMD3406
	Formed-in-place (cold applied)	
Silicone sealant	Non-sag (toolable) or self-leveling (low or ultralow modulus)	
Nitrile rubber sealant	Self-leveling (no tooling), non sag, toolable	
Polysulfide sealant	Self-leveling (no tooling), low modulus	SS-S-200A
	Preformed compression seals	
Preformed compression seals with lubricant adhesive	20–50% allowable strain	ASTM D2628 with ASTM D2835

When using silicone sealants, a minimum shape factor (ratio of sealant depth to width) of 1:2 is recommended. The maximum shape factor should not exceed 1:1. For best results, the minimum width of the sealant should be [9.5 mm] 3/8 inch. The surface of the sealant should be recessed [6.5 to 9.5 mm] 1/4 to 3/8 inch below the pavement surface to prevent abrasion caused by traffic. The use of a backer rod is necessary to provide the proper shape factor and to prevent the sealant from bonding to the bottom of the joint reservoir. This backer rod should be a closed-cell polyurethane foam rod having a diameter approximately 25 percent greater than the width of the joint to ensure a tight fit. When using preformed compression seals, the joint should be designed so that the seal will be in 20 to 50 percent compression at all times. The surface of the seal should be recessed [3 to 9.5 mm] 1/8 to 3/8 inch to protect it from traffic.

(FHWA 1990a)

The ACPA course manual (ACPA 1996a: VII-45) lists the following critical factors for sawing and sealing joints in new pavements:

- saw equipment selection;
- locating joints;
- saw timing, considering weather effects on the sawing window;

- sawing straight and to proper width and depth;
- selection of sealant materials that meet project specifications and requirements;
- preparing the joint reservoir; and
- sealing longitudinal joints before widening transverse joints.

Preparation of the joint reservoir is cited as the single most important element. A detailed discussion of joint sawing and sealing operations, with photographs, is provided in the AASHTO/FHWA/Industry Joint Training Participant's Manual, *Construction of Portland Cement Concrete Pavements* (ACPA 1996a: VII-1–VIII-52). A CD with PowerPoint presentations that accompanies this manual may be purchased separately from ACPA.

The American Concrete Pavement Association has published *Joint and Crack Sealing and Repair for Concrete Pavements, Technical Bulletin TB–012.P* (ACPA 1993a). This 32-page guide addresses reasons for joint and crack sealing and provides additional information about techniques for both new construction and rehabilitation. Joint sealing is particularly important if the pavement is not well drained. With drainable bases, joint sealing provides a second level of protection against moisture-related pavement damage.

The ACPA guide provides details about the selection of joint sealing materials. Important properties include elasticity, stiffness (modulus), adhesion, cohesion, compatibility with other materials, resistance to weathering, and jet fuel resistance for airfield pavements. Many types are governed by ASTM, AASHTO, or military specifications (ACPA 1993a: 4–5).

The joint reservoir should be sized to accommodate the expected movement of the joints. The expected joint movement ΔL in mm or inches is provided by:

$$\Delta L = CL(\alpha \ \Delta T + \varepsilon) \tag{15.4}$$

where C = subbase/slab frictional coefficient, 0.65 for stabilized material and 0.80 for granular material;
L = slab length in mm or inches;
α = coefficient of thermal expansion, in microstrain per degree C or F;
ΔT = expected temperature change in degrees C or F (provided in figure 3 of ACPA 1993a). This is "generally the temperature of the concrete at the time of placement minus the average daily minimum temperature in January" (FHWA 1990a). This range varies from 59 °C (106 °F) in the southeast of the United States to 71 °C (128 °F) in the West and upper midwest; and
ε = concrete shrinkage coefficient, in microstrain (ACPA 1993a: 8).
Typical thermal coefficients (α) based on coarse aggregate type are shown in Table 15.2, and typical shrinkage coefficients (ε) based on concrete strength are shown in Table 15.3.

Table 15.2 Typical values for PCC CTE (α) (modified from AASHTO 1993: II-29)

Type of coarse aggregate	PCC CTE α (10^{-6}/degree)	
	Per degree C	Per degree F
Quartz	3.7	6.6
Sandstone	3.6	6.5
Gravel	3.3	6.0
Granite	2.9	5.3
Basalt	2.7	4.8
Limestone	2.1	3.8

Table 15.3 Typical values for PCC coefficient of shrinkage (ε) (modified from AASHTO 1993: II-29)

Indirect tensile strength of concrete		PCC coefficient of shrinkage (mm/mm or in/in)
MPa	psi	
2.1 (or less)	300 (or less)	0.0008
2.8	400	0.0006
3.4	500	0.00045
4.1	600	0.0003
4.8	700 (or greater)	0.0002

For example, a 6.1 m (20 ft) slab with sandstone aggregate ($\alpha = 11.7\ \mu\varepsilon/°C$ or $6.5\ \mu\varepsilon/°C$), with 3.45 MPa (500 psi) splitting tensile strength concrete ($\varepsilon = 0.00045$) built on granular base in California ($\Delta T = 99°C$ or $179°F$) will open by:

$$\Delta L = CL(\alpha\ \Delta T + \varepsilon) = 0.80\ (6, 100)\ (11.7 \times 10^{-6} \times 99 + 0.00045)$$

$$= 7.9 \text{ mm or}$$

$$\Delta L = CL(\alpha\Delta T + \varepsilon) = 0.80\ (240)\ (6.5 \times 10^{-6} \times 179 + 0.00045)$$

$$= 0.31 \text{ in}$$

Therefore, if the joint material can accommodate 100 percent strain the reservoir must be 15.8 mm (0.62 in) wide. With longer slabs, the joints move more, which in part accounts for the frequent poor performance of JRCP. Also, the use of joint materials with lower deformation capacity requires larger joint reservoirs which may be more difficult to maintain. Tied joints between lanes and between lanes and shoulders do not move as much.

A 20-page checklist is available from the FHWA for Pavement Preservation Checklist Series 6 (2002). The checklist addresses general preparation items as well as items specific to different types of joint sealants, and may be downloaded from the web at http://www.fhwa.dot. gov/pavement/pub_listing.cfm.

Hot and cold weather paving and curing precautions

Special precautions are generally necessary for hot and cold weather paving. Adverse weather conditions are rarely a complete surprise, so it is advisable to have planning meetings and discussions about probable weather conditions during construction, what methods may need to be taken to protect the freshly placed concrete, and the conditions under which paving must be suspended. For example, during hot weather, it is important to ensure an orderly traffic flow so that concrete trucks do not wait and build up heat. In cold weather, materials to protect the concrete in case of freezing temperatures must be readily available, preferably stockpiled at or near the job site.

Hot weather precautions

Hot weather conditions elevate the risk of early-age cracking for concrete pavements. Hall et al. (2005: 10) note

> When ambient temperatures are in excess of 90 °F (32 °C), the risk of early cracking is significantly elevated. Hot-weather concreting practices must be followed when paving under these conditions. Further, if the concrete is being placed on a dark-colored base, such as an ATB or ATPB, care needs to be taken to cool the surface of the base prior to PCC placement. If hot-weather paving precautions are ignored, excessive drying shrinkage can lead to warping and axial deformations. The effect of warping is similar to that of a negative thermal gradient. Axial deformations lead to stress build-up at locations of restraint—slab/base interface, tie bars, etc.

Therefore, hot weather precautions need to be included in specifications and enforced if these conditions may occur during paving.

A useful general reference on hot weather concrete construction is *ACI 305R-99, Hot Weather Concreting* (ACI Committee 305 1999). This document discusses the effects of hot weather on concrete properties and construction operations. Practices intended to minimize the effects of hot weather include:

- selecting materials and proportions;
- precooling ingredients;
- special batching;
- length of haul;
- considering the as-placed concrete temperature; and
- placing and curing equipment and techniques.

"Hot weather may create problems in mixing, placing, and curing hydraulic cement concrete...Most of these problems relate to the increased rate of cement hydration at higher temperature and increased evaporation of moisture from the freshly mixed concrete" (ACI Committee 305 1999: 2). Consequently, practices need to focus on managing temperature effects and moisture loss.

The ACI 305R-99 report (ACI Committee 305 1999: 2) defines hot weather as any combination of high ambient temperature, high concrete temperature, low relative humidity, wind speed, and solar radiation that will impair the quality of freshly mixed or hardened concrete. Problems may occur year round in tropical climates, and during the summer in temperate climates. Paving at night may address these issues – ambient and concrete temperatures are lower, and winds often die down at night. However, with night paving, issues of safety and quality control need to be addressed.

Fresh concrete problems are likely to include increases in (ACI Committee 305 1999: 2)

- water demand;
- rate of slump loss;
- rate of setting;
- tendency for plastic shrinkage cracking; and
- difficulty in controlling air content.

Hardened concrete problems are likely to include, among others (ACI Committee 305 1999: 2–3),

- lower 28-day and later strength;
- increased drying shrinkage;
- increased tendency for thermal cracking;
- decreased durability (due to cracking); and
- increased permeability.

Some of the considerations for selecting materials and mixture proportions are discussed in Chapter 2 of Poole (2005: 3–16) and sections 2.4–2.9 of ACI 305R-99 (ACI Committee 305 1999: 7–11). In general, concrete mixtures in hot weather should avoid type III (high early strength cements), reduce cement content when possible, and use larger substitutions of fly ash or slag. When paving projects extend over several months, it is not uncommon to have problems as spring moves into summer or as summer moves into fall if the mixture is not adjusted for the various environmental conditions. Contracting procedures that award bonuses on the basis of concrete strength may encourage contractors to add unnecessary cement and exacerbate hot weather problems.

Kohn and Tayabji (2003: 56) note that in addition to using less cement, concrete should use class F fly ash, calcined clay, or slag in hot weather. Some class C fly ashes can be used, but others will cause problems in hot weather. Trial batches for hot-weather concreting should include retarders to verify dosages and setting times.

Because of low slump and slump loss in hot weather, there is often a strong temptation to add water without otherwise adjusting the mixture. This will, of course, adversely affect strength and durability. It is preferable to instead add chemical admixtures at the job site, particularly high range water reducers (ACI 305 1999: 10).

One effective method of cooling the concrete mixture is to use cold water for mixing. This is more effective than cooling cement or aggregates, because water has four or five times the specific heat of these materials. ACI 305R-99 provides a formula for calculating concrete temperature based on the temperature of the mixture constituents. Ice may also be added. Although not as effective in reducing concrete temperature as cold water, aggregates may be cooled with water or blown air (ACI Committee 305 1999: 7, 19–20).

Liquid nitrogen may also be used to cool either mixing water or concrete. Water may be chilled with liquid nitrogen in an insulated holding tank, and may even be turned into ice slush, reducing the temperature by as much as 11 °C (20 °F). To cool concrete directly, an injection nozzle is inserted into a ready-mixed concrete truck. Generally, an injection nozzle cannot reduce the temperature to lower than 10 °C (50 °F) or the concrete closest to the injection nozzle may form a frozen lump. The injection nozzles may be used at the job site, where the cooling will be most effective. These methods may be costly (ACI Committee 305 1999: 19–20).

Chemical admixtures play an important role in hot weather concrete. They lower mixing water demand and extend the period of time for concrete placement. Some water-reducing and high range water-reducing admixtures also have set retarding properties, such as ASTM C 494, types D and G. Mid-range water-reducing admixtures are not ASTM classified but comply with ASTM C494 type A and in some cases type F. Extended set control admixtures stop hydration and may be particularly useful with long haul distances or placement delays (ACI 305 1999: 8–10, ASTM C494 2005).

For most pavement projects, moist curing is not practical. White pigmented curing compounds can reflect sunlight and help cure concrete, in addition to making it easier to determine if coverage is adequate (ACI 305 1999: 15–16).

Cold weather precautions

Cold weather concreting is addressed in two ACI Committee 306 documents, *ACI 306R-88, Cold Weather Concreting* and *ACI 306.1-90, Standard Specification For Cold Weather Concreting* (ACI Committee 306 1988). The main objective is to prevent damage due to early freezing. In cooler weather, concrete pavement has an advantage over hot mix asphalt pavement because it is difficult to compact asphalt to proper density at low temperatures.

"Cold weather is defined as a period when, for more than 3 consecutive days, the following conditions exist: 1) the average daily air temperature is less than 40 °F (5 °C) and 2) the air temperature is not greater than 50 °F (10 °C) for more than one-half of any 24-hr period" (ACI Committee 306 1988: 2). In temperate climates this will generally be from some time in fall until some time in spring. The rapid temperature drops that may occur in these seasons are addressed in the next section.

As a general rule, concrete should be protected against freezing until it has attained a compressive strength of at least 3.5 MPa (500 psi). Corners and edges are particularly susceptible to freezing. The required minimum temperature for concrete as mixed for sections less than 300 mm (12 in) thick is 16 °C (60 °F) for air temperatures above −1 °C (30 °F), 18 °C (60 °F) for air temperatures between −18 and −1 °C (0 and 30 °F), and 21 °C (70 °F) below −18 °C (0 °F) (ACI Committee 306 1988: 2–4).

Concrete temperature may be increased by heating mixing water, heating aggregates, or steam heating aggregates. However, very hot water, above about 80 °C (140 °F), may cause flash set and cement balls in the mixer. It is important also that all snow and ice be removed from the subgrade or subbase. The concrete should never be placed on a frozen surface – the subgrade or subbase may be thawed by covering it with insulating material for a few days before concrete placement (ACI Committee 306 1988: 5–7).

Types of insulating blankets that may be used to protect freshly placed concrete include polystyrene foam sheets, urethane foam, foamed vinyl blankets (with or without embedded electrical wires), mineral wool or cellulose fibers, straw, or blanket or batt insulation. Insulation needs to be anchored so that it cannot be removed by wind. At exposed edges and corners, triple thicknesses of insulation should be used (ACI Committee 306 1988: 13–14).

In cold weather, the strength gain of concrete pavement slows considerably. This has three implications. First, the proper time for joint sawing may be delayed. This is chiefly a matter of convenience, although additional expenses may be incurred for laborer overtime. Secondly, it may be necessary to cure the pavement longer, although this is only an issue if water retentive or water added methods are used. Finally, the time that the pavement may be opened to traffic is delayed, which may be an issue for accelerated construction. The strength gain of concrete in cold weather may be accelerated by adjusting the mixture (with more cement or with type III cement), by using accelerating admixtures, or by using curing blankets to retain heat of hydration.

Large ambient temperature drops

Large ambient temperature drops following paving may occur in late fall or early spring in northern climates due to cold fronts or rain and snow. Hall et al. (2005: 10) note

> A drop in ambient temperature of 25 °F (14°C) or greater, shortly after initial set of the concrete, is sufficient to elevate the risk of early cracking significantly. This is particularly true when the temperature falls to a level where the strength gain of the PCC is affected significantly. For example, a temperature drop from 70 to 45 °F (21 to 7 °C) causes a greater risk of early cracking than a drop from 100 to 75 °F (38 to 24 °C).
>
> A large ambient temperature drop imposes a negative thermal gradient through the slab (where the top is cooler than the bottom). If the slab is sufficiently hardened, this can lead to tensile stresses at the top of the slab and a potential for top-down cracking. Large temperature swings are typically prevalent in northern and northeastern climates during late fall or spring construction. A sudden rain or snow event shortly after PCC placement can also cause these swings.

When concrete pavement is protected by blankets, the protection should be removed carefully and slowly. "At the end of the protection period, concrete should be cooled gradually to reduce crack-inducing differential strains between the interior and exterior of the structure" (ACI Committee 306 1988: 9).

HIPERPAV

The FHWA developed the HIPERPAV software package to provide concrete pavement design and construction guidelines. In 2005, an enhanced version, HIPERPAV II, was developed (Ruiz et al. 2005a,b). This program may be downloaded free from the FHWA website (http://www.fhwa.dot.gov/pavement/pccp/hipemain.cfm) along with the technical documentation. The software developer provides additional information on its website (http://www.hiperpav.com/). HIPERPAV II may use either US customary or SI units.

The original version of HIPERPAV used concrete pavement design, materials and mix design, construction, and environment to predict the likelihood of early-age cracking and the optimum time for sawing joints. This program applied to the early-age behavior and performance of JPCP only. HIPERPAV II added long-term performance of JPCP as well as the early-age behavior and performance of CRCP. A detailed user's guide has been published by the FHWA (Ruiz et al. 2005b).

The use of the software may best be illustrated through an example, using the default values from the software package for early-age JPCP analysis. The two input modules are project info and strategies. Project info provides the location of the project, and links to a climate data base. For one project, a number of strategies may be analyzed.

For each strategy, input module sections are provided for strategy information, design, materials and mix design, construction, and environment. The main input in strategy information is the desired reliability, which defaults to 90 percent.

In the design section, inputs are required for geometry, dowels, and slab support. Geometry includes the pavement thickness, width, and transverse joint spacing. The default values are pavement thickness of 254 mm (10 in), slab width of 3.66 m (12 ft) and joint spacing of 4.57 m (15 ft).

If dowels are included in the early-age analysis, the dowel diameter is required. Default values may be used for other dowel properties. For the default JPCP case, the dowel diameter is 38 mm ($1\frac{1}{2}$ in).

Eight different subbase or subgrade types may be defined within the slab support module. These include asphalt concrete or asphalt-stabilized subbase (both of which may be defined as smooth or rough), cement-stabilized subbase, lime treated clay subgrade, untreated clay subgrade, or unbound aggregate subbase. The stiffness and the friction of each subbase or subgrade are provided in the program, and may be modified. Figure 15.4 shows the properties of rough asphalt concrete subbase. This stiff subbase, with high friction, would be expected to contribute to a substantial risk of mid-slab cracking, particularly in hot weather.

Within the materials and mix design section, inputs are required for cement type, PCC mix, PCC properties, and maturity data. Type III cement is used. An example of concrete mixture proportions is shown in Figure 15.5. In addition to the concrete constituents, it is important to characterize the aggregate type, the admixtures used, and the type of fly ash (if any). PCC properties encompasses the strength, modulus of elasticity, shrinkage, and coefficient of thermal expansion. Shrinkage may be

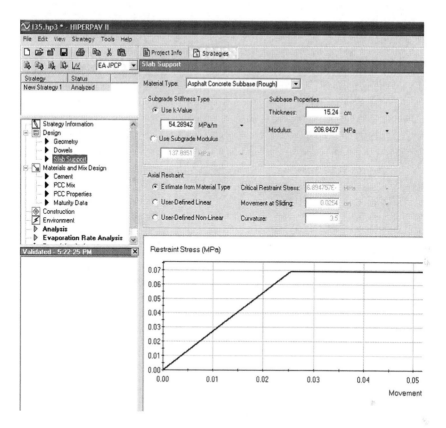

Figure 15.4 Properties of rough asphalt concrete subbase.

estimated from the mixture proportions, and the thermal coefficient may be estimated from the aggregate type. The maturity data may rely on default strength gain, or may be input from laboratory tests.

An example of the construction inputs is shown in Figure 15.6. These include the temperatures of the concrete and the support layer, the type of curing and time of application, and the sawcutting strategy.

Sample environment inputs are shown in Figure 15.7. Either temperature, humidity, wind speed, or cloud cover may be displayed. These are determined based on the location of the project – in this case Round Rock, Texas, near Austin – and on the date of construction. For July 8, the high temperatures are near 33 °C (91 °F) and the low temperatures are near 23 °C (73 °F).

The construction time is an important input, and in this case it makes the difference between the pavement cracking and not cracking. For this example, paving starts at 8 am.

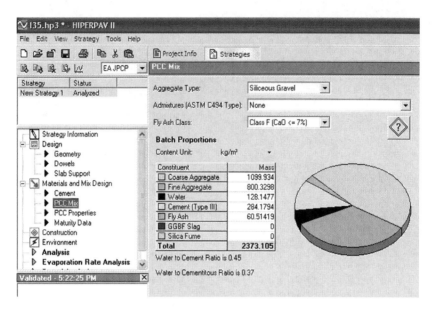

Figure 15.5 Concrete mixture proportions.

Finally, the results of the analysis are shown in Figure 15.8. The upper line shows the strength development of the concrete, while the lower line shows the stress build up in the concrete.

At noon, 4 hours after paving, the stress exceeds the strength, indicating a substantial risk of early-age cracking. The reason is that on a hot day, the heat of the afternoon and the heat of hydration of the concrete combine. In contrast, if paving is delayed until 2 pm, the stress does not exceed the pavement strength and it is less likely to crack, if all other variables remain the same.

HIPERPAV should be used with experience and engineering judgment, because at 2 pm the concrete and subbase may both be warmer, and that should be taken into account. It is, therefore, necessary to understand how the different variables interact with each other. One way to avoid cracking would be to use a subbase with lower stiffness and friction.

HIPERPAV also provides predictions of the concrete evaporation rate, and thus the risk of plastic shrinkage cracking. This is shown in Figure 15.9. With the 8 am time of paving, the limit is exceeded before 9 am with a wind speed of 48 kph (30 mph) and at 9:30 am with a wind speed of 24 kph (15 mph). Fortunately, the predicted wind speed is less than 8 kph (5 mph).

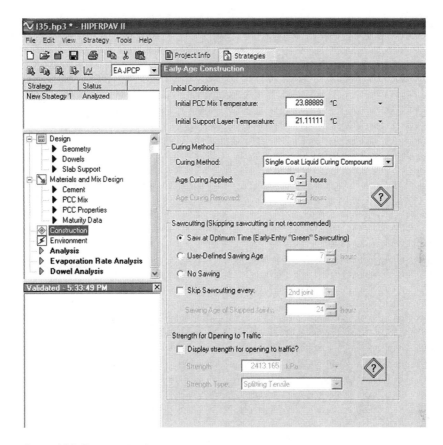

Figure 15.6 Construction inputs.

HIPERPAV is a very useful tool for concrete pavement construction, and allows engineers and contractors to realistically asses the risk of pavement cracking under certain conditions. The ODOT has published a Supplement entitled "HIPERPAV Requirements for Concrete Pavement" addressing HIPERPAV use by the agency. A copy of the ODOT Supplement may be downloaded from http://www.hiperpav.com/. If the Supplement is in force, the contractor is required to provide the results of the HIPERPAV analysis to the engineer at least 24 hours before concrete placement. HIPERPAV has also been used for airfield concrete pavement construction at Airborne Airpark (Wilmington, Ohio) and Cincinnati-Northern Kentucky Airport (Peshkin et al. 2006b).

In a similar fashion, Hall et al. (2005: 28–29) discuss the importance of various factors on early-age cracking and provide threshold values beyond

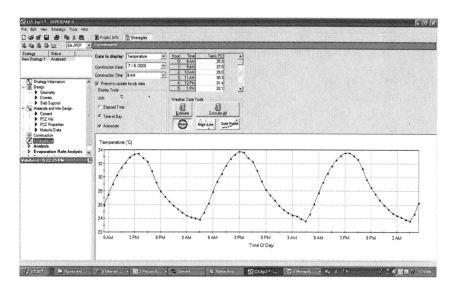

Figure 15.7 Environmental inputs.

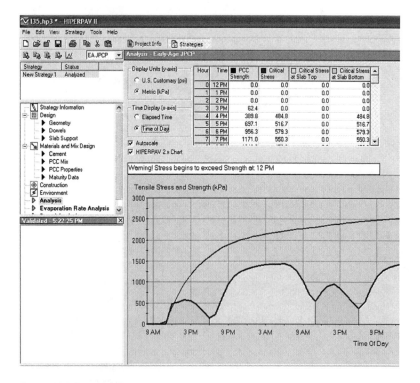

Figure 15.8 Results of analysis.

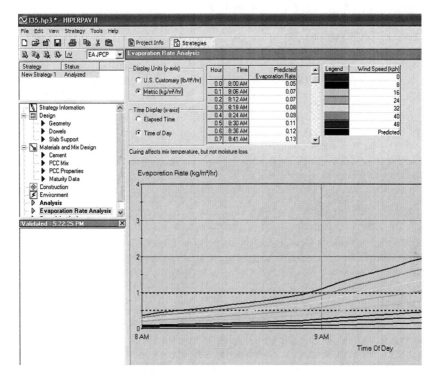

Figure 15.9 Evaporation rate predictions.

which the risk of cracking is increased. Ranked in order of importance, they are:

- "Base strength/stiffness.
- Sawing.
- Panel sizes and aspect ratios.
- PCC/base interface friction.
- PCC cement factor.
- Presence or absence of bond-breaker.
- PCC curing.
- Shrinkage susceptibility of PCC mixes.
- Base thickness.
- Presence of shrinkage cracking in base.
- Internal slab restraint (dowel bars, tie bars, etc.)."

Chapter 16

Concrete pavement maintenance

Well-built concrete pavements generally require less maintenance than asphalt pavements under similar traffic and environmental conditions, but they are rarely maintenance free. Timely maintenance helps preserve the pavement investment and generally extends the time until more costly rehabilitation or reconstruction strategies are necessary. One important consideration is that the pavement condition will deteriorate between the time the decision is made to apply the maintenance and the time that the maintenance is performed, as shown in Figure 16.1.

Many of the concrete pavement distresses, as well as the maintenance strategies that may be used to address them, have been discussed in Chapter 3. Joints and cracks may need to be sealed or resealed. Proper maintenance will also prevent blowups by keeping incompressible materials out of joints. Expansion joints must be able to expand freely without building up stress. Spalls and other small areas of surface deterioration require patches.

Lane-to-shoulder dropoff and lane-to-shoulder separation generally occur with shoulders made of asphalt or aggregate. The shoulders may be maintained or replaced, and any joints that open between the shoulder and main line pavement may be sealed. In one instance, the State of Georgia used RCC to replace deteriorated asphalt shoulders on the heavily traveled Interstate 285 around Atlanta (Bacon 2005). This technique shows promise for heavy duty pavements where shoulder maintenance is costly as well as disruptive to traffic.

Different agencies may define different repair techniques as either maintenance or rehabilitation, based on the circumstances and on agency practices. Arguably, partial depth patching, discussed in Chapter 17, may also be considered a maintenance technique. It is important to carefully prepare distressed areas to ensure good bond between the patch and the existing pavement, and thus a durable patch. Some agencies use asphalt as a patching material for concrete – asphalt is fast and cheap, but these patches generally do not last long and should be replaced with proper partial depth patches as soon as possible.

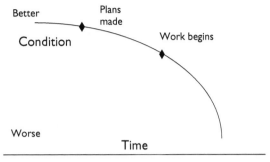

Condition then ≠ condition now

Figure 16.1 Effect of delay in applying a maintenance treatment.

Patch deterioration is also an issue. As an engineer from the US Army Corps of Engineers once said (in reference to locks and dams) "We spend too much time repairing the repairs." Patches are often placed under less than ideal conditions – with short closure windows and/or at night – and may not be cured well. Other causes of premature patch failure may be neglecting to remove all of the distressed concrete, and not preparing the hole well enough to ensure bond. Agencies can avoid premature patch failure through worker training and enforcement of proper patching procedures, as outlined in Chapter 17.

Methods to restore pavement surface friction might also be considered maintenance. These include diamond grinding, discussed in detail in Chapter 17, or thin overlays. The thin overlays may be asphalt-based chip seals, in which case durability is likely to be an issue, or they may be epoxy based (and generally expensive).

Maintenance is particularly important for airfield pavements from the standpoint of preventing foreign object damage (FOD). "When an item that shouldn't have been there – a foreign object – causes aircraft or support equipment damage, it's classified as FOD." (FOD News 2006). Concrete spalls that would merely be a nuisance on a highway can be sucked into a jet engine and cause extensive damage or even loss of an aircraft and crew.

Popouts from unsound aggregates, which generate potential FOD hazards, were observed on military airfields as early as 1959. As a result, military engineers developed strict limitations on deleterious material in concrete aggregates (Rollings 2005: 171). Rollings (2005: 176) also discusses construction problems that lead to early spalling and FOD, such as problems with mixture proportions and curing.

Pavement management

Maintenance, rehabilitation, overlays, and inlays, discussed in this and the following two chapters, all fall under this heading. Garber and Hoel (2002: 1,065) note

> A major problem that faces highway and transportation agencies is that the funds they receive are usually insufficient to adequately repair and rehabilitate every roadway section that deteriorates. The problem is further complicated in that roads may be in poor condition but still usable, making it easy to defer repair projects until conditions become unacceptable. Roadway deterioration usually is not the result of poor design and construction practices but is caused by the inevitable wear and tear that occurs over a period of years. The gradual deterioration of a pavement occurs due to many factors including variations in climate, drainage, soil conditions, and truck traffic.

Pavement management requires the collection and analysis of pavement condition data in order to establish project priorities, establish options, and forecast performance. Roadway condition measurements include roughness, distress, structural condition, and skid resistance.

Methods for measuring pavement condition include (Garber and Hoel 2002: 1,067–1,083):

- Response-type road roughness measuring systems (RTRRMS), as described in Chapter 3. These measure pavement smoothness.
- Inertial road profiling systems (IRPS), also described in Chapter 3. These devices also measure pavement smoothness.
- Locked-wheel trailers with ribbed or smooth tires to measure skid resistance, generally expressed as a skid number (abbreviated SK or SN). These devices are also discussed in Chapter 3.
- Manual methods of distress evaluation, which are slow and disrupt traffic, and often also involve risk to inspectors. Manual methods are, in general, accurate, but different inspectors may classify the same distress with different degrees of severity. Consequently, if pavement distress increases in severity from one report to another, it may mean that the pavement is deteriorating, or it may be due to a difference of interpretation. To reduce these differences, standardized distress manuals such as the *Distress Identification Manual for the Long-Term Pavement Performance Program (Fourth Revised Edition)*, (Miller and Bellinger 2003) have been developed.
- To speed distress data collection, vehicles such as the PASCO ROAD-RECON system have been developed. These vehicles use film and video devices to collect pavement condition data at speeds of up to 80 kph (50 mph) for subsequent manual or automated analysis.

- Falling weight deflectometers (FWDs) are trailer mounted devices that apply an impulse load to a pavement and collect deflections with an array of about six sensors placed at intervals of about 300 mm (12 in).

One example of an agency testing program that uses a van for distress data collection, a locked-wheel trailer, and an FWD is provided by the Washington State Department of Transportation website.

"The distress identification van collects pavement distress, wheelpath rutting, and roughness (IRI) every year on the state route system. These condition measures are processed into pavement performance measures and include PSC (Pavement Structural Condition), PRC (Pavement Rutting Condition), and PPC (Pavement Profile Condition – IRI)" (WSDOT 2006). This van

> records pavement profile (ride, faulting, and rutting) and video images of the pavement surface, ahead view, and shoulder view. This data can be collected at highway speeds and will significantly enhance the accuracy of the data collection process as well as provide a variety of research and analysis options concerning pavement performance. In years previous to 1999, this data was collected with a "windshield" survey. The raters would ride in a van over portions of the state routes, drive approximately [16 kph] 10 mph, and rate the roadway from what they could see. The collection van greatly improves the accuracy and quickness of the data collection process.
>
> (WSDOT 2006)

> Surface friction is measured on the complete WSDOT route system every two years. In essence, a coefficient of friction is measured via a locked-wheel towed trailer between a tire and the pavement surface (the actual value is called Friction Number). Vehicle speed is maintained while water is applied to the pavement surface in front of the test wheel and a brake is applied to the test wheel. When the test wheel stops rotating (locked-wheel state), the drag and load (horizontal and vertical force) are measured.
>
> The friction of most dry pavements is high. Wet pavements are the problem. Thus, the Friction Number testing process involves application of water to the pavement surface prior to determination of the friction value.
>
> (WSDOT 2006)

"The FWD produces a dynamic impulse load that simulates a moving wheel load, rather than a static, semi-static or vibratory load" (Dynatest 2006). The FWD is shown in Figure 16.2.

For concrete pavements, the FWD may be used to measure overall pavement stiffness and structural integrity, or joint load transfer. To assess

Figure 16.2 Falling weight deflectometer (photo by the author).

overall pavement stiffness, "Variations in the applied load may be achieved by altering either the magnitude of the mass or the height of drop. Vertical peak deflections are measured by the FWD in the center of the loading plate and at varying distances away from the plate to draw what are known as 'deflection basins'" (Garber and Hoel 2002: 1,079).

To measure joint load transfer, the load is applied near a joint with sensors on either side of the joint to measure the deflections. If the deflections are nearly the same on both sides, the joint approaches 100 percent load transfer. If the deflection of the loaded slab is considerably higher than that of the unloaded slab, load transfer efficiency is poor.

For airfield pavements, the load applied by an FWD of 7–120 kN (1,500–27,000 lb) is not sufficient to assess the pavement capacity. Therefore, the heavy weight deflectometer (HWD), which can apply loads in the range of 30–240 kN (6,500–54,000 lb), has been developed. "With an expanded loading range, simulating up to a heavy aircraft such as the Boeing 747 (one wheel), the HWD can properly introduce anticipated load/deflection measurements on even heavy pavements such as airfields and very thick highway pavements" (Dynatest 2006).

Other elements of airport pavement evaluation are addressed in Chapter 6 of FAA AC 150/5320-D (FAA 2004: 135–139).

Airport pavement evaluations are necessary to assess the ability of an existing pavement to support different types, weights, or volumes of aircraft traffic. The load-carrying capacity of existing bridges, culverts, storm drains, and other structures should also be considered in these evaluations. Evaluations may also necessary to determine the condition of existing pavements for use in the planning or design of improvements to the airport. Evaluation procedures are essentially the reversal of design procedures

(FAA 2004: 135)

Various maintenance, rehabilitation, overlay, and inlay options are discussed in Chapters 16, 17, and 18. These strategies are generally appropriate for different levels of pavement condition and distress.

When the pavement is in good condition, generally maintenance is sufficient to address minor distresses. If the pavement is in worse condition, various rehabilitation strategies can be applied. Various types of overlays are appropriate as pavement condition deteriorates – bonded overlays for pavements in good condition, and unbonded overlay for pavements in poor condition, as discussed in Chapter 18. If the pavement is in very poor condition, it is generally necessary to resort to a complete reconstruction. This is much more expensive and time consuming than other alternatives.

One challenge is the selection of an appropriate maintenance and rehabilitation strategy. An alternative is to use expert systems.

Expert systems (ES) are computer models that exhibit, within a specific domain, a degree of expertise in problem solving that is comparable to that of a human being. The knowledge required to build the expert system is obtained by interviewing pavement engineers who have long experience and knowledge about pavement management. The acquired knowledge is stored in the expert system, which can then be used to recommend appropriate maintenance or rehabilitation strategies. Since the information is stored in a computer knowledge base, diagnostic advice is available to those users who may not be experienced.

(Garber and Hoel 1,096)

In many agencies, the scope and difficulty of maintenance challenges increases as the pavement network ages, at the same time that the engineers who built that network are retiring and being replaced by a smaller number of inexperienced engineers. Expert systems provide a way to capture the knowledge of the experienced engineers before they depart.

The pavement condition data and the maintenance and rehabilitation strategy selection methods provide the necessary inputs to pavement rehabilitation programming. Some of the methods used include condition assessment, priority assessment models, and optimization:

- Condition assessment is used for short-term programming, on the order of a year. A trigger criterion is established for pavement rehabilitation, such as a PSI of 2.5, and pavements that fall below the trigger are selected for action. If the needs exceed the budget, this must be supplemented with a ranking.
- Priority assessment models extend short-term programming to multiyear programming. Performance models are used to project when different pavement sections will reach trigger points.
- Optimization models such as linear programming or dynamic and integer programming are able to simultaneously evaluate entire networks. The purpose is to either maximize total network benefits or minimize total network cost, within budget constraints and performance standards (Garber and Hoel 2002: 1,098–1,101).

For airfield pavements, the FAA makes use of a pavement condition index (PCI) to numerically rate the surface condition of the pavement. The PCI values range from 100 for a pavement with no defects to 0 for a pavement with no remaining functional life (FAA 2004: 135–136).

Joint and crack resealing

One important maintenance task is resealing joints and cracks. Hot-pour sealants typically need to be replaced at 3–5 years, although some low-modulus or PVC coal tars can last more than 8 years. Silicone sealants last 8–10 years or more and compression seals may provide 15–20 years of service. Narrow, hairline cracks less than 3 mm (1/8 in) wide do not need sealing. Sealing is appropriate for 3–12 mm (1/8–1/2 in) cracks. Wider cracks require patching. Resealing may also be done in conjunction with patching or surface grinding as part of a restoration strategy (ACPA 1996a: VIII-12). In extreme cases, poorly maintained joints may lead to blowups, as discussed in Chapter 3.

As a general rule, nearly all joint sealants fail long before the concrete pavements do. Therefore, there will be a need to periodically reseal joints. Cracks that form in pavement slabs may also need to be sealed after dowel bar retrofit or cross-stitching, as discussed in Chapter 17.

The American Concrete Pavement Association publication *Joint and Crack Sealing and Repair for Concrete Pavements, Technical Bulletin TB – 012.P* discusses how to assess the condition of existing joint sealants to determine if resealing is necessary. Distresses include adhesion loss (sealant not sticking to concrete), cohesion loss (internal disintegration), and oxidation/hardening. A thin metal blade may be inserted into the joint to check adhesion. Joint spalling may also indicate joint sealant failure (ACPA 1993a: 10–13).

Joint resealing

Resealing operations should generally be carried out in the spring and fall to take advantage of moderate temperatures, so that joints will not be either completely open or completely closed. The five steps of resealing are (ACPA 1996a: VIII-83):

1 removing old sealant;
2 shaping the sealant reservoir;
3 cleaning the reservoir;
4 installing the backer rod; and
5 placing the sealant.

Compression seals may be manually pulled out of the joint. Other types of seals may be removed with diamond-bladed saws, or with a metal plow attached to construction equipment (ACPA 1996a: VIII-83–VIII-84).

Once the old seal has been removed, the reservoir should be sawn to clean and widen it. The same equipment and blades used to saw new joint reservoirs may be used. This step may be eliminated for compression seals removed by hand, or if the reservoir is in good shape after sawing out the old sealant. If any spalls adjacent to joints need to be patched, they should be repaired before replacing the sealant. It is important to have the reservoir clean enough to promote bond of the new sealant. Installation of backer rods and new sealant then follows the steps for joint sealing during new construction as outlined in Chapter 14 (ACPA 1996a: VIII-85–VIII-87).

Expansion and isolation joints only require removing the sealant down to the compressible filler, and then replacing it. Contraction joints within 30 m (100 ft) of an expansion joint should be checked to determine if they have opened more widely than usual, particularly if the expansion joint has closed. Lane and shoulder joints should have a 25 mm (1 in) reservoir sawn out, which is then filled with sealant in the usual manner. Asphalt shoulders may settle and joints between concrete pavement and asphalt shoulders present a difficult sealing challenge. The reservoir may need to be 25 mm (1 in) or wider to accommodate the separation between the main line pavement and the shoulder. Propane torches should not be used for joint drying and cleaning, because they can cause concrete spalling and raveling (ACPA 1996a: VIII-88–VIII-90).

Crack sealing

Crack sealing presents a challenge because, unlike joints, cracks are neither straight nor of uniform width. They are, therefore, more difficult to clean, shape, and seal. The first step is to saw the crack out with a small diameter (175–200 mm, 7–8 in) blade that is flexible enough to follow the crack wander. Large diameter blades may overheat and lose segments and endanger the saw operator. Routers do not provide good results either. Once the sawing is complete, the other steps – backer rod and sealant installation – are the same as those for joint resealing (ACPA 1996a: VIII-91–VIII-93).

The American Concrete Pavement Association publication *Joint and Crack Sealing and Repair for Concrete Pavements, Technical Bulletin TB – 012.P* provides a short discussion of cross-stitching and crack sealing/resealing. It also notes that thin hairline cracks do not generally require special treatment or sealing (ACPA 1993a: 24–27).

A detailed discussion of joint and crack resealing operations, with photographs, is provided in the AASHTO/FHWA/Industry Joint Training Participant's Manual, *Construction of Portland Cement Concrete Pavements* (ACPA 1996a: VIII-83–VIII-93). A CD with PowerPoint presentations that accompanies this manual may be purchased separately from ACPA.

Maintenance of drainage systems

It is important to maintain surface and subsurface drainage, or the benefits of the drainage system will be lost. Surface drainage is relatively easy to maintain, because it is simply a matter of keeping side ditches mowed and relatively clear and removing surface obstructions.

Subsurface drainage systems are difficult to maintain because for the most part they are inaccessible. For this reason, it is particularly important to design these systems to resist clogging, as discussed in Chapter 4. Pipes may be cleaned out if they are large enough, so it may make sense to install larger drain pipes than required by hydraulic considerations to make them easy to inspect and clean. Retrofitting edge-drains fall in the category of rehabilitation, not maintenance.

Chapter 17

Rehabilitation

Concrete pavement rehabilitation is necessary when maintenance is no longer sufficient to restore serviceability. One category of rehabilitation is termed concrete pavement restoration (CPR) operations. This encompasses slab stabilization, full and partial depth patching, dowel bar retrofit, diamond grinding, and joint and crack resealing. Joint and crack resealing is also a maintenance technique, as discussed in Chapter 16. In some cases, such as dowel bar retrofit and diamond grinding, the CPR strategies fix load transfer or smoothness deficiencies in the original pavement and make it better than new.

A detailed discussion of CPR operations, with photographs, is provided in the AASHTO/FHWA/Industry Joint Training Participant's Manual, *Construction of Portland Cement Concrete Pavements* (ACPA 1996a). A CD with PowerPoint presentations that accompanies this manual may be purchased separately from ACPA. Sections in the manual include:

- general overview of CPR (pages VIII-1–VIII-13);
- slab stabilization operations (pages VIII-14–VIII-25);
- full-depth patching operations (pages VIII-26–VIII-44);
- partial-depth patching operations (pages VIII-45–VIII-58);
- dowel bar retrofit (pages VIII-59–VIII-68);
- diamond grinding operations (pages VIII-69–VIII-82);
- joint and crack resealing operations (pages VIII-83–VIII-93). Joint and crack resealing is discussed in Chapter 15.

Another pavement management acronym is CPR^3, which stands for concrete pavement restoration, resurfacing, and reconstruction. Restoration encompasses the techniques discussed above. Resurfacing techniques are overlays that are used to improve the structural and/or functional characteristics of pavements. These are discussed in Chapter 18. Some form of preoverlay repair is nearly always necessary to prevent distress in the existing pavement from reflecting up into the overlay. The third technique, reconstruction, is the most expensive and time consuming, and involves the complete removal of the pavement structure. The pavement may be rebuilt as an inlay, without changing the surface elevation. Recycling of the old pavement materials should be considered, both to preserve scarce natural resources and to save time and energy by reducing the need to haul out old materials and haul in fresh materials.

One useful rehabilitation strategy for all types of pavements is retrofitting edge-drains. Edge-drains are discussed in Chapter 4. Although it is not possible to retrofit a drainable subbase under the pavement, edge-drains will help drain the subbase and subgrade and lower the water table in the vicinity of the pavement, which is likely to improve performance.

Selecting maintenance and overlay strategies

The selection of pavement rehabilitation strategies is a complicated topic. Meeting expanding pavement rehabilitation needs with limited resources is difficult. Because of the importance of addressing deteriorating infrastructure, the technology of pavement rehabilitation is changing rapidly. One useful reference is the ACPA document *Pavement Rehabilitation Strategy Selection* (ACPA 1993d).

Many factors affect choices between rehabilitation alternatives. Among these factors are:

- level of pavement distress;
- geometry;
- traffic;
- constructability; and
- future serviceability requirements for the project (ACPA 1993d: 1).

Project information that needs to be collected falls into five categories. These are the design data for the existing pavement, the construction data, traffic data (past, current, and projected), environmental data (precipitation, temperature, and freeze-thaw conditions), and distress and condition data. The type and amount of data that needs to be collected often depends on the scale and complexity of the planned rehabilitation activities.

It is important in pavement rehabilitation to address the causes of deterioration and not just the symptoms. The engineer should determine the structural and functional condition of the pavement. Structural problems may be manifested by corner breaks, pumping, faulted joints, or shattered slabs for concrete pavements. Functional distresses affect ride quality or safety but not the pavement structure, and include smoothness and skid resistance. The evaluation should include extraction and testing of concrete samples to determine if ASR or D-cracking are present. Problems caused by drainage deficiencies should be obvious. It is also useful to note whether there are significant differences in pavement condition between lanes – often the outside (truck) lane has more distress, and the other lanes will not need as much work (ACPA 1993d: 3–4).

Timing of strategies is important. As a pavement deteriorates, the type of rehabilitation which is most appropriate also changes. Rehabilitation applied earlier generally costs less, while deferred treatments cost more and may not be as effective. The concept of pavement preservation refers to early maintenance (e.g. sealing joints and cracks) which delays subsequent pavement deterioration (e.g. moisture-related distress). Furthermore, because there are inevitable time lags between gathering pavement condition data, planning and programming the rehabilitation strategy, and executing the strategy, the effect of continuing deterioration must be considered. This lag is shown in Figure 16.1.

Full and partial depth repairs

Full and partial depth patching or repair entails removing and replacing a portion of the slab, to restore areas of deterioration. Full depth patching extends to the bottom of the concrete and is often used to address joint deterioration, corner breaks, or multiple cracks in slabs. Full depth patching is shown in Figure 17.1. Punchouts in CRCP are also addressed with full depth patching. Partial depth patching is usually no more than 1 sq m in area (11 sq ft) and 50–75 mm (2–3 in) deep. Partial depth repairs are

Figure 17.1 Hole prepared for a full depth repair at a joint (ACPA 1996a: VIII-28).

Figure 17.2 Partial depth repair (ACPA 1996a: VIII-57).

generally used to address spalling or small areas with severe scaling. Partial depth patching is shown in Figure 17.2 (ACPA 1996a: VIII-7–VIII-9).

One important consideration is when to use full depth repairs rather than partial depth repairs. With partial depth repairs, more extensive preparation of the patch area is usually necessary and there is a risk of not removing all of the deteriorated concrete. If extensive partial depth patching is necessary, it may be faster and more economical to provide a smaller number of full depth patches instead. Although material costs may be higher, the labor cost may be lower and performance will probably be better with fewer, larger patches.

In many cases, a small amount of surface damage is the visible effect of considerable internal pavement damage. The extent of damage may be determined from cores extracted from the pavement. If this is the case, partial depth patches cannot address the problem and full depth repairs will be necessary.

Full depth repair patches

If designed and constructed well, full depth patches should last as long as the surrounding pavement. It is important that patches be sized to go beyond the deterioration and encompass any below surface spalling. The minimum length of patches should be 2 m (6 ft), and patches that are close together should be combined. Patches should extend past doweled transverse joints by at least 0.3 m (1 ft) and past CRCP cracks by at least 0.15 m (6 in) (ACPA 1996a: VIII-26–VIII-28). Full depth patches are generally across the entire width of the pavement.

A diagram of a typical patch geometry for JPCP, JRCP, and CRCP is provided in the ACPA course manual on page VIII-27 (ACPA 1996a) and in *Guidelines for Full-Depth Repair, Technical Bulletin TB002.02P* on page 4 (ACPA 1995). Bear in mind that if the patching becomes extensive, it may be cheaper and faster to reconstruct the pavement or to build an unbonded overlay. This is because it is simpler from the point of construction to pave continuously rather than in discrete patches, and because if extensive patching is necessary the remaining pavement in good condition may only have a limited life remaining. One important item is that patches should be as close to square as possible, with no more than a 1.5–1 aspect ratio.

Before removing the deteriorated concrete, full depth saw cuts should be used to isolate the area from adjacent concrete and shoulder materials. It is important to remove the deteriorated concrete without damaging adjacent pavement. Diamond-bladed saws should be used. Saw blades may bind due to pavement thermal expansion in hot weather. This may be addressed by sawing at night or by sawing pressure relief cuts at 180–360 m (600–1,200 ft) intervals before starting any boundary sawing. Large carbide toothed wheel or rock saws may be used to make interior, but not perimeter, cuts. Cuts in CRCP should be only 1/4 to 1/3 of the way through the slab thickness at the perimeter, with full depth cuts further in to allow for bar splicing (ACPA 1996a: VIII-29–VIII-32).

The existing concrete is then removed by lifting it out or by breaking it up in place. Lifting is preferred because lifting causes less damage to the subbase. Lift pins are inserted into the slab and chained together, then attached to lifting equipment. If the pavement is too badly deteriorated, it must be removed in small pieces. Mechanized breakers may also be used. In any case it may be necessary to repair the subbase as well as chipping of adjacent concrete. The hole should be dry and well compacted (ACPA 1996a: VIII-32–VIII-35).

Except for pavements that only carry light traffic, it is necessary to use dowels to provide load transfer between the patch and the existing pavement. Dowels may be drilled into the adjacent concrete and grouted in place to provide load transfer between the full depth patch and the existing pavement. Automated dowel drill rigs can make this process more rapid and accurate. Next, the drill holes are cleaned, and then dowels are installed with grout and grout retention disks. Deformed reinforcing bars are used to restore longitudinal joints. A bond breaking board may be installed along the edge of the full depth patch, as shown in Figure 17.3 (ACPA 1996a: VIII-35–VIII-41).

Figure 17.3 Full depth repair with dowels and fiberboard for isolation (ACPA 1996a: VIII-41).

Concrete is placed in the repair area from ready-mix trucks or other mobile batch vehicles. Concrete must be distributed evenly without excessive shoveling, and vibrated uniformly. Vibrating screeds or 3 m (10 ft) straight edges may be used to strike off and finish the repair. The finished patch should be textured in a similar manner to the surrounding pavement, and cured with a heavy application of membrane-forming curing compound. Insulating mats may be used to accelerate strength gain so that the patch may be opened to traffic earlier. As a final step, it may be necessary to seal joints (ACPA 1996a: VIII-41–VIII-44).

Additional information about full depth repairs is provided by the ACPA publication *Guidelines for Full-Depth Repair, Technical Bulletin TB002.02P* (ACPA 1995). This 20-page document provides further details on the full-depth repair technique, including recommendations for minimum strength for opening patches to traffic.

The FHWA has published a 16-page checklist entitled *Full-Depth Repair of Portland Cement Concrete Pavements* (Pavement Preservation Checklist Series 10 2005). This checklist covers the preparation and execution of full depth repairs, and may be downloaded from the web at http://www.fhwa.dot.gov/pavement/pub_listing.cfm.

Similar issues are raised by utility cuts in concrete streets. It is, unfortunately, often necessary to make cuts to repair or install utilities. The ACPA has published *Utility Cuts and Full-Depth Repairs in Concrete Streets, Information Series IS235.02P* (ACPA 1988). Overall techniques are similar to those for conventional full depth patches.

It is generally not necessary to use dowels, but cuts should be extended to slab joints and edges when possible. This avoids sympathy cracks in the repair induced by existing joints. Either compacted backfill or flowable fill may be used to fill the utility trench (ACPA 1988). Flowable fill is a mixture of Portland cement, fly ash, water, and fine aggregate, and is also termed controlled low strength material or CLSM, and is discussed in detail in ACI Committee Report *Controlled Low Strength Materials, 229R-99* (ACI Committee 229 1999). Flowable fill does not require hand compaction and will not subside over time like compacted backfill. It is generally delivered by ready-mixed concrete trucks.

A different type of full depth repair is used to replace individual UTW panels. Selective panel replacement has been found to be a useful rehabilitation method for UTW. This technique is discussed in *Repair of Ultra-Thin Whitetopping, Publication PA397P* (ACPA 2000a).

Partial depth repair patches

Partial depth patches address spalling and shallow delaminations. Delaminations and unsound concrete may not be visible from the surface. Therefore, a critical first step is accurately locating and mapping the delaminations and unsound concrete, using hammer tapping, steel rods, or steel chains. Repairs should be at least 300 by 100 mm (12 by 4 in), and extend beyond the problem area by 75–100 mm (3–4 in). Combine patches if they are closer than 0.6 m (2 ft), and repair entire joints if there are more than two patches. Patch boundaries should be square or rectangular (ACPA 1996a: VIII-45–VIII-47).

Two common methods for removing concrete before patching are sawing and chipping or carbide milling. For sawing and chipping, the patch is first sawn along the perimeter with a diamond-bladed saw to a depth of 50 mm (2 in). Within the patch areas, concrete should be removed to a minimum depth of 35 mm ($1\frac{1}{2}$ in). Lighter pneumatic hammers are easier to control during this process, and spade bits are preferred to gouge bits. Carbide milling machines are particularly effective at removing concrete for patches that extend across an entire pavement lane (ACPA 1996a: VIII-48–VIII-51).

After removing the existing concrete, the patch area should be checked for weak spots and then carefully cleaned so that the patch will bond well. If the deterioration goes completely through the slab, then a full depth repair may be necessary. Sandblasting or high-pressure waterblasting may be used to clean the patch, followed by air blowing. Patches that abut working joints or cracks require compressible inserts such as Styrofoam or asphalt-impregnated fiberboard. Cementitious bonding agents may be applied in thin coats to promote bond between the patch and the existing concrete (ACPA 1996a: VIII-52–VIII-55).

Patching materials may be mixed in small batches in small mobile drums or paddle mixers, or supplied in ready-mixed concrete trucks if there are enough patches to justify the quantity of material. The patch area should be slightly overfilled. Vibration with small spud vibrators to consolidate the patches is important. The patches are then carefully finished to match the surrounding pavement elevation, and cured with liquid membrane forming curing compounds. Standard Portland cement-based patching materials should not be used if the air temperature falls below 4 °C (40 °F). Minimum temperatures may be different for other patching materials (ACPA 1996a: VIII-56–VIII-58).

The ACPA has published *Guidelines for Partial-Depth Repair, Technical Bulletin TB003P* (ACPA 1989). This document provides additional information about the technique, and stresses the importance of the preliminary survey and delineation of the repair boundaries. The pavement may be struck with a steel rod or hammer, with a metallic ring indicating sound concrete and a dull or hollow sound indicating delaminated areas.

The FHWA has published a 16-page checklist entitled *Partial-Depth Repair of Portland Cement Concrete Pavements* (Pavement Preservation Checklist Series 9 2005). This checklist covers the preparation and execution of partial depth repairs, and may be downloaded from the web at http://www.fhwa.dot.gov/pavement/pub_listing.cfm.

Dowel bar retrofit and cross-stitching

Dowel bar retrofit and cross stitching establish or re-establish load transfer across deteriorated joints or cracks. The key difference is that dowel bar retrofit allows joints to slide open and closed, as with dowels in new construction, while cross-stitching locks joints and cracks together. Cross-stitching, therefore, performs the same function as tie bars in a new construction.

For dowel bar retrofit, slots are cut across each joint to house new dowel bars. If the joints are already badly cracked or spalled, full or partial depth patching should be done instead (ACPA 1996a: VIII-10). Dowel bar retrofit is shown in Figure 17.4.

When dowels are installed in pavements during new construction, they are generally evenly spaced across the width of the joint at approximately 300 mm (12 in) intervals, as shown in Figure 14.6. In contrast, retrofitted dowel bars are installed only in vehicle wheel paths, with three bars per path. This limits the number of dowels required and positions them where they are most needed to provide load transfer.

Figure 17.4 Dowel bar retrofit (ACPA 1996a: VIII-66).

Retrofitting dowel bars

Dowel bar retrofit may be used where dowels were never installed, such as JPCP designed to rely on aggregate interlock or at mid-slab cracks. It may also be used where dowels have failed because of small diameter and high bearing stress wearing away at the concrete, causing the dowels to become loose. Dowels may also fail due to corrosion.

The four main operations of dowel bar retrofitting are:

1 cutting the slots for the dowels;
2 cleaning and preparing the slots to ensure bond;
3 placing the dowel bars; and
4 filling the slots with repair material (ACPA 1996a: VIII-59).

The slots may be cut using either standard saws or slot sawing machines that cut three to six slots at the same time. Standard saws are more difficult to control and do not allow for high production rates. It is vitally important that the slots be cut parallel to the pavement centerline, which also favors the use of slot sawing machines. Slot sawing machines use multiple diamond blades, leaving a piece of concrete between each pair of saw cuts. Slots are cut 60–65 mm (2.4–2.6 in) wide, about 20 mm (3/4 in) wider than the dowel diameter. The slot must provide enough clearance around each dowel to provide for both accurate placement and backfilling of patch material or grout (ACPA 1996a: VIII-59–VIII-62).

Preparation of the slot is another important step. Light pneumatic hammers (7–13.5 kg, 15–30 lb) are used to remove the concrete between the slots. Next, rocks and burrs are knocked of the slot bottom with a small hammerhead. As with all patching operations, it is necessary to provide a clean surface for repair material bonding. Generally, sand blasting followed by air blowing is used to clean out the slot walls and bottom. The cleanliness of the slot is checked by wiping with a hand in a dark glove. Finally, the joint or crack is caulked to keep patch material from entering the joint (ACPA 1996a: VIII-63–VIII-65).

Dowels are carefully prepared. It makes sense to use corrosion-resistant dowels because dowel bar retrofit is a relatively costly operation, and therefore the added cost of using epoxy dowels is not high. Each dowel has a non-metallic expansion cap, two non-metallic chairs, and a compressible joint reformer added. Dowels must also be lubricated. The chairs should provide 12 mm (1/2 in) clearance around the dowel. They are placed carefully and firmly into the slots, parallel to the pavement centerline, with the joint reformer in the joint or crack (ACPA 1996a: VIII-65–VIII-66).

The material used for backfilling the slot is often the same as that used for partial depth patching. Aggregate size should be small enough to allow the material to completely fill the slot and embed the dowel. The slot should be slightly overfilled and then vibrated. Finally, the backfill material is cured with a liquid membrane forming curing compound, and the joint or crack is sawn (ACPA 1996: VIII-67–VIII-68).

The FHWA has published a 16-page checklist entitled *Dowel-Bar Retrofit for Portland Cement Concrete Pavements* (Pavement Preservation Checklist Series 8 2005). This checklist covers the preparation and execution of dowel bar retrofits, and may be downloaded from the web at http://www.fhwa.dot.gov/pavement/pub_listing.cfm.

Cross-stitching

Cross-stitching is used to restore load transfer at longitudinal joints in much the same way that dowel bar retrofit restores load transfer at transverse joints. Cross stitching replaces missing or damaged tie bars.

Holes are drilled at a shallow angle (35–45°) from each side of the joint to form an "x" pattern. Reinforcing bars are then inserted into the holes and epoxied in place, locking the joint together. 20 mm (3/4 in or #6) bars are sufficient, with a typical spacing of 600–900 mm (24–36 in) (ACPA 2001).

Slot stitching is similar, but slots are cut into the pavement and tie bars are epoxied into the slots. It is carried out much like dowel bar retrofit, except that alignment is less critical because the tie bars do not have to slip. Slot stitching is more expensive than cross-stitching, but is less likely to damage the concrete slabs (ACPA 2001).

Specific applications of cross and slot stitching are:

- providing tie bars if they were omitted from the original construction due to design or construction error;
- strengthening longitudinal joints, preventing slab migration and maintaining aggregate interlock;
- tying together lanes and shoulders that are separating;
- tying longitudinal centerlines and preventing faulting; and
- strengthening key joints (ACPA 2001: 1).

One important item to consider is whether the crack to be cross-stitched has been providing stress relief to the slab. If so, then cross-stitching the crack will lead to a stress buildup and the formation of a new crack just next to the cross-stitched crack.

Grinding and grooving

Diamond grinding removes bumps and restores the surface of concrete pavements. It can remove roughness from warped and curled slabs and ruts from studded tires. Specially designed equipment for diamond grinding uses gang-mounted diamond saw blades. In some cases, new pavements are diamond ground to improve initial ride quality or to earn agency smoothness bonuses (ACPA 1996a: VIII-11). Diamond grinding is shown in Figure 17.5.

Figure 17.5 Diamond grinding (ACPA 1996a: VIII-11).

The pavement shown in Figure 17.4 was also diamond ground before the dowel bar retrofit.

Typically, the average depth of removal from diamond grinding is about 5 mm (1/5 in). The four keys to success include (ACPA 1996a: VIII-69–VIII-70):

1   understanding the pavement type and condition;
2   setting up the grinding head properly;
3   operating the machine properly; and
4   monitoring the operation.

There are some important parameters about the pavement that should be investigated when planning the diamond grinding project, including the year the pavement was built, type (JPCP, JRCP, or CRCP), transverse joint spacing, aggregate source and hardness, aggregate size and exposure, depth of studded tire ruts (if any), average planned depth of removal, existing pavement profile, amount of joint faulting, and amount of patching. Aggregate hardness varies from soft (limestone, dolomite, coral, and river gravel) to medium (river gravel, trap rock, and granite) to hard (granite, flint, chert, quartz, and river gravel) (ACPA 1996a: VIII-70). Some types of aggregate, such as river gravel and granite, span multiple hardness categories, so some further investigation may be useful.

The grinding head is made of multiple diamond saw blades attached to a shaft or arbor about 1 or 1.25 m (3.3 or 4.1 ft) wide. Key setup parameters are the depth of groove, the height of groove, and the width between grooves or land area. These vary by aggregate hardness – hard aggregates require a tighter spacing. Saw blades must be selected with the proper bond hardness and diamond concentration to optimize cutting life and provide an even final surface. Blade manufacturers provide information to help select the correct blade for a particular aggregate type and hardness (ACPA 1996a: VIII-71–VIII-73).

The three most important aspects of grinding machine operations are the machine weight, grinding head horsepower, and blade setup. Operation is analogous to a wood plane, with the grinding machine frame spanning between leading and trailing bogies and the grinding head positioned in between. The machine weight keeps the grinding head from riding up over the bumps, and the operator controls the machine speed, head depth, and down pressure to cut through bumps (ACPA 1996a: VIII-73–VIII-74).

The overlap between side to side patches should ensure a vertical match of <3mm in a 3 m span (<1/8 in in 10 ft). Successive paths should overlap by 25–50 mm (1–2 in) to avoid unground areas or "dogtails." Unground areas or "holidays" that result from isolated low areas should be kept to 0.25 m² (2.7 ft²) and not more than 5 percent of the total area. If there are too many holidays, the grinding head should be lowered and the area should be ground again. Wide expansion joints on the order of 75–125 mm (3–5 in) will allow the leading or trailing bogies to dip, lowering the cutting head. One potential problem is if the slabs defect substantially under the load and do not allow much surface removal. This indicates that the slabs may need to be stabilized (ACPA 1996a: VIII-75–VIII-79).

The progress of the diamond grinding project should be evaluated by measuring rideability, using a tool such as a California profilograph. Devices for measuring smoothness and rideability are discussed in Chapter 3. Generally, grinding can improve the previous profile index by at least 65 percent. After grinding, the pavement will have thin concrete "fins" from the area between saw blades, but these generally break away quickly under normal traffic (ACPA 1996a: VIII-78–VIII-81).

The FHWA has published a 12-page checklist entitled *Diamond Grinding of Portland Cement Concrete Pavements* (Pavement Preservation Checklist Series 7 2005). This checklist covers the preparation and execution of diamond grinding, and may be downloaded from the web at http://www.fhwa.dot.gov/pavement/pub_listing.cfm.

Additional information on diamond grinding and grooving is available from the International Grooving & Grinding Association website (International Grooving & Grinding Association 2006), http://www.igga.net/. This website also provides links to a number of state specifications for diamond grinding as well as other maintenance and CPR activities.

Slab stabilization

In slab stabilization, grout is pumped through holes drilled through the slab surface to fill small voids that have developed below the slabs. These may be caused by heavy truck loads and pumping or consolidation of subbase or subgrade material, and are generally less than 3 mm (1/8 in) deep. "Slab stabilization should not be confused with slab jacking, which is done to raise depressed slabs" (ACPA 1996a: VIII-7). Slab stabilization adjacent to a joint is shown in Figure 17.6.

The ACPA has published *Slab Stabilization Guidelines for Concrete Pavements, Technical Bulletin TB–018P* (ACPA 1994b) to provide additional information about this CPR strategy. This document notes that voids may be caused by:

- pumping of water and soil through pavement joints under heavy, fast moving traffic;
- consolidation of materials under the slab;
- subgrade failure due to bearing failure of saturated subgrade; or
- bridge approach failure caused by consolidation and washout of fill material.

Figure 17.6 Grout injection for slab stabilization (ACPA 1996a: VIII-23).

As the voids progress, more of the slab becomes unevenly supported or unsupported, leading to higher flexural stresses. This leads to corner and mid-slab cracking and eventual widespread structural failure of the pavement.

The success of slab stabilization depends on accurately finding voids, designing and mixing a good grout, using good drilling and injection practices, and testing the completed work. Voids may be found through visual inspection by looking for faulted joints, stains from pumping, shoulder blow holes, corner breaks, large shoulder drop-offs, and depressed areas. Static or dynamic deflection testing may also be used to find voids – high deflections indicate a probable void that requires stabilization if the deflection under a truck load is greater than about 0.5 to 0.6 mm (0.02 to 0.025 in) (ACPA 1996a: VIII-14–VIII-21).

Voids may also be detected using ground penetrating radar (GPR) or pulsed electromagnetic wave (PEW) technology. Another technique to determine the extent of a known void is to drill a hole and pouring in a colored epoxy, and then drill down in other locations and examine the drill bit for evidence of the colored epoxy (ACPA 1994b: 5–6).

Pozzolan-cement grout is the most common stabilizing material. The grout must be flowable and dense enough to displace free water. Holes must be carefully drilled, with downward drill pressure monitored so as not to spall the bottom of the slab (ACPA 1996a: VIII-14–VIII-21). Other materials used include polyurethane, limestone-dust cement, cement only, asphalt cement only, and sand–cement. Material tests for this grout are primarily concerned with flowability (ACPA 1994b: 6–10).

During injection, the upward movement of the slab must be measured so that it is not lifted too high. The grout should be injected using a positive displacement injection pump or a non-pulsing progressive cavity pump, at a rate of about 5.5 liters (1.5 gallons) per minute. Grout pressure should be between 0.15 and 1.4 MPa (20 and 200 psi) during injection. Injection should be stopped when the slab begins to rise, grout no longer pumps below the maximum allowable pressure, or grout flows up through adjacent holes. After injection, the holes may be closed with tapered wooden plugs (ACPA 1996a: VIII-22–VIII-25).

A variety of drills and injection devices are available, depending on required production rates. The location of holes, and the order in which the holes are injected, are important to ensure that the slabs rise evenly (ACPA 1994b: 10–15).

The stabilized slabs should be tested using the deflection method 24–48 hours after finishing injection. If high deflections remain, the area will require a second attempt. Slab stabilization should not be carried out if the air temperature drops below 4 °C (40 °F) or the subgrade freezes (ACPA 1996a: VIII-25).

Chapter 18

Overlays and inlays

Concrete overlays are discussed in detail in *ACI 325.13R-06 Concrete Overlays for Pavement Rehabilitation* (ACI Committee 325 2006) and Smith et al. (2002). The Smith et al. report was used as the basis for ACI 325.13R-06. Concrete overlays may be placed over either asphalt or concrete pavements.

Information about concrete overlays and recommendations for determining which concrete overlay alternatives should be used are provided in Chapter 3 of ACI Committee 325 report 325.13R-06. Factors to consider include user costs, lane-closure requirements, traffic control considerations, desired performance life, duration of construction, and local experience with rehabilitation alternatives (ACI Committee 325 2006: 11–13). For example, the US States of Iowa and Texas have developed considerable experience with bonded concrete overlays, and are thus more likely to consider them and use them under the appropriate conditions than other agencies. Many other agencies, however, are very reluctant to use them.

In practice, it is almost always necessary to consider asphalt overlay alternatives as well as concrete overlays. Advantages and disadvantages of different types of pavement overlays, as well as typical life estimates, are reviewed in ACI 325.13R. As an example, structurally bonded concrete overlays are similar to asphalt overlays, although with a longer life (15–25 years versus 10–15 years) (ACI Committee 325 2006: 12). It has been found that concrete overlays deteriorate much more slowly than asphalt overlays, and are likely to last significantly longer (Packard 1994).

Concrete overlays of concrete pavements

Concrete overlays of concrete pavements may be classified as unbonded or bonded. A third category, partially bonded overlays, is rarely used now.

Unbonded overlays

Unbonded overlays are discussed in Chapter 5 of ACI report 325.13R-06 (ACI Committee 325 2006: 18–25) as well as Chapter 4 of Smith et al. (2002: 4-1–4-28). Unbonded overlays are distinguished from bonded and partially bonded overlays in that specific measures are taken to break the bond with the existing pavement, so that damage to the existing pavement does not cause reflective cracking in the overlay. The ACPA published *Guidelines for Unbonded Concrete Overlays, Technical Bulletin TB–005P* (ACPA 1990a). This document recommends minimum unbonded overlay thickness of 125 mm (5 in) for JPCP overlays and 150 mm (6 in) for CRCP overlays.

Because the overlay and the base pavement are structurally isolated, it is difficult for the damage (cracking and joint faulting) in the existing pavement to reflect through the overlay. Therefore, unbonded overlays may be used to rehabilitate concrete pavements with extensive deterioration, including material related distresses such as D-cracking or reactive aggregate. However, the condition of the existing pavement still needs to be evaluated in order to determine the effective thickness (D_{eff}) for designing the overlay, and very weak areas that may lead to localized failures of the overlay may need to be repaired. Weak areas include JPCP or JRCP shattered slabs and CRCP punchouts. A key consideration is ensuring adequate and uniform support for the overlay. By far the most common type of unbonded overlay constructed is JPCP, although a significant number of CRCP overlays have also been built (ACI Committee 325 2006: 18).

A list of distresses and recommended methods of repair is provided in ACI 325.13R-06. The existing pavement may also be fractured into smaller pieces, so that movements at individual cracks will be smaller. This technique has been used in Europe. For example, in Germany, one standard technique used for unbonded concrete overlays is to crack and seat (compact) the existing pavement, place a 100 mm (4 in) lean-concrete separator layer, cut notches in the separator layer to prevent random cracking, and then place the overlay (FHWA 1993).

The design procedures for concrete overlays date back to pioneering work by the US Army Corps of Engineers for bonded, unbonded, and partially bonded overlays that was documented in previous ACI Committee 325 reports (ACI Committee 325 1958, 1967). The National Cooperative Highway Research Program syntheses 99 and 204 also reviewed concrete overlays (Hutchinson 1982, McGhee 1994). In highway and airfield applications, unbonded overlays have been much more widely used than either bonded or partially bonded overlays (ACI Committee 325 2006: 6).

Typical thickness of an unbonded overlay for a highway is 150–300 mm (6–12 in). Generally, there are no special materials or mixture proportions for the concrete other than those for conventional concrete pavement discussed above in Chapters 5 and 6. A separator layer, typically hot-mixed asphalt at least 25 mm (1 in) thick, is placed over the existing pavement as a bond breaker. The intent is that this layer prevents crack reflection from the existing pavement into the overlay.

Unbonded concrete overlay thickness design for highways uses the structural deficiency approach of the AASHTO Pavement Design Guide (AASHTO 1993). The thickness of the overlay, D_{OL}, in mm or inches, is given by:

$$D_{OL} = \sqrt{D_f^2 - D_{eff}^2} \qquad (18.1)$$

The required thickness of a new concrete pavement to carry the projected traffic, D_f, is determined using any thickness design procedure such as AASHTO (1993, 1998) or PCA (1984). D_{eff} is the effective thickness of the existing pavement, which is the actual thickness multiplied by a condition factor CF (≤ 1). The condition factors may be determined based on the product of joint condition, durability, and fatigue adjustment factors, or based on an estimate of remaining life. Tables and charts for determining condition factors are provided in Section 5.9.5 of the AASHTO Design Guide (1993: III-146–III-152).

For example, consider an unbonded concrete overlay where a 250-mm (10-in) pavement would be needed to carry future projected traffic, and the existing pavement is 200 mm (8 in) thick. If the condition factor is 0.8, then the existing pavement is equivalent to 163 mm (6.4 in) and the overlay should be 196 mm (7.7 in) thick. Therefore, the final composite pavement would be 424 mm (16.7 in) thick, counting the 25 mm (1 in) asphalt separator layer. The final grade of the pavement would be raised by 225 mm (8.9 in), which might affect the feasibility of the project, as discussed below.

The FAA unbonded concrete overlay design procedure also uses the structural deficiency approach from equation 18.1. The condition factor is 0.75 for pavements with initial corner cracks due to loading without progressive cracking or joint faulting. A condition factor of 0.35 applies to pavements in poor structural condition which are badly cracked or crushed or have faulted joints. Figures illustrating these two pavement condition factors are provided in the FAA overlay design procedure. Unbonded overlays for airports must be at least 127 mm (5 in) thick. A modified formula applies if the flexural strength of the overlay and existing pavement are different by more than 700 kPa (100 psi). The separator layer must be a highly stable hot-mixed asphalt (FAA 2004: 112–116).

Available FAA software packages (R805faa.xls and LEDFAA) have options for overlay design. In LEDFAA, an internal friction coefficient is used to model the interaction between the existing pavement and the overlay. This coefficient is a fixed parameter that cannot be modified by the

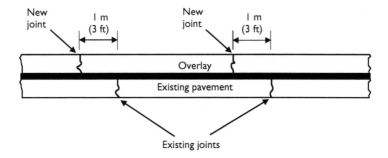

Figure 18.1 Joint mismatching (Smith et al. 2002: 4–9).

designer. The structural condition index (SCI) is used to characterize the existing pavement (FAA 2004: 145).

The key design and construction considerations are achieving separation between the two concrete layers, and mismatching the joints when jointed overlays, typically JPCP, are placed over existing JPCP or JRCP. Joint mismatching is shown in Figure 18.1. The purpose for joint mismatching is to provide good load transfer in the overlay by locating it over relatively sound existing pavement, as opposed to a possibly faulted and deteriorated joint. The existing pavement therefore forms a "sleeper slab" under the joint. Doweled joints are recommended, although load transfer in unbonded overlays is likely to be much better than that in new JPCP construction because of the support provided by the existing pavement slabs (ACI Committee 325 2006: 22).

For unbonded overlays, attention should be paid to the requirements for the separator layer. In addition to ensuring that the old and new slabs act independently, the layer also provides a sound working platform for constructing a smooth overlay. In the past, polyethylene sheeting, wax-based curing compounds, liquid asphalts, and thin chip seal or slurry seals have been used as separators, but these are not recommended because they do not effectively separate the layers. Generally, a thin densely graded hot-mix asphalt layer has provided the best results (ACI Committee 325 2006: 9–10). An open graded separator layer might trap moisture, leading to stripping in the asphalt.

In hot weather, the black asphalt separator layer may become hot, increasing the risk of thermal cracking in the concrete overlay. If the temperature of the layer is expected to exceed 43°C (110°F) at the time of overlay placement, whitewashing the asphalt layer with a lime slurry mixture or white-pigmented curing compound will reduce the surface temperature (ACI Committee 325 2006: 10).

Factors that that may affect the feasibility of unbonded concrete overlays include traffic control, shoulders, and overhead clearance (ACI Committee

325 2006: 18). A TRB workshop identified management of detour traffic during construction as a critical issue in urban areas (TRB 1998). The accelerated construction techniques discussed above in Chapters 6 and 14 and in ACI 325.11R-01 (ACI Committee 325 2001) may be useful in these circumstances.

Unbonded overlay construction will typically require new shoulders because the pavement grade will be raised as much as 250–300 mm (10–12 in). This may also require adjusting access ramps, and clearances under bridge overpasses may prove to be a critical issue (ACI Committee 325 2006: 18–19). If necessary, it is possible to raise bridges and provide concrete or steel spacers on top of bridge piers (ACPA 1990a).

As noted above, the change in grade elevation for an unbonded concrete overlay will require new shoulders. Edge-stresses in the overlay may be reduced either by widening the pavement or by providing tied shoulders. Since the widened slab would extend across the old shoulder, there would be a higher risk of longitudinal cracking. For this reason, tied shoulders are preferred. If it is necessary to widen the traffic lane onto existing shoulder in order to improve traffic flow, the widened portion should be provided with a pavement cross-section that closely matches the underlying pavement (ACI Committee 325 2006: 22). Lane widening options are shown in Figure 18.2.

Figure 18.3 shows three lanes being paved simultaneously on an interstate highway. Two lanes are unbonded overlay on existing concrete pavement, with the asphalt separator visible. The third lane is a new full depth concrete pavement.

A transition detail for bridge approaches and overpasses is shown in Figure 18.4. A recommended length for this transition is 90–150 m (300–500 ft) (ACI Committee 325 2006: 23).

ACI 325.13R-06 (ACI Committee 325 2006: 20) points out some of the significant limitations of current unbonded overlay design procedures. These include:

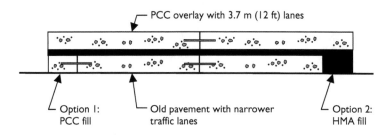

Figure 18.2 Lane widening options (Smith et al. 2002: 4–12).

Figure 18.3 Unbonded overlay on two lanes with a third lane paved at the same time (photo courtesy: Dale Crowl).

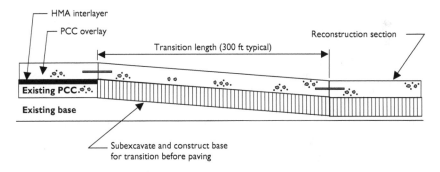

Figure 18.4 A transition detail for bridge approaches and overpasses for unbonded overlays (Smith et al. 2002: 4–13).

- lack of consideration of layer interaction – the separator layer provides a structural contribution to the overlay that is ignored, which would tend to reduce the stresses in the overlay and prolong its life;
- excessive credit given to existing pavement and possible unconservative results if the existing pavement is relatively thick; and

- lack of consideration of the effects of curling and warping on joint stresses. As discussed in Chapter 7, joint spacing should be reduced with high k-values, particularly for thin overlays. This is likely to be an issue for unbonded overlays with joint spacings longer than 4.6 m (15 ft).

The first two issues cancel out, and therefore unbonded overlays are likely to perform well if the joint spacing is not excessive.

For unbonded overlays up to 230 mm (9 in) thick, a joint spacing of no more than 21 times the slab thickness seems to be reasonable, although the joint spacing should not be less than the pavement lane width. For thicker pavements, a maximum L/ℓ ratio of 4.5 to 5.5 is reasonable, and a maximum spacing of 4.6 m (15 ft) greatly reduces the risk of premature cracking (ACI Committee 325 2006: 21–22). Figure 7.3 may be used, with a k-value of 135 MPa/m (500 psi/in) to represent the high level of support provided by the existing pavement.

Smith et al. (2002: 4-17–4-27) provides a detailed discussion on the performance of unbonded overlays, including some details of specific projects. Overall the field performance of unbonded overlays has been very good, with the exception of some thin 150–175 mm (6–7 in) overlays. Performance is also discussed in ACI 325.13R-06 (ACI Committee 325 2006: 24–25). The ACPA *Guidelines for Unbonded Concrete Overlays, Technical Bulletin TB–005P* (ACPA 1990a) documents the successful performance of several dozen projects.

Bonded concrete overlays

Bonded concrete overlays (BCO) are discussed in Chapter 4 of ACI 324.13R-06 (ACI Committee 325 2006: 13–18) as well as Chapter 3 of Smith et al. (2002: 3-1–3-16). Bonded overlays, in contrast to unbonded overlays, are most appropriate when the existing pavement is in good condition. The ACPA has published *Guidelines for Bonded Concrete Overlays, Technical Bulletin TB–007P* (ACPA 1990c).

The design intent of a bonded concrete overlay is to ensure that the overlay and the existing concrete behave monolithically (ACI Committee 325 2006: 13). Typical bonded concrete overlay thicknesses are 75–100 mm (3–4 in) for highways and up to 150 mm (6 in) for airports, although one overlay 165 mm (6½ in) thick was constructed on Interstate Highway I-10 in El Paso, Texas in 1996 (Delatte et al. 1996b). A practical minimum thickness for slipforming is 50 mm (2 in) (ACPA 1990c: 3).

Bonded concrete overlays may be used to address structural deficiencies of concrete pavements, if the existing pavement is in good condition. They may also be used to address functional deficiencies, including

- poor surface friction;
- surface roughness not caused by faulting;
- surface rutting caused by studded tires; and
- excessive noise levels (ACI Committee 325 2006: 11).

Most bonded concrete overlays have been JPCP on JPCP, with care taken to match joints (unlike unbonded overlays). If the joints are not matched, then the old joint will cause a reflection crack in the overlay. Because bonded overlays are thin, dowels are not placed in the overlay and the dowels in the existing pavement provide the load transfer. CRCP overlays have also been constructed on CRCP, primarily in Texas (ACI Committee 325 2006: 14–15).

ACI Committee 325 (2006: 12) notes that

> when considering a bonded concrete overlay, it is imperative that the existing structural capacity of the underlying pavement not already be compromised. Where structural-related distresses are present, such as pumping, faulting, mid-panel cracks, or corner breaks, the load-carrying capabilities of the underlying pavement are already compromised, and a bonded concrete overlay is not an appropriate rehabilitation technique. Furthermore, the presence of D-cracking or other materials-related distresses in the underlying concrete suggest conditions where the effectiveness of a bonded overlay may be limited.

Therefore, an effective evaluation of the existing pavement is a key step for determining whether a bonded concrete overlay is an appropriate rehabilitation solution. The evaluation should include a visual survey, deflection testing using a FWD, and coring (ACI Committee 325 2006: 14). The visual survey may be carried out using a distress identification manual such as Miller and Bellinger (2003).

Examples of distresses that indicate structural deterioration include:

- deteriorated transverse cracks;
- corner breaks;
- pumping of joints;
- faulting of joints; and
- punchouts of CRCP.

If these are present, the severity and extent should be evaluated. If they are not widespread, they should be corrected with full-depth repairs before placing the overlay. If the distresses are extensive, then a bonded overlay will not perform well. Also, pavements with material-related distresses are not good candidates for bonded overlays (ACI Committee 325 2006: 14).

Other possible preoverlay repairs include partial depth repair of joint spalls, slab stabilization to fill voids and prevent future pumping and loss of support, and load transfer restoration across working cracks or non-doweled joints (ACI Committee 325 2006: 17). Load transfer may be restored using dowel bar retrofit, as discussed in Chapter 16. Details are provided by Hoerner et al. (2001).

ACI Committee 325 (2006: 16) suggests that drainage should be part of the evaluation and design process for a bonded concrete overlay. Drainage may need to be improved if pumping, faulting, and corner breaks are present and related to high moisture levels under the pavement. Edge-drains may be retrofitted to correct drainage problems.

Bonded concrete overlay thickness design for highways uses the structural deficiency approach of the AASHTO Pavement Design Guide (AASHTO 1993). The thickness of the overlay, D_{OL}, in mm or inches, is given by:

$$D_{OL} = D_f - D_{eff} \qquad (18.2)$$

The required thickness of a new concrete pavement to carry the projected traffic, D_f, is determined using any thickness design procedure such as AASHTO (1993, 1998) or PCA (1984). D_{eff} is the effective thickness of the existing pavement, which is the actual thickness multiplied by a condition factor CF (≤ 1). The condition factors may be determined based on the product of joint condition, durability, and fatigue adjustment factors, or based on an estimate of remaining life. Tables and charts for determining condition factors are provided in Section 5.8.5 of the AASHTO Design Guide (1993: III-124, III-137–III-144).

For example, consider a bonded concrete overlay where a 250-mm (10-in) pavement would be needed to carry future projected traffic, and the existing pavement is 200 mm (8 in) thick. If the condition factor is 0.8, then the existing pavement is equivalent to 163 mm (6.4 in) and the overlay should be 87 mm (3.4 in) thick. Therefore, the final composite pavement would be 287 mm (11.4 in) thick.

The pavement grade would only be raised by 87 mm (3.4 in) as opposed to the 225 mm (8.9 in) required for an unbonded overlay in the example discussed previously. However, the situations are not directly comparable, since most pavements that are good candidates for unbonded overlays would be poor candidates for bonded overlays.

Bonded concrete overlays on JPCP do not use reinforcement. CRCP overlays bonded to CRCP may have reinforcement of approximately 40 percent of the underlying pavement (ACPA 1990c: 4). The reinforcement may be placed directly on the old pavement after surface preparation. If the additional reinforcement is not provided, the composite section may be under-reinforced and may have a greater risk of developing punchouts.

The FAA minimum thickness for a bonded concrete overlay is 75 mm (3 in) (FAA 2004: 112). The FAA AC 150/5320-D states

> Concrete overlays bonded to existing rigid pavements are sometimes used under certain conditions. By bonding the concrete overlay to the existing rigid pavement, the new section behaves as a monolithic slab. The thickness of bonded overlay required is computed by subtracting the thickness of the existing pavement from the thickness of the required slab thickness determined from design curves... Bonded overlays should be used only when the existing rigid pavement is in good condition. Defects in the existing pavement are more likely to reflect through a bonded overlay than other types of concrete overlays. The major problem likely to be encountered with bonded concrete overlays is achieving adequate bond. Elaborate surface preparation and exacting construction techniques are required to ensure the bond.
>
> (FAA 2004: 114)

The FAA method uses equation 18.2 without any condition factor to modify the existing pavement thickness.

The FAA provides design procedures for airport unbonded and partially bonded overlays using R805faa.xls and LEDFAA software, but not bonded concrete overlays (FAA 2004: 145). The design procedures for partially bonded overlays or, in the FAA terminology, a concrete overlay without a leveling course (FAA 2004: 113–114), could conservatively be used for bonded concrete overlays.

The USACE is currently working to update its BCO design procedures for military airfield pavements.

> New research on failure mechanisms, surface preparation, bonding agent use, and interface texture has produced fresh insight... Integrating contemporary findings into recommended practice should improve the oft-times variable nature of achieving the best bonding conditions for bonded overlays and ultimately improve their reliable long-term performance.
>
> (Semen and Rollings 2005: 900)

At present, BCO use for military airfields is limited to correcting the functional surface deficiencies of a structurally adequate pavement.

Concrete mixtures for bonded overlays should be proportioned for rapid strength gain, minimal thermal expansion and contraction, and minimum shrinkage (Delatte et al. 1998). Although conventional paving mixtures have performed well in many instances, high early strength mixtures have also been used successfully and should be considered when the pavement must be opened quickly to traffic. Water-reducing admixtures are often used to reduce the w/cm ratio (ACI Committee 325 2006: 15–16). The aggregate with the lowest available thermal coefficient of expansion should be used, because this will reduce the thermal stresses produced between the overlay and the base concrete. Specific examples of bonded concrete overlay mixture proportions are provided by Smith et al. (2002). For early opening to traffic, mixtures such as those shown in Table 6.1 are likely to work well.

After the existing pavement condition assessment and preoverlay repair, the preparation of the existing surface for achieving bond is probably the most important step. Loose concrete, paint, and other materials that could inhibit bond must be removed, and a clean, coarse macrotexture for mechanical bonding must be provided. Many surface preparation techniques have been used, including sandblasting, steel shotblasting, high pressure waterblasting, and other techniques. Today, the most common and effective technique is steel shotblasting followed by air blasting immediately before placing the overlay. Waterblasting also provides a suitable surface if water is dried before placing the overlay (ACI Committee 325 2006: 16).

Warner et al. (1998) note that while steel shotblasting and waterblasting do not bruise the existing surface, other techniques such as scabblers and drum-type carbide pick mills create a weak layer immediately below the bond interface. This can lead to subsequent delamination. The *ACI 546R Concrete Repair Guide* provides additional guidance on surface preparation (ACI Committee 546 2004). The texture depth following surface preparation may be measured using ASTM E965 (2001).

Two other considerations are the surface moisture condition at the time of overlay placement, and whether or not a bonding agent is required. Although standing water is clearly detrimental, a surface that is too dry may inhibit hydration of the paste at the interface. Therefore, a near SSD condition is recommended. Bonding agents are not necessary, although they have been used in the past. In fact, some bonding agents have the potential to act as bond breakers if they are allowed to dry or set before the overlay is placed. Vehicles should not be allowed to drip oil or otherwise contaminate the prepared surface (ACI Committee 325 2006: 16–17).

As noted previously, curing of thin concrete layers is both difficult and important. With bonded overlays, high shrinkage due to poor curing practices may lead directly to debonding. Wet burlap or mat curing is preferable, but if curing compounds are used they should be applied at least at twice the manufacturer's recommended rate (ACI Committee 325 2006: 17).

For JPCP overlays bonded to JPCP, the location, timing, and depth of joint sawing are important. It is important to accurately locate the new joints over the old joints, and to cut them completely through the overlay plus an additional 12 mm (1/2 in). Otherwise, there is a risk of debonding and spalling caused by a stress concentration in the uncut overlay. Joints should be cut as soon as the pavement can support the sawing equipment and the cut may be made without spalling or raveling the joint, typically within 4–12 hours after paving (ACI Committee 325 2006: 17–18). It should be noted that the ACPA publication *Guidelines for Bonded Concrete Overlays, Technical Bulletin TB–007P* has an incorrect sawing recommendation for bonded overlays more than 100 mm (4 in) thick (ACPA 1990c: 12). Despite the recommendation in this document, in all cases the sawcut should go completely through the overlay, regardless of overlay thickness. At least one BCO failure has been attributed to not cutting the joint all the way through the overlay.

Bonded concrete overlays have had an uneven performance history. Some projects, particularly in Iowa and Texas, have performed very well. Others have suffered reflective cracking and early debonding in areas where bond was lost (Delatte et al. 1996a,b, ACI Committee 325 2006: 12). Some problems have been attributed to using bonded overlays on pavements that were excessively cracked (McGhee 1994) or inadequate or ineffective preoverlay repair (Peshkin and Mueller 1990). Some detailed case studies of bonded concrete overlays are provided by Smith et al. (2002: 3-12-3-16).

Concrete overlays of asphalt pavements

Concrete overlays of asphalt pavements are divided by overlay thickness into UTW 50–100 mm (2–4 in) thick, thin whitetopping 100–200 mm (4–8 in) thick, and whitetopping 200 mm (8 in) or more in thickness. UTW and thin whitetopping are further distinguished from whitetopping in that the concrete is deliberately bonded to the asphalt.

Whitetopping overlays

As noted in Chapter 2, the oldest type of concrete overlay over an existing asphalt pavement is termed whitetopping. These overlays are placed directly on the existing asphalt pavement, and no special measures are taken to either achieve or prevent bond between the new concrete and the old asphalt. They are also called conventional whitetopping to distinguish them from UTW and thin whitetopping.

Whitetopping overlays may be used to rehabilitate asphalt pavements that exhibit a wide range of structural distresses, including rutting, shoving, and alligator cracking. In general, there is no need to repair the existing asphalt pavement before placing the overlay, although some milling may be necessary to correct severe rutting or profile deficiencies. Although whitetopping overlays have been constructed with JPCP, JRCP, or CRCP, the most common is JPCP (ACI Committee 325 2006: 13). The ACPA has published two documents on whitetopping overlays, *Guidelines for Concrete Overlays of Existing Asphalt Pavements (Whitetopping), Technical*

Bulletin TB-009P (ACPA 1991) and *Whitetopping – State of the Practice, Engineering Bulletin EB210.02P* (ACPA 1998).

Thickness design of whitetopping follows conventional concrete pavement design as discussed in previous chapters, with a relatively high k-value assigned to the existing asphalt. Typical overlay thickness has been 200–300 mm (8–12 in) for primary interstate highways and 125–175 mm (5–7 in) for secondary roads (ACPA 1998). The FAA considers whitetopping overlays to be essentially new concrete pavements, with engineering judgment required to evaluate the structural contribution of the existing pavement. The k-value assigned to the existing flexible pavement may be based on a plate bearing test of the existing pavement or on non-destructive testing, but should not exceed 135 MPa/m (500 psi/in). The computer programs R805faa.xls and LEDFAA may be used for airport whitetopping overlay design (FAA 2004: 113, 144).

Because of this high k-value and potentially high curling and warping stresses, shorter joint spacings should be considered. A reasonable rule of thumb is 21 times the pavement thickness for joint spacing (ACPA 1998). This would lead to a 4.3 m (14 ft) joint spacing for a 200-mm (8-in) thick pavement. However, for pavements 240 mm (9.5 in) and thicker, the 21 rule of thumb may result in excessive joint spacing. Therefore, joint spacing should be considered directly in the design procedure. Figure 7.3 may be used, assuming a k-value of 145 MPa/m (500 psi/in).

Joints should be doweled to prevent faulting, and the asphalt layer will help ensure good load transfer. Some whitetopping overlays built on interstate highways in Wyoming and Utah without dowels quickly developed significant faulting (McGhee 1994). However, it should also be noted that some highway pavements in California without dowels were reported to be performing well after 20 years of service (Hutchinson 1982).

Generally, no preoverlay repairs are required for any cracking, raveling, or bleeding. Ruts of more than 50 mm (2 in) may be milled or filled in with a leveling layer of asphalt. Shallower ruts may be left alone or milled. Shoving may be addressed by milling. Potholes should be filled in with crushed stone cold mixture or a hot-asphalt mixture. The most important issue is to provide uniform support for the concrete overlay (ACI Committee 325 2006: 26).

Surface preparation for whitetopping overlays is not an important issue because there is no need to ensure bond with the existing pavement. Therefore, the overlay may be placed directly on the asphalt surface after sweeping, if ruts do not exceed 25 mm (1 in). For deeper ruts or to avoid raising the pavement grade excessively, 25–75 mm (1–3 in) of the asphalt may be milled off. Alternatively, a 25–50 mm (1–2 in) asphalt leveling course may be placed, although milling is often less expensive (ACI Committee 325 2006: 26–27).

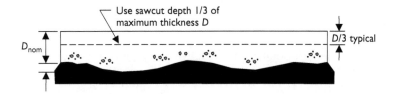

Figure 18.5 Saw cutting for whitetopping with ruts (figure courtesy: ACPA, Smith et al. 2002: 5–12).

When the overlay is placed directly on the existing asphalt pavement, it will be thickest where the ruts are deepest, as shown in Figure 18.5. This provides additional pavement section where the existing pavement is weakest, which will be beneficial for performance. It will, however, create difficulties in estimating the amount of concrete for the project. Also, the depth of saw cutting must be at least one-third the greatest thickness of the overlay.

As for unbonded concrete overlays, there are generally no special materials or mixture proportions for the concrete other than those for conventional concrete pavement discussed above in Chapters 5 and 6. In areas where congestion or heavy traffic limit available construction time, the accelerated paving techniques discussed in chapters 6 and 14 and in ACI 325.11R (ACI Committee 325 2001) should be considered.

The traffic control, shoulder, and overhead clearance considerations for whitetopping overlays are also similar to those for unbonded concrete overlays, as discussed previously. Shoulders will need to be provided, and tied shoulders will reduce pavement edge-stresses. A transition detail for bridge approaches and overpasses is shown in Figure 18.6. A recommended length for this transition is 90–150 m (300–500 ft) (ACI Committee 325 2006: 28).

In hot weather, the existing asphalt pavement may become hot, increasing the risk of thermal cracking in the concrete overlay. If the air temperature

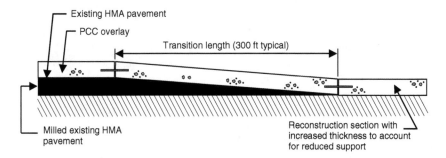

Figure 18.6 A transition detail for bridge approaches and overpasses for whitetopping (figure courtesy: ACPA, Smith et al. 2002: 5–9).

is expected to exceed 32°C (90°F) at the time of overlay placement, water fogging or whitewashing with a lime slurry mixture or white-pigmented curing compound will reduce the surface temperature (ACI Committee 325 2006: 28).

Whitetopping performance on US highways has generally been good, possibly in part due to the conservatism of the design in neglecting the bond to the existing asphalt. The performance of whitetopping is discussed in detail in ACI 325.13R-06 (ACI Committee 325 2006: 29–30) and Smith et al. (2002: 5-12–5-14). The existing asphalt also provides a strong, uniform base of support for the overlay. ACPA reported excellent performance of whitetopping overlays on Iowa county roads. These overlays reduced the crown of the pavement and ranged from 125–150 mm (5–6 in) thick in the center of the road to 150–175 mm (6–7 in) thick at the edges, with 4.6 m (15 ft) joint spacings (ACI Committee 325 2006: 29–30).

Ultrathin whitetopping (UTW) overlays

UTW overlays are suited to asphalt pavements on local roads and intersections with severe rutting, shoving, and pothole problems (ACI Committee 325 2006: 13). Use of UTW began in the early 1990s and has included parking lots, residential streets, low-volume roads, general aviation airports, and intersections, with extension to highway applications in several states (ACI Committee 325 2006: 7). The ACPA has published two documents on UTW, *Construction Specification Guideline for Ultra-Thin Whitetopping, Information Series IS120P* (ACPA 1999a) and *Ultra-Thin Whitetopping, Information Series IS100.02* (ACPA 1999b).

Typically, UTW is 50–100 mm thick (2–4 in) with joint spacing of 0.6–1.8 m (2-6 ft) (ACI Committee 325 2006: 7). One key consideration is the type and amount of asphalt in the existing pavement. Suggested minimums range from 75 to 150 mm (3 to 6 in). The condition of the asphalt is also important, because UTW relies on bond with a competent existing pavement to reduce flexural stresses in the concrete. In particular, very severely distressed or stripped asphalt pavements are poor candidates for UTW. A thorough evaluation of the existing pavement and of the need for preoverlay repair is important. In particular, if the poor performance of the existing asphalt pavement is due to poor drainage, then a UTW overlay is unlikely to be a long-term solution unless the drainage is also improved (ACI Committee 325 2006: 13, 30–31, 33).

Thickness design of UTW is based on the concrete pavement strength, thickness of asphalt remaining after surface preparation, the k-value of the layers under the asphalt, and the vehicle traffic. The ACPA has developed charts for designing UTW and also provides a web UTW thickness design calculator (http://www.pavement.com/Concrete_Pavement/Technical/UTW_Calculator/index.asp). The ACPA website also provides documentation of the theory and assumptions behind the online calculator. The charts are provided in ACI 325.13R.

For design, traffic is classified as either low or medium truck volume. Low truck volume, or axle load category A, assumes a distribution with a maximum single axle load of 80 kN (18,000 lb) and a maximum tandem axle load of 160 kN (36,000 lb). Axle load category B, medium truck volume, uses a distribution with the single axles increased to 116 kN (26,000 lb) and the tandem axles increased to 196 kN (44,000 lb). The ACPA UTW calculator on the web provides a detailed breakdown of the number of axles per 1,000 trucks assumed in the two axle load categories.

As an example of the design procedure, consider a UTW overlay to carry 75 trucks per day in Category B, with a design life of 15 years. The total number of trucks over that time would be approximately 411,000. The existing pavement will be 100 mm (4 in) thick after milling and the subgrade k-value underlying the asphalt is 54 MPa/m (200 psi/in). If the flexural strength of the concrete is 4.8 MPa (700 psi) and the joint spacing is 0.9 m (3 ft), a 75 mm (3 in) overlay would allow 284,000 trucks, which is insufficient. The output from the web calculator is shown in Figure 18.7. Increasing the overlay thickness to 100 mm (4 in) and increasing the joint spacing to 1.2 m (4 ft) would increase the allowable number of trucks to 578,000, which is sufficient.

One advantage of using the UTW calculator is that it allows the use of input values other than those listed in the design tables. There are, however, limits on the ranges that may be input into the calculator. These are:

- UTW thickness between 50 and 100 mm (2 and 4 in);
- UTW concrete flexural strength between 4.8 and 5.6 MPa (700 and 810 psi);
- joint spacing between 0.6 and 1.9 m (2 and 6.2 ft);
- asphalt thickness between 75 and 150 mm (3 and 6 in) – if the remaining asphalt is thicker, then 150 mm (6 in) is used;
- modulus of subgrade reaction (k-value) between 24 and 57 MPa/m (90 and 210 psi/in).

Joint spacing is typically 12–18 times pavement thickness, and the pavement is sawcut into squares. The joint cuts are made with thin early entry saws, typically 3–6 hours after concrete placement, and are not sealed. UTW is too thin to allow for dowels or reinforcing steel, so the joints rely on aggregate interlock.

Because bond between the concrete and asphalt is critical for the performance of the UTW system, there is generally a need for extensive surface preparation. The most common technique is to mill the asphalt surface, followed by a final cleaning of the surface before UTW placement. The surface cleaning may use air blasting, power brooming, water blasting, or

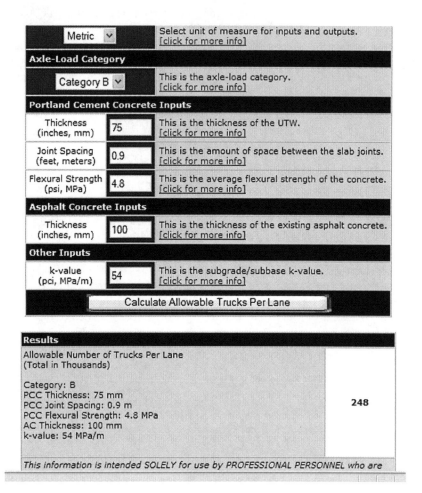

Figure 18.7 Output of ACPA UTW calculator (http://www.pavement.com/Concrete_Pavement/Technical/UTW_Calculator/index.asp).

washing. If water is used, then the prepared surface should be allowed to dry before the overlay is placed, so that the dampness does not inhibit bond (ACI Committee 325 2006: 34).

Figure 18.8 shows UTW placed as an inlay at the outer truck lane of an intersection on an US highway. Milling out an inlay for the full depth of the UTW allows the use of the existing asphalt pavement as a form and makes it easier to build a smooth final pavement surface. However, it is necessary to ensure that enough asphalt will be in place after the inlay milling, and that drainage problems are not created.

Figure 18.8 UTW placed as an inlay (photo by the author).

Joint spacings of less than 1.8 m (6 ft) can cause problems because, with this geometry, joints are located in or near a vehicle wheel path. Heavy truck tire loads near the joints may break down the aggregate interlock and initiate failure.

Figure 18.9 illustrates the condition of a 100-mm (4-in) thick overlay with 1.2 m (4 ft) joint spacing, which breaks the overlay up into three slabs across the lane. This photograph was taken approximately 3½ years after the construction of the UTW overlay and illustrates a premature failure. Several of the outer panels have been displaced laterally by trucks braking at the intersection, and some faulting of the joints is visible. Evidence of asphalt stripping, which would have contributed to an early failure, was found nearby.

Therefore, for the design example discussed previously, it would be desirable to consider increasing the joint spacing to 1.8 m (6 ft). The required number of trucks cannot be carried with 4.8 MPa (700 psi) flexural strength concrete, so the strength must be increased to 5.5 MPa (800 psi). With the higher strength concrete, a 100 mm (4 in) overlay can carry 502,000 trucks in Category B with 1.8 m (6 ft) joint spacing.

UTW often makes use of high early strength, fiber reinforced concrete. Typical flexural strengths are in the range of 4.8–5.5 MPa (700–800 psi), which is higher than for typical concrete paving. Synthetic fibers, principally polypropylene or polyolefin, are often used to help control plastic shrinkage

Figure 18.9 Failure of UTW by panel sliding (photo by the author).

cracking. ACI 544.1R-96 (ACI Committee 544 1996) discusses the use of fibers in concrete. ACI 325.13R-06 (ACI Committee 325 2006: 32) and Smith et al. (2002: 6–8) provide examples of sample concrete mixture designs that have been used for UTW projects.

High early strength concrete is often used when UTW is constructed during weekend closures, as at Selma and Jasper in Alabama (Delatte et al. 2001). Typically, the concrete is required to achieve a compressive strength of 21 MPa (3,000 psi) within 24 hours (Mack et al. 1998). Accelerated concrete paving techniques are discussed in Chapters 6 and 14 and in ACI 325.11R-01 (ACI Committee 325 2001), and the durability of these mixtures is addressed by Van Dam et al. (2005).

As with bonded concrete overlays, the high surface to volume ratios of UTW overlays makes curing the pavement particularly important. High early strength UTW concrete is also likely to generate more heat of hydration than conventional concrete, exacerbating the problem, although this is offset to some degree by the thinness of the overlay. Also, the short duration of closure to traffic permitted for many UTW projects precludes the use of wet mats or curing blankets. Double applications of curing compound are often used, but particular attention should be paid to hot weather concreting (ACI

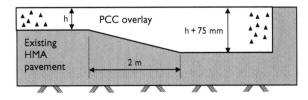

Base or Subgrade

Figure 18.10 UTW-thickened end transition detail (figure courtesy: ACPA, Smith et al. 2002: 6–7).

Committee 305 1999) and curing guidelines (ACI Committee 308 2001) if warranted by the weather conditions anticipated for the project.

A transition detail for connecting to asphalt pavement at the beginning and end of the UTW project is shown in Figure 18.10. The transition increases the UTW thickness by 75 mm (3 in) where it abuts the asphalt (ACI Committee 325 2006: 33). This transition is typically one slab length plus 1.8 m (6 ft). The transition detail helps reduces the effect of impact loads from vehicles approaching and leaving the pavement, particularly if the asphalt adjacent to the UTW ruts. Also, the detail addresses the sort of panel sliding at the pavement end implied in Figure 18.9.

Because UTW is a relatively new pavement rehabilitation technique, relatively little long-term performance information is available. Furthermore, because UTW tends to be used on the low volume roads, agencies have generally not investigated their performance in great detail. Cole (1997) surveyed a number of four- to five-year old UTW pavements and found generally good performance, although his observations of the problems at the beginning and end of sections led to the development of the transition detail shown in Figure 18.10. Shorter joint spacings and thicker remaining asphalt layers improved performance. The performance of UTW is discussed in detail in ACI 325.13R-06 (ACI Committee 325 2006: 34–35) and Smith et al. (2002: 6-12–6-17).

Thin whitetopping overlays

Thin whitetopping overlays fill the gap between UTW and conventional whitetopping and are 100–200 mm (4–8 in) thick, with 1.8–3.7 m (6–12 ft) joint spacing. Like UTW, thin whitetopping relies on bond with the underlying asphalt pavement for performance (ACI Committee 325 2006: 7). One example is a 150-mm (6-in) thin whitetopping overlay placed in October 2001 on the US Highway 78 in Jasper, Alabama. This pavement carries a substantial number of heavy industrial and logging trucks. This project was inspected 2 years after construction, and no distress was observed at that time (Delatte et al. 2001).

At present no design recommendations for thin whitetopping are available, although the conventional whitetopping recommendations in this thickness range developed by ACPA (1998) would be conservative. Tables 9.4 and 9.5 with a k-value of 135 MPa/m (500 psi/in) could also be used. Construction of thin whitetopping follows UTW guidelines, generally with a 1.8 m (6 ft) joint spacing.

Concrete inlays

Inlays provide a rehabilitation option between overlays and complete pavement removal or reconstruction. They may be used to replace one or more deteriorated lanes of a concrete or asphalt pavement. An important advantage of inlays is that they do not change the final grade of the pavement. Also, existing pavement shoulders may be re-used. In general, inlays would be designed as either conventional concrete pavement, as UTW, or as thin whitetopping.

The American Concrete Pavement Association publication *Reconstruction Optimization Through Concrete Inlays, Technical Bulletin TB – 013.P* (ACPA 1993b) discusses inlays in detail, including three pages of sample cross-sections. For roadways and streets, lane-specific (single lane) and multilane inlays are used. Runway keel sections may be replaced as inlays. Partial depth inlays, which are bonded concrete overlays using some of the existing pavement, are also possible.

In order to avoid creating a "bathtub" effect that traps moisture, it is necessary to evaluate and possibly correct drainage as part of an inlay project. Retrofitted edge-drains offer a useful solution.

Bibliography

American Association of State Highway and Transportation Officials (AASHTO) (1993) *Guide for Design of Pavement Structures*, Washington, DC: American Association of State Highway and Transportation Officials.

American Association of State Highway and Transportation Officials (AASHTO) (1998) *Supplement to the AASHTO Guide for Design of Pavement Structures*, Washington, DC: American Association of State Highway and Transportation Officials.

American Association of State Highway and Transportation Officials (AASHTO) (2000) *Highway Capacity Manual 2000*, Washington, DC: American Association of State Highway and Transportation Officials.

American Association of State Highway and Transportation Officials (AASHTO) (2004) *Distribution of the Recommended Mechanistic-Empirical Pavement Design Guide, NCHRP Project 1-37A*, Washington, DC: American Association of State Highway and Transportation Officials.

American Association of State Highway and Transportation Officials (AASHTO) (2005) *Standard Specifications for Transportation Materials and Methods of Sampling and Testing, 25th Edition, and AASHTO Provisional Standards, 2005 Edition*, Washington, DC: American Association of State Highway and Transportation Officials.

American Concrete Institute (ACI) Committee 116 (2000) *Cement and Concrete Terminology, 116R-00*, Farmington Hills, MI: American Concrete Institute.

American Concrete Institute (ACI) Committee 201 (1992) *Guide for Making a Condition Survey of Concrete in Service, 201.1R-92*, Farmington Hills, MI: American Concrete Institute.

American Concrete Institute (ACI) Committee 201 (2001) *Guide to Durable Concrete, 201.2R-01*, Farmington Hills, MI: American Concrete Institute.

American Concrete Institute (ACI) Committee 211 (1991) *Standard Practice for Selecting Proportions for Normal Heavyweight, and Mass Concrete, 211.1-91*, Farmington Hills, MI: American Concrete Institute.

American Concrete Institute (ACI) Committee 211 (1998) *Standard Practice for Selecting Proportions for Structural Lightweight Concrete (Re-Approved 2004), 211.2-98*, Farmington Hills, MI: American Concrete Institute.

American Concrete Institute (ACI) Committee 211 (2002) *Guide for Selecting Proportions for No-Slump Concrete, 211.3R-02*, Farmington Hills, MI: American Concrete Institute.

American Concrete Institute (ACI) Committee 213 (2003) *Guide for Structural Lightweight-Aggregate Concrete, 213R-03*, Farmington Hills, MI: American Concrete Institute.

American Concrete Institute (ACI) Committee 221 (1998) *Report on Alkali-Aggregate Reactivity, 221.1R-98*, Farmington Hills, MI: American Concrete Institute.

American Concrete Institute (ACI) Committee 229 (1999) *Controlled Low Strength Materials, 229R-99*, Farmington Hills, MI: American Concrete Institute.

American Concrete Institute (ACI) Committee 304 (2000) *Guide for Measuring, Mixing, Transporting, and Placing Concrete, ACI 304R-00*, Farmington Hills, MI: American Concrete Institute.

American Concrete Institute (ACI) Committee 305 (1999) *Hot Weather Concreting, ACI 305R-99*, Farmington Hills, MI: American Concrete Institute.

American Concrete Institute (ACI) Committee 306 (1988) *Cold Weather Concreting (Reapproved 2002), ACI 306R-88*, Farmington Hills, MI: American Concrete Institute.

American Concrete Institute (ACI) Committee 308 (1998) *Standard Specification for Curing Concrete, ACI 308.1-98*, Farmington Hills, MI: American Concrete Institute.

American Concrete Institute (ACI) Committee 308 (2001) *Guide to Curing Concrete, ACI 308R-01*, Farmington Hills, MI: American Concrete Institute.

American Concrete Institute (ACI) Committee 325 (1958) *Recommended Practice for Design of Concrete Pavements, ACI Standards 1958, ACI 325-58*, Farmington Hills, MI: American Concrete Institute.

American Concrete Institute (ACI) Committee 325 (1967) *Design of Concrete Overlays for Pavements, ACI 325-1R-67*, Farmington Hills, MI: American Concrete Institute.

American Concrete Institute (ACI) Committee 325 (1995) *State-of-the-Art Report on Roller-Compacted Concrete Pavements, ACI 325.10R-95*, Farmington Hills, MI: American Concrete Institute.

American Concrete Institute (ACI) Committee 325 (2001) *Accelerated Techniques for Concrete Paving, ACI 325.11R-01*, Farmington Hills, MI: American Concrete Institute.

American Concrete Institute (ACI) Committee 325 (2002) *Guide for Design of Jointed Concrete Pavements for Streets and Local Roads, ACI 325.12R-02*, Farmington Hills, MI: American Concrete Institute.

American Concrete Institute (ACI) Committee 325 (2006) *Concrete Overlays for Pavement Rehabilitation, ACI 325.13R-06*, Farmington Hills, MI: American Concrete Institute.

American Concrete Institute (ACI) Committee 330 (2001) *Guide for Design and Construction of Concrete Parking Lots, ACI 330.1R-01*, Farmington Hills, MI: American Concrete Institute.

American Concrete Institute (ACI) Committee 330 (2003) *Specification for Unreinforced Concrete Parking Lots, ACI 330.1-03*, Farmington Hills, MI: American Concrete Institute.

American Concrete Institute (ACI) Committee 522 (2006) *Pervious Concrete, ACI 522R-06*, Farmington Hills, MI: American Concrete Institute.

American Concrete Institute (ACI) Committee 544 (1996) *Report on Fiber Reinforced Concrete (Reapproved 2002), ACI 544.1R-96*, Farmington Hills, MI: American Concrete Institute.

American Concrete Institute (ACI) Committee 546 (2004) *Concrete Repair Guide, ACI 546R-04*, Farmington Hills, MI: American Concrete Institute.

American Concrete Pavement Association (ACPA) (1988) *Utility Cuts and Full-Depth Repairs in Concrete Streets, Information Series IS235.02P*, Skokie, IL: American Concrete Pavement Association.

American Concrete Pavement Association (ACPA) (1989) *Guidelines for Partial-Depth Repair, Technical Bulletin TB003P*, Arlington Heights, IL: American Concrete Pavement Association.

American Concrete Pavement Association (ACPA) (1990a) *Guidelines for Unbonded Concrete Overlays, Technical Bulletin TB-005P*, Arlington Heights, IL: American Concrete Pavement Association.

American Concrete Pavement Association (ACPA) (1990b) *Constructing Smooth Concrete Pavements, Technical Bulletin TB-006.0-C*, Arlington Heights, IL: American Concrete Pavement Association.

American Concrete Pavement Association (ACPA) (1990c) *Guidelines for Bonded Concrete Overlays, Technical Bulletin TB-007P*, Arlington Heights, IL: American Concrete Pavement Association.

American Concrete Pavement Association (ACPA) (1991) *Guidelines for Concrete Overlays of Existing Asphalt Pavements (Whitetopping), Technical Bulletin TB-009P*, Arlington Heights, IL: American Concrete Pavement Association.

American Concrete Pavement Association (ACPA) (1992a) *Scale-Resistant Concrete Pavements, Information Series IS117.02P*, Skokie, IL: American Concrete Pavement Association.

American Concrete Pavement Association (ACPA) (1992b) *Design of Concrete Pavements for City Streets, Information Series IS184P*, Skokie, IL: American Concrete Pavement Association.

American Concrete Pavement Association (ACPA) (1992c) *Design and Construction of Joints for Concrete Streets, Information Series IS061P*, Skokie, IL: American Concrete Pavement Association.

American Concrete Pavement Association (ACPA) (1993a) *Joint and Crack Sealing and Repair for Concrete Pavements, Technical Bulletin TB-012.P*, Skokie, IL: American Concrete Pavement Association.

American Concrete Pavement Association (ACPA) (1993b) *Reconstruction Optimization through Concrete Inlays, Technical Bulletin TB-013.P*, Skokie, IL: American Concrete Pavement Association.

American Concrete Pavement Association (ACPA) (1993c) *Recycling Concrete Pavement, Technical Bulletin TB-014.P*, Skokie, IL: American Concrete Pavement Association.

American Concrete Pavement Association (ACPA) (1993d) *Pavement Rehabilitation Strategy Selection, Technical Bulletin TB-015.P*, Skokie, IL: American Concrete Pavement Association.

American Concrete Pavement Association (ACPA) (1994a) *Fast-Track Concrete Pavements, Technical Bulletin TB004.02*, Skokie, IL: American Concrete Pavement Association.

American Concrete Pavement Association (ACPA) (1994b) *Slab Stabilization Guidelines for Concrete Pavements, Technical Bulletin TB–018P*, Skokie, IL: American Concrete Pavement Association.

American Concrete Pavement Association (ACPA) (1994c) *Cement-Treated Permeable Base for Heavy-Traffic Concrete Pavements, Information Series IS404.01P*, Skokie, IL: American Concrete Pavement Association.

American Concrete Pavement Association (ACPA) (1995) *Guidelines for Full-Depth Repair, Technical Bulletin TB002.02P*, Skokie, IL: American Concrete Pavement Association.

American Concrete Pavement Association (ACPA) (1996a) *Construction of Portland Cement Concrete Pavements, Participant's Manual, FHWA HI-96-027*, AASHTO/FHWA/INDUSTRY Joint Training, Washington, DC: Federal Highway Administration, United States Department of Transportation.

American Concrete Pavement Association (ACPA) (1996b) *Intersection Joint Layout, IS006.01P*, Skokie, IL: American Concrete Pavement Association.

American Concrete Pavement Association (ACPA) (1997) *Concrete Intersections: A Guide for Design and Construction, Technical Bulletin TB019P*, Skokie, IL: American Concrete Pavement Association.

American Concrete Pavement Association (ACPA) (1998) *Whitetopping – State of the Practice, Engineering Bulletin EB210.02P*, Skokie, IL: American Concrete Pavement Association.

American Concrete Pavement Association (ACPA) (1999a) *Construction Specification Guideline for Ultra-Thin Whitetopping, Information Series IS120P*, Skokie, IL: American Concrete Pavement Association.

American Concrete Pavement Association (ACPA) (1999b) *Ultra-Thin Whitetopping, Information Series IS100.02*, Skokie, IL: American Concrete Pavement Association.

American Concrete Pavement Association (ACPA) (1999c) *1999 Database of State DOT Concrete Pavement Practices*, Skokie, IL: American Concrete Pavement Association. http://www.pavement.com/PavTech/Tech/StPract/Query.asp.

American Concrete Pavement Association (ACPA) (2000a) *Repair of Ultra-Thin Whitetopping, Publication PA397P*, Skokie, IL: American Concrete Pavement Association.

American Concrete Pavement Association (ACPA) (2000b) *Concrete Pavement Surface Textures, Special Report SR902P*, Skokie, IL: American Concrete Pavement Association.

American Concrete Pavement Association (ACPA) (2000c) *Traffic Management – Handbook for Concrete Pavement Reconstruction and Rehabilitation, Engineering Bulletin EB213P*, Skokie, IL: American Concrete Pavement Association.

American Concrete Pavement Association (ACPA) (2001) *Stitching Concrete Pavement Cracks and Joints, Special Report SR903P*, Skokie, IL: American Concrete Pavement Association.

American Concrete Pavement Association (ACPA) (2002) *Concrete Pavement for General-Aviation and Commuter Aircraft, IS202.02P*, Skokie, IL: American Concrete Pavement Association.

American Concrete Pavement Association (ACPA) (2006) *Concrete Pavement (About, Research, Technical)*, Skokie, IL: American Concrete Pavement Association. http://216.25.88.215/Concrete_Pavement/.

American Society of Civil Engineers (ASCE) (2006) *American Society of Civil Engineers Code of Ethics*, Reston, VA: American Society of Civil Engineers. www.asce.org.

Applied Research Associates (2006) *About the Mechanistic-Empirical Pavement Design Guide*. http://www.ara.com/transportation/mepdg/mepdg_about.htm.

ASTM C33 (2003) *Standard Specification for Concrete Aggregates, C33-03*, West Conshohocken, PA: ASTM International.

ASTM C39 (2005) *Standard Test Method for Compressive Strength of Cylindrical Concrete Specimens, C39/C39M-05*, West Conshohocken, PA: ASTM International.

ASTM C78 (2002) *Standard Test Method for Flexural Strength of Concrete (Using Simple Beam with Third-Point Loading), C78-02*, West Conshohocken, PA: ASTM International.

ASTM C150 (2005) *Standard Specification for Portland Cement, C150-05*, West Conshohocken, PA: ASTM International.

ASTM C293 (2002) *Standard Test Method for Flexural Strength of Concrete (Using Simple Beam With Center-Point Loading), C293-04*, West Conshohocken, PA: ASTM International.

ASTM C494 (2005) *Standard Specification for Chemical Admixtures for Concrete, C494/C494M-05a*, West Conshohocken, PA: ASTM International.

ASTM C496 (2004) *Standard Test Method for Splitting Tensile Strength of Cylindrical Concrete Specimens, C496/C496M-04*, West Conshohocken, PA: ASTM International.

ASTM C618 (2005) *Standard Specification for Coal Fly Ash and Raw or Calcined Natural Pozzolan for Use in Concrete, C618-05*, West Conshohocken, PA: ASTM International.

ASTM C989 (2005) *Standard Specification for Ground Granulated Blast-Furnace Slag for Use in Concrete and Mortars, C989-05*, West Conshohocken, PA: ASTM International.

ASTM C1074 (2004) *Standard Practice for Estimating Concrete Strength by the Maturity Method, C1074-04*, West Conshohocken, PA: ASTM International.

ASTM D698 (2000) *Standard Test Methods for Laboratory Compaction Characteristics of Soil Using Standard Effort (12,400 ft-lbf/ft³ (600 kN-m/m³)), D698-00ae1*, West Conshohocken, PA: ASTM International.

ASTM D1557 (2002) *Standard Test Methods for Laboratory Compaction Characteristics of Soil Using Modified Effort (56,000 ft-lbf/ft³ (2,700 kN-m/m³)), D1557-02e1*, West Conshohocken, PA: ASTM International.

ASTM D2487 (2000) *Standard Classification of Soils for Engineering Purposes (Unified Soil Classification System), D2487-00*, West Conshohocken, PA: ASTM International.

ASTM E274 (2006) *Standard Test Method for Skid Resistance of Paved Surfaces Using a Full-Scale Tire, E274-06*, West Conshohocken, PA: ASTM International.

ASTM E965 (2001) *Standard Test Method for Measuring Pavement Macrotexture Depth Using a Volumetric Technique, E965-96(2001)*, West Conshohocken, PA: ASTM International.

ASTM E1170 (2001) *Standard Practices for Simulating Vehicular Response to Longitudinal Profiles of Traveled Surfaces, Standard E1170-97(2001)*, West Conshohocken, PA: ASTM International.

Bacon, S. (2005) "Roller-Compacted Concrete: New Application Used Along Shoulders of Atlanta Perimeter," *Southeast Construction Magazine*, McGraw-Hill Construction, Features–May 2005. http://www.southeast.construction.com/features/archive/0505_feature1.asp.

Barber, E.S. and Sawyer, C.L. (1952) *Highway Subdrainage*, Proceedings, Highway Research Board, pp. 643–666.

Barksdale, R.D. (1991) *The Aggregate Handbook*, Washington, DC: National Stone Association.

Bentz, D.P., Lura, P., and Roberts, J.W. (2005) "Mixture Proportioning for Internal Curing," *Concrete International*, Vol. 27, No. 2, pp. 35–40.

Bradbury, R.D. (1938) *Reinforced Concrete Pavements*, Washington, DC: Wire Reinforcement Institute.

Buch, N., Vongchusiri, K., Meeker, B., Command, M., and Ardani, A. (2005) *Accelerated Repair of Jointed Concrete Pavements (JCP) Using Precast Concrete Panels – Colorado Experience*, Proceedings, 8th International Conference on Concrete Pavements, Volume II, Colorado Springs, CO: International Society for Concrete Pavements, pp. 840–856.

Burke, M.P., Jr. (2004a) "Reducing Bridge Damage Caused by Pavement Forces, Part 1: Some Examples," *Concrete International*, Vol. 26, No. 1, pp. 53–57.

Burke, M.P., Jr. (2004b) "Reducing Bridge Damage Caused by Pavement Forces, Part 2: The Phenomenon," *Concrete International*, Vol. 26, No. 2, pp. 83–89.

Bury, M.A. and Nmai, C. (2005) *Innovative Admixture Technology Facilitates Rapid Repair of Concrete Pavements*, Proceedings, 8th International Conference on Concrete Pavements, Volume II, Colorado Springs, CO: International Society for Concrete Pavements, pp. 441–452.

Byrum, C.R. (2000) "Analysis by High-Speed Profile of Jointed Concrete Pavement Slab Curvature," *Transportation Research Record 1730*, Washington, DC: Transportation Research Board.

Byrum, C.R. (2001) "The Effect of Locked-In Curvature on PCC Pavement," *Long-Term Performance Program: Making Something of It*, Reston, VA: ASCE Press, pp. 1–20.

Canadian Portland Cement Association (CPCA) (1999) *Thickness Design for Concrete Highway and Street Pavements (Canadian Edition)*, EB209P, Ottawa, ON: Canadian Portland Cement Association.

Casagrande, A. and Shannon, W.L. (1952) *Base Course Drainage for Airport Pavements*, Proceedings of the American Society of Civil Engineers, Volume 77, pp. 792–814.

Cedergren, H.R., Arman, J.A., and O'Brien, K.H. (1973) *Development of Guidelines for the Design of Subsurface Drainage Systems for Highway Pavement Structural Sections, FHWA-RD-73-14*, Washington, DC: Federal Highway Administration, United States Department of Transportation.

Cole, L.W. (1997) *Pavement Condition Surveys of Ultrathin Whitetopping Projects*, Proceedings of the Sixth International Purdue Conference on Concrete Pavement Design and Materials for High Performance, West Lafayette, IN: Purdue University.

Concrete Reinforcing Steel Institute (CRSI) (1983) *Construction of Continuously Reinforced Concrete Pavements*, Schaumburg, IL: Concrete Reinforcing Steel Institute.

Covarrubias, J.P. (2007) *Concrete Pavements with Small-Size Slabs for Heavier Loads or Thinner Slabs (paper number 07-2358)*, 86th Annual Meeting, Washington, DC: Transportation Research Board.

Darter, M. (1994) "Performance of CRCP in Illinois," *Concrete Reinforcing Steel Institute Research Series – 3*, Schaumburg, IL: Concrete Reinforcing Steel Institute.

Delatte, N.J. (2001) "Lessons from Roman Cement and Concrete," *ASCE Journal of Professional Issues in Engineering Education and Practice*, Vol. 127, No. 3, pp. 109–115.

Delatte, N.J. (2002) "Using the LTPP Database in a Pavement Design Course," *ASCE Journal of Professional Issues in Engineering Education and Practice*, Vol. 128, No. 4, October 2002.

Delatte, N.J. (2004) *Simplified Design of Roller-Compacted Composite Pavement, Pavement Design and Accelerated Testing 2004, Transportation Research Record 1896*, Washington, DC: Transportation Research Board, pp. 57–65.

Delatte, N.J., Fowler, D.W., and McCullough, B.F. (1996a) *High-Early Strength Bonded Concrete Overlay Designs and Construction Methods, Research Report 2911-4*, Austin, TX: Center for Transportation Research.

Delatte, N.J., Gräter, S., Treviño, M., Fowler, D., and McCullough, B. (1996b) *Partial Construction Report of a Bonded Concrete Overlay on IH-10, El Paso and Guide for Expedited Bonded Concrete Overlay Design and Construction, Report 2911-5F*. Austin, TX: Texas Department of Transportation.

Delatte, N.J., Fowler, D.W., McCullough, B.F., and Gräter, S.F. (1998) "Investigating Performance of Bonded Concrete Overlays," *Journal of Performance of Constructed Facilities*, Vol. 12, No. 2, New York: American Society of Civil Engineers.

Delatte, N.J., Safarjalani, M., and Zinger, N.B. (2000) *Concrete Pavement Performance in the Southeastern United States, Report No. 99247*, Tuscaloosa, AL: The University Transportation Center for Alabama. http://utca.eng.ua.edu/projects/final_reports/99247report.pdf.

Delatte, N.J., Chen, S.-E., Davidson, J., Sehdev, A., Amer, N., and Endfinger, M. (2001) *Design and Quality Control of Concrete Overlays, Report No. 01220*, Tuscaloosa, AL: The University Transportation Center for Alabama. http://utca.eng.ua.edu/projects/final_reports/01220-RPT.pdf.

Delatte, N.J., Amer, N.H., and Storey, C. (2003) *Improved Management of RCC Pavement Technology, Report No. 01231*, Tuscaloosa, AL: The University Transportation Center for Alabama. http://utca.eng.ua.edu/projects/final_reports/01231-rpt.pdf.

Drinkard, J.L. (1991) *A Concrete Alternative to Runway Reconstruction, Aircraft/Pavement Interaction: An Integrated System*, Proceedings of the Specialty Conference held in Kansas City, Missouri, September 4–6, 1991, ASCE.

Dynatest (2006) *Dynatest FWD/HWD Test Systems.* http://www.dynatest.com/hardware/fwd_hwd.htm.

Environmental Council of Concrete Organizations (ECCO) (2006) *Why Concrete?* www.ecco.org.

Environmental Protection Agency (EPA) (2006) *Heat Island Effect.* http://www.epa.gov/hiri/index.html.

ERES Consultants (1999) *Evaluation of Unbonded Portland Cement Concrete Overlays, NCHRP Report 415*, Washington, DC: National Cooperative Highway Research Program, Transportation Research Board.

EverFE (2006) *EverFE: Software for the 3D Finite Element Analysis of Jointed Plain Concrete Pavements*. http://www.civil.umaine.edu/EverFE/.

Federal Aviation Administration (FAA) (1970) *Airport Drainage, Advisory Circular (AC) 150/5320-5B*, Washington, DC: Federal Aviation Administration. http://www.faa.gov/airports_airtraffic/airports/resources/advisory_circulars/.

Federal Aviation Administration (FAA) (1981) *Engineering Brief No. 24, Prestressed Concrete Overlay at Chicago O'Hare International Airport*, Washington, DC: Federal Aviation Administration.

Federal Aviation Administration (FAA) (1989) *Airport Design, Advisory Circular (AC) 150/5300-13, with Change 38*, Washington, DC: Federal Aviation Administration. http://www.faa.gov/airports_airtraffic/airports/resources/advisory_circulars/.

Federal Aviation Administration (FAA) (2004) *Airport Pavement Design, Advisory Circular (AC) 150/5320-6D, with Change 3*, Washington, DC: Federal Aviation Administration. http://www.faa.gov/airports_airtraffic/airports/resources/advisory_circulars/.

Federal Aviation Administration (FAA) (2005) *Standards for Specifying Construction of Airports, AC 150/5370-10B*, Washington, DC: Federal Aviation Administration. http://www.faa.gov/airports_airtraffic/airports/regional_guidance/central/construction/aip_development/.

Federal Highway Administration (FHWA) (1989) *Experimental Project No. 12, Concrete Pavement Drainage Rehabilitation*, Washington, DC: Federal Highway Administration.

Federal Highway Administration (FHWA) (1990a) *Concrete Pavement Joints, Technical Advisory, T 5040.30*, Washington, DC: Federal Highway Administration. http://www.fhwa.dot.gov/legsregs/directives/techadvs/t504030.htm.

Federal Highway Administration (FHWA) (1990b) *Continuously Reinforced Concrete Pavement, Technical Advisory, T 5080.14*, Washington, DC: Federal Highway Administration. http://www.fhwa.dot.gov/legsregs/directives/techadvs/t508014.htm.

Federal Highway Administration (FHWA) (1993) *Report on the 1992 U.S. Tour of European Concrete Highways, FHWA-SA-93-012*, Washington, DC: Federal Highway Administration. http://www.fhwa.dot.gov/pavement/ltpp/.

Federal Highway Administration (FHWA) (2003) *Manual on Uniform Traffic Control Devices*, Washington, DC: Federal Highway Administration, United States Department of Transportation. http://mutcd.fhwa.dot.gov/.

Federal Highway Administration (FHWA) (2004) *Transportation Applications of Recycled Concrete Aggregate, FHWA State of the Practice National Review, Publication No. FHWA-IF-05-013*, Washington, DC: Federal Highway Administration.

Federal Highway Administration (FHWA) DataPave Online (FHWA DataPave) (2006). http://www.datapave.com/.

Federal Highway Administration (FHWA) Long Term Pavement Performance Program (FHWA LTPP). (2006) http://www.fhwa.dot.gov/pavement/ltpp/.

Federal Highway Administration Techbrief (FHWA Techbrief) (2006) *Protocol to Identify Incompatible Combinations of Concrete Materials, FHWA-HRT-06-082*, Washington, DC: Federal Highway Administration.

Ferguson, B.K. (2005) *Porous Pavements*, Boca Raton, FL: CRC Press.

FOD News (2006). http://www.fodnews.com/.

Garber, N.J. and Hoel, L.A. (2002) *Traffic & Highway Engineering, 3rd edition*, Pacific Grove, CA: Brooks-Cole.

Gharaibeh, N.G. and Darter, M.I. (2002) *Technical Guide for the Construction of Continuously Reinforced Concrete Pavement, Contract Number DTFH61-00-P-00421*, Washington, DC: Federal Highway Administration, United States Department of Transportation.

Goldbeck, A.T. (1919) "Thickness of Concrete Slabs," *Public Roads*, pp. 34–38.

Gomez-Dominguez, J. (1978) *Mortar Mixtures for Thin, Skid Resistant Concrete Surfaces*, Purdue and Indiana State Highway Commission JHRP.

Griffiths, S., Bowmaker, G., Bryce, C., and Bridge, R. (2005) *Design and Construction of Seamless Pavement on Westlink M7, Sydney, Australia*, Proceedings of the 8th International Conference on Concrete Pavements, Colorado Springs, CO: International Society for Concrete Pavements, August 14–18, 2005, pp. 48–65.

Grogg, M.G., and Smith, K.D. (2002) *PCC Pavement Smoothness: Characteristics and Best Practices for Construction, FHWA-IF-02-025*, Washington, DC: Federal Highway Administration, United States Department of Transportation.

Hachiya, Y., Akamine, F., Takahashi, O., and Miyaji, Y. (2001) *Rapid Repair With Precast Prestressed Concrete Slab Pavements Using Compression Joint System*, Proceedings of the 7th International Conference on Concrete Pavements, Volume II, Orlando, FL: International Society for Concrete Pavements, pp. 891–905.

Hall, K.T., Darter, M.I., Hoerner, T.E., and Khazanovich, L. (1997) *LTPP Data Analysis Phase I: Validation of Guidelines for k-Value Selection and Concrete Pavement Performance Prediction, FHWA-RD-96-198*, Washington, DC: Federal Highway Administration, United States Department of Transportation.

Hall, J.W., Mallela, J., and Smith, K.L. (2005) *Stabilized and Drainable Base for Rigid Pavement: A Design and Construction Guide, Report IPRF-01-G-002-021(G)*, Skokie, IL: Innovative Pavement Research Foundation. http://www.iprf.org/products/main.html.

Hanson, D.I. and Waller, B. (2005) *Evaluation of the Noise Characteristics of Minnesota Pavements*, Sponsored by the Minnesota Department of Transportation. http://www.igga.net/downloads/noise.html.

Hoel, L.A. and Short, A.J. (2006) "The Engineering of the Interstate Highway System," *TR News*, Washington, DC: Transportation Research Board.

Hoerner, T.E., Smith, K.D., Yu, H.T., Peshkin, D.G., and Wade, M.J. (2001) *PCC Pavement Evaluation and Rehabilitation, Reference Manual, NHI Course 131062*, Arlington, VA: National Highway Institute.

Huang, Y.H. (2004) *Pavement Analysis and Design, 2nd Edition*, Upper Saddle River, NJ: Prentice-Hall.

Hutchinson, R.L. (1982) *Resurfacing with Portland Cement Concrete, NCHRP Synthesis of Highway Practice 99*, Washington, DC: National Cooperative Highway Research Program, Transportation Research Board.

International Grooving & Grinding Association (2006). http://www.igga.net/.

Ioannides, A.M., Thompson, M.R., and Barenberg, E.J. (1985) "Westergaard Solutions Reconsidered," *Transportation Research Record 1043*, Washington, DC: Transportation Research Board, pp. 13–23.

Karamihas, S.M. and Cable, J.K. (2004) *Developing Smooth, Quiet, Safe Portland Cement Concrete Pavements, FHWA Project DTFH61-01-X-0002*, Ames, IA: Center for Portland Cement Concrete Pavement Technology, Iowa State University. http://www.igga.net/downloads/noise.html.

Khazanovich, L., Darter, M., Bartlett, R., and McPeak, T. (1998) *Common Characteristics of Good and Poorly Performing PCC Pavements, FHWA-RD-97-131*, Washington, DC: Federal Highway Administration, United States Department of Transportation.

Keith, F.R., Bentley, C.L., Sr., Walker, W.W., and Holland, J.A. (2006) "Shrinkage-Compensating Concrete Pavements Perform Well," *Concrete International*, Vol. 28, No. 1, January 2006, Farmington Hills, MI: American Concrete Institute, pp. 47–51.

Kohn, S.D. and Tayabji, S.D. (2003) *Best Practices for Airport Portland Cement Concrete Pavement Construction, IPRF-01-G-002-1*, Washington, DC: Innovative Pavement Research Foundation. http://www.iprf.org/products/main.html.

Kosmatka, S.H., Kerkhoff, B., and Panarese, W.C. (2002) *Design and Control of Concrete Mixtures, 14th Edition, Engineering Bulletin EB001.14*, Skokie, IL: Portland Cement Association.

Kreuer, B. (2006) *Bond Shear Strength of a Rigid Pavement System with a Roller Compacted Concrete Base*, Master's Thesis, Cleveland, OH: Cleveland State University.

Kuemmel, D.A., Sontag, R.C., Crovetti, J.A., Becker, Y., Jaeckel, J.R., and Satanovsky, A. (2000) *Noise and Texture on PCC Pavements: Results of a Multi-State Study, Report No. WI/SPR-08-99*, Madison, WI: Wisconsin Department of Transportation. http://www.igga.net/downloads/noise.html.

Lam, H. (2005) *Effects of Internal Curing Methods on Restrained Shrinkage and Permeability, PCA R&D Serial No. 2620*, Skokie, IL: Portland Cement Association.

Lane, B. and Kazmierowski, T. (2005) *Use of Innovative Pre-Cast Concrete Slab Repair Technology in Canada*, Proceedings, 8th International Conference on Concrete Pavements, Volume II, Colorado Springs, CO: International Society for Concrete Pavements, pp. 771–788.

Larson, R.M., Scofield, L., and Sorenson, J. (2005) *Providing Durable, Safe, and Quiet Highways*, Proceedings of the 8th International Conference on Concrete Pavements, Volume II, Colorado Springs, CO: International Society for Concrete Pavements, pp. 500–522.

LAW PCS (1999) *Introduction to LTPP Data*. Office on Infrastructure Research and Development, US DOT, FHWA, McLean, VA.

Lee, E.-B., Harvey, J.T., Choi, K., and Thomas, D. (2005) *Innovative Approach to Rapid Rehabilitation of Concrete Pavement on Urban Highway*, Proceedings of the 8th International Conference on Concrete Pavements, Volume III, Colorado Springs, CO: International Society for Concrete Pavements, pp. 1066–1086.

Mack, J.W., Hawbaker, L.D., and Cole, L.W. (1998) "Ultrathin Whitetopping: State-of-the-Practice for Thin Concrete Overlays of Asphalt," *Transportation Research Record 1610*, Washington, DC: Transportation Research Board.

Mack, E.C. (2006) *Using Internal Curing to Prevent Concrete Bridge Deck Cracking*, Master's Thesis, Cleveland, OH: Cleveland State University.

Malhotra, V.M. (2006) "Reducing CO_2 Emissions," *Concrete International*, Vol. 28, No. 9, pp. 42–45.

Mallela, J., Larson, G., Wyatt, T., Hall, J., and Barker, W. (2002) *User's Guide for Drainage Requirements in Pavements – DRIP 2.0 Microcomputer Program*, Washington, DC: Federal Highway Administration, United States Department of Transportation. http://www.fhwa.dot.gov/pavement/software.cfm.

McCullough, B.F. and Cawley, M.L. (1981) *CRCP Design Based on Theoretical and Field Performance*, Proceedings, 2nd International Conference on Concrete Pavement Design, West Lafayette, IN: Purdue University, pp. 239–251.

McCullough, B.F. and Elkins, G.E. (1979) *CRC Pavement Design Manual*, Austin, TX: Austin Research Engineers.

McGhee, K.H. (1994) *Portland Cement Concrete Resurfacing, NCHRP Synthesis of Highway Practice 204*, Washington, DC: National Cooperative Highway Research Program, Transportation Research Board.

McGovern, M.S. (1998) *Smooth Landing for Runway Intersection Replacement*, Publication #R980124, Publication Three, The Aberdeen Group.

Merritt, D.K., McCullough, B.F., and Burns, N.H. (2002) *Construction and Preliminary Monitoring of the Georgetown, Texas, Precast Prestressed Concrete Pavement, FHWA/TX-03-1517-01-IMP-1*, Austin, TX: Texas Department of Transportation.

Merritt, D.K., Stahl, K., Tyson, S.P., and McCullough, B.F. (2005) *Expedited Construction Using Precast Prestressed Concrete Pavement in California*, Proceedings 8th International Conference on Concrete Pavements, Volume II, Colorado Springs, CO: International Society for Concrete Pavements, pp. 808–823.

Miller, J.S. and Bellinger, W.Y. (2003) *Distress Identification Manual for the Long-Term Pavement Performance Program (Fourth Revised Edition), FHWA-RD-03-031*, Washington, DC: Federal Highway Administration, United States Department of Transportation.

Mindess, S., Young, J.F., and Darwin, D. (2003) *Concrete, 2nd Edition*, Upper Saddle River, NJ: Pearson Education.

Moulton, L.K. (1980) *Highway Subsurface Design, Report No. FHWA-TS-80-224*, Washington, DC: Federal Highway Administration, United States Department of Transportation.

Murray, T.M. (2000) "Floor Vibrations: 10 Tips for Designers of Office Buildings," *Structure*, Fall 2000, pp. 26–30.

National Ready Mixed Concrete Association (NRMCA) (2004) *Freeze Thaw Resistance of Pervious Concrete*, Silver Spring, MD: NRMCA.

Neville, A.M. (1997) *Properties of Concrete, 4th Edition*, Essex, England, Prentice-Hall.

Niethelath, N., Garcia, R., Weiss, J., and Olek, J. (2005) *Tire-Pavement Interaction Noise: Recent Research on Concrete Pavement Surface Type and Texture*, Proceedings of the 8th International Conference on Concrete Pavements, Volume II, Colorado Springs, CO: International Society for Concrete Pavements, pp. 523–540.

Nussbaum, P.J. and Lokken, E.C. (1978) *Portland Cement Concrete Pavements, Performance Related to Design-Construction-Maintenance, Report No. FHWA-TS-78-202*, Prepared by PCA for Federal Highway Administration, Washington, DC: Federal Highway Administration, United States Department of Transportation.

ODOT (2005) *Construction and Material Specifications*, Columbus, OH: State of Ohio Department of Transportation.

Older, C. (1924) "Highway Research in Illinois," *Transactions of ASCE*, Vol. 87, pp. 1180–1222.

Packard, R.G. (1973) *Design of Concrete Airport Pavement, EB050.03P*, Skokie, IL: Portland Cement Association.

Packard, R.G. (1994) "Pavement Costs and Quality," *Concrete International*, August 1994, Farmington Hills, MI: American Concrete Institute. Vol. 16, No. 8, pp. 36–38.

Packard, R.G. and Tayabji, S.D. (1985) "New PCA Thickness Design Procedure for Concrete Highway and Street Pavements," *Third International Conference on Concrete Pavement Design and Rehabilitation*, West Lafayette, IN: Purdue University, pp. 225–236.

Pasko, T.J. Jr. (1998) "Concrete Pavements – Past, Present, and Future," *Public Roads*, Vol. 62, No. 1, Washington, DC: Federal Highway Administration, United States Department of Transportation.

Pavement Preservation Checklist Series 6 (2002) *Joint Sealing Portland Cement Concrete Pavements, FHWA-IF-03-003*, Washington, DC: Federal Highway Administration, United States Department of Transportation.

Pavement Preservation Checklist Series 7 (2005) *Diamond Grinding of Portland Cement Concrete Pavements, FHWA-IF-03-040*, Washington, DC: Federal Highway Administration, United States Department of Transportation.

Pavement Preservation Checklist Series 8 (2005) *Dowel-Bar Retrofit for Portland Cement Concrete Pavements, FHWA-IF-03-041*, Washington, DC: Federal Highway Administration, United States Department of Transportation.

Pavement Preservation Checklist Series 9 (2005) *Partial-Depth Repair of Portland Cement Concrete Pavements, FHWA-IF-03-042*, Washington, DC: Federal Highway Administration, United States Department of Transportation.

Pavement Preservation Checklist Series 10 (2005) *Full-Depth Repair of Portland Cement Concrete Pavements, FHWA-IF-03-043*, Washington, DC: Federal Highway Administration, United States Department of Transportation.

Peshkin, D.G. and Mueller, A.L. (1990) "Field Performance and Evaluation of Thin Bonded Overlays," *Transportation Research Record 1296*, Washington, DC: Transportation Research Board.

Peshkin, D.G., Hoerner, T.E., and Zimmerman, K.A. (2004) *Optimal Timing of Pavement Preventive Maintenance Treatment Applications, NCHRP Report No. 523*, Washington, DC: National Cooperative Highway Research Program, Transportation Research Board. http://trb.org/news/blurb_browse.asp?id=2.

Peshkin, D.G., Bruinsma, J.E., Wade, M.J., and Delatte, N.J. (2006a) *Accelerated Practices for Airfield Concrete Pavement Construction – Volume I: Planning Guide, Report IPRF-01-G-002-02-3*, Skokie, IL: Innovative Pavement Research Foundation.

Peshkin, D.G., Bruinsma, J.E., Wade, M.J., and Delatte, N.J. (2006b) *Accelerated Practices for Airfield Concrete Pavement Construction – Volume II: Case Studies, Report IPRF-01-G-002-02-3*, Skokie, IL: Innovative Pavement Research Foundation.

Piggott, R.W. (1999) *Roller Compacted Concrete Pavements – A Study of Long Term Performance, PCA RP366*, Skokie, IL: Portland Cement Association.

Poole, T.S. (2005) *Guide for Curing of Portland Cement Concrete Pavements, Volume I, FHWA-RD-02-099*, Washington, DC: Federal Highway Administration, United States Department of Transportation.

Portland Cement Association (PCA) (1966) *Thickness Design for Concrete Pavements*, Skokie, IL: Portland Cement Association.

Portland Cement Association (PCA) (1984) *Thickness Design for Concrete Highway and Street Pavements, Engineering Bulletin EB109P*, Skokie, IL: Portland Cement Association.

Portland Cement Association (PCA) (1987) *Structural Design of Roller-Compacted Concrete for Industrial Pavements, Information Series IS233.01*, Skokie, IL: Portland Cement Association.

Portland Cement Association (PCA) (1988) *Design of Heavy Industrial Concrete Pavements, Information Series IS234.01P*, Skokie, IL: Portland Cement Association.

Portland Cement Association (PCA) (1992) *Soil-Cement Laboratory Handbook, Engineering Bulletin EB052.07S*, Skokie, IL: Portland Cement Association.

Portland Cement Association (PCA) (2005) *Roller-Compacted Concrete Pavements for Highways and Streets, IS328*, Skokie, IL: Portland Cement Association.

Portland Cement Association (PCA) (2006) *Norfolk International Terminal Selects RCC for Port Facility Expansion, PL620*, Skokie, IL: Portland Cement Association.

Portland Cement Association/American Concrete Pavement Association (PCA/ACPA) (1991) *Subgrades and Subbases for Concrete Pavements, IS029.03P/TB-011.0D*, Skokie and Arlington Heights, IL: Portland Cement Association and American Concrete Pavement Association.

Rasmussen, R.O. and Rozycki, D.K. (2004) *Thin and Ultra-Thin Whitetopping, NCHRP Synthesis 338*, Washington, DC: National Cooperative Highway Research Program. http://www4.trb.org/trb/synthsis.nsf.

Ray, G.K. (1981) *35 Years of Pavement Design and Performance*, Proceedings, 2nd International Conference on Concrete Pavement Design, West Lafayette, IN: Purdue University, pp. 3–8.

Ridgeway, H.H. (1976) "Infiltration of Water Through the Pavement Surface," *Transportation Research Record 616*, Washington, DC: Transportation Research Board, pp. 98–100.

RMC Research Foundation (2005) *Ready Mixed Concrete Industry LEED Reference Guide*, Silver Spring, MD: NRMCA. www.rmc-foundation.org.

Rollings, R.S. (2001) *Concrete Pavement Design: It's More Than a Thickness Design Chart*, Proceedings of the 7th International Conference on Concrete Pavements, Volume I, Orlando FL: International Society for Concrete Pavements, pp. 281–295.

Rollings, R.S. (2005) *Why Do Our Concrete Pavements Still Fail?* Proceedings of the 8th International Conference on Concrete Pavements, Volume I, Colorado Springs, CO: International Society for Concrete Pavements, pp. 167–180.

Rollings, R.S, Rollings, M.P., Poole, T., Wong, G.S., and Gutierrez, G. (2006) "Investigation of Heaving at Holloman Air Force Base, New Mexico," *ASCE Journal of Performance of Constructed Facilities*, Vol. 20, No. 1, Reston, VA: American Society of Civil Engineers, pp. 54–63.

Ruiz, J.M., Rasmussen, R.O., Chang, G.K., Dick, J.C., Nelson, P.K., and Farragut, T.R. (2005a) *Computer-Based Guidelines for Concrete Pavements: Volume I – Project Summary, FHWA-HRT-04-121*, Washington, DC: Federal Highway Administration, United States Department of Transportation.

Ruiz, J.M., Rasmussen, R.O., Chang, G.K., Dick, J.C., and Nelson, P.K. (2005b) *Computer-Based Guidelines for Concrete Pavements: Volume II – Design and Construction Guidelines and HIPERPAV II User's Manual, FHWA-HRT-04-122*, Washington, DC: Federal Highway Administration, United States Department of Transportation.

Sarkar, S.L. (1999) "Durability of Lightweight Aggregate Pavement," *Concrete International*, Vol. 21, No. 5, pp. 32–36.

Sarkar, S.L. and Godiwalla, A. (2003) "Airport Runway Concrete," *Concrete International*, Vol. 16, No. 2, pp. 61–68.

Sarkar, S.L., Little, D.N., Godiwalla, A., and Harvey, G.G. (2001) "Evaluation of Runway Distress and Repair Strategy at Hobby Airport, Houston, Texas," *Concrete Repair Bulletin*, March/April 2001, pp. 10–13.

Semen, P.M. and Rollings, R.S. (2005) *A Review and Discussion of Current Developments Involving Bonded Concrete Overlays of Airfield Pavements*, Proceedings of the 8th International Conference on Concrete Pavements, Volume III, Colorado Springs, CO: International Society for Concrete Pavements, pp. 900–910.

Shilstone, J.M. Sr. (1990) "Concrete Mixture Optimization," *Concrete International*, Vol. 12, No. 6, pp. 33–39.

Shilstone, J.M. Sr. and Shilstone, J.M. Jr. (2002) "Performance-Based Concrete Mixtures and Specifications for Today," *Concrete International*, Vol. 24, No. 2, pp. 80–83.

Shilstone Companies (undated) *Guide Specification for Concrete Paving Mixture (A, B, Cs of Concrete Mixes)*, Dallas, TX: The Shilstone Companies.

Smith, K.D. and Hall, K.T. (2001) "Concrete Pavement Design Details and Construction Practices — Technical Digest," *NHI Training Course 131060*, Arlington, VA: National Highway Institute.

Smith, T. and Jolly, R. (2005) *Concrete Pavement a Sustainable Choice*, Proceedings of the 8th International Conference on Concrete Pavements, Colorado Springs, CO: International Society for Concrete Pavements, pp. 585–606.

Smith, K.D., Peshkin, D.G., Darter, M.I., Mueller, A.L., and Carpenter, S.H. (1990) *Performance of Jointed Concrete Pavements, Vol. 1, Evaluation of Concrete Pavement Performance and Design Features, Report No. FHWA-RD-89-136*, Washington, DC: Federal Highway Administration, United States Department of Transportation.

Smith, K.D., Yu, H.T., and Peshkin, D.G. (2002) *Portland Cement Concrete Overlays: State of the Technology Synthesis, FHWA-IF-02-045*, Washington, DC: Federal Highway Administration, United States Department of Transportation.

Snell, L.M. and Snell, B.G. (2002) "Oldest Concrete Street in the United States," *Concrete International*, Vol. 24, No. 3, pp. 72–74.

Snyder, M.B. (2005) *An Evaluation of Cathodically Protected Dowels for Concrete Pavements*, Proceedings of the 8th International Conference on Concrete Pavements, Colorado Springs, CO: International Society for Concrete Pavements.

Tayabji, S.D., Stephanos, P.J., and Zollinger, D.G. (1995) "Nationwide Field Investigation of Continuously Reinforced Concrete Pavements," *Transporta-*

tion Research Record 1482, Washington, DC: Transportation Research Board, pp. 7–18.

Taylor, P.C., Graf, L.A., Zemajtis, J.Z., Johansen, V.C., Kozikowski, R.L., and Ferraris, C.L. (2006a) *Identifying Incompatible Combinations of Concrete Materials: Volume I, Final Report, FHWA-HRT-06-079*, Washington, DC: Federal Highway Administration, United States Department of Transportation.

Taylor, P.C., Graf, L.A., Zemajtis, J.Z., Johansen, V.C., Kozikowski, R.L., and Ferraris, C.L. (2006b) *Identifying Incompatible Combinations of Concrete Materials: Volume I, Test Protocol, FHWA-HRT-06-080*, Washington, DC: Federal Highway Administration, United States Department of Transportation.

Titus-Glover, L., Mallela, J., Darter, M.I., Voigt, G., and Waalkes, S. (2005) *Enhanced Portland Cement Concrete Fatigue Model for StreetPave, Transportation Research Record No. 1919*, Washington, DC: Transportation Research Board, pp. 29–37.

Transportation Research Board (TRB) (1998) *Get In Get Out Stay Out!* Proceedings of the Workshop on Pavement Renewal for Urban Freeways, Washington, DC: Transportation Research Board.

Transportation Research Board (TRB) (2006) *Long-Term Pavement Performance Studies*, Washington, DC: Transportation Research Board. http://www4. trb.org/trb/dive.nsf/web/long-term_pavement_performance_studies.

Tyson, S.S. and Merritt, D.K. (2005) "Pushing the Boundaries," *Public Roads*, Vol. 68, No. 4, January/February 2005, pp. 28–33.

Unified Facility Criteria (UFC) (2001a) 3-250-02 *Standard Practice Manual for Rigid Pavements*, US Army Corps of Engineers (Preparing Activity), Naval Facilities Engineering Command, Air Force Civil Engineer Support Agency.

Unified Facility Criteria (UFC) (2001b) 3-260-02 *Pavement Design for Airfields*, US Army Corps of Engineers (Preparing Activity), Naval Facilities Engineering Command, Air Force Civil Engineer Support Agency.

Van Dam, T.J., Peterson, K.R., Sutter, L.L., Panguluri, A., Sysma, J., Buch, N., Kowli, R., and Desaraju, P. (2005) *Guidelines for Early-Opening-to-Traffic Portland Cement Concrete for Pavement Rehabilitation, NCHRP Report No. 540*, Washington, DC: National Cooperative Highway Research Program, Transportation Research Board. http://trb.org/news/blurb_browse.asp?id=2.

Warner, J., Bhuhan, S., Smoak, W.G., Hindo, K.R., and Sprinkel, M.M. (1998) "Surface Preparation for Overlays," *Concrete International*, Vol. 20, No. 5, pp. 43–46.

Washington, State Department of Transportation (WSDOT) (2006) *Pavement Testing*. http://www.wsdot.wa.gov/biz/mats/pavement/testing.htm.

Wayson, R.L. (1998) *Relationship Between Pavement Surface Texture and Highway Traffic Noise, NCHRP Synthesis of Highway Practice No. 268*, Washington, DC: National Cooperative Highway Research Program, Transportation Research Board. http://www.igga.net/downloads/noise.html.

Westergaard, H.M. (1926) "Stresses in Concrete Pavements Computed by Theoretical Analysis," *Public Roads*, Vol. 7, pp. 225–235.

Won, M., Hankins, K., and McCullough, B.F. (1989) *A Twenty-four Year Performance Review of Concrete Pavement Sections Made with Siliceous and Lightweight Coarse Aggregates, Interim Report*, Center for Transportation Research, Austin, TX: The University of Texas.

Woodstrom, J.H. (1990) *Measurements, Specifications, and Achievement of Smoothness for Pavement Construction, NCHRP Synthesis of Highway Practice No. 167*, Washington, DC: National Cooperative Highway Research Program, Transportation Research Board.

Yoder, E.J. and Witczak, M.W. (1975) *Principles of Pavement Design, 2nd Edition*, New York, NY: John Wiley & Sons.

Yu, H.T., Smith, K.D., Darter, M.I., Jiang, J., and Khazanovich, L. (1998) *Performance of Concrete Pavements, Volume III – Improving Concrete Pavement Performance, FHWA-RD-95-111*, McLean, VA: Federal Highway Administration.

Index

eBooks – at www.eBookstore.tandf.co.uk

A library at your fingertips!

eBooks are electronic versions of printed books. You can store them on your PC/laptop or browse them online.

They have advantages for anyone needing rapid access to a wide variety of published, copyright information.

eBooks can help your research by enabling you to bookmark chapters, annotate text and use instant searches to find specific words or phrases. Several eBook files would fit on even a small laptop or PDA.

NEW: Save money by eSubscribing: cheap, online access to any eBook for as long as you need it.

Annual subscription packages

We now offer special low-cost bulk subscriptions to packages of eBooks in certain subject areas. These are available to libraries or to individuals.

For more information please contact webmaster.ebooks@tandf.co.uk

We're continually developing the eBook concept, so keep up to date by visiting the website.

www.eBookstore.tandf.co.uk